Water Governance for Sustainable Development

Edited by Sylvain Perret, Stefano Farolfi and Rashid Hassan

earthscan
from Routledge

First published by Earthscan in the UK and USA in 2006
Moved to digital printing 2010

Cirad Editions
TA 283/04
Avenue Agropolis
34938 Montpellier Cedex 5
France

ISBN-13: 978-1-84407-319-1 (hbk)
ISBN-13: 978-0-41585-202-9 (pbk)

Typesetting by Composition and Design Services
Cover design by Mike Fell

For a full list of publications please contact:

Earthscan

2 Park Square, Milton Park, Abingdon, Oxon OX14 4RN

Simultaneously published in the USA and Canada by Earthscan

711 Third Avenue, New York, NY 10017

Earthscan is an imprint of the Taylor & Francis Group, an informa business

First issued in paperback 2013

Earthscan publishes in association with the International Institute for Environment and Development

A catalogue record for this book is available from the British Library

Library of Congress Cataloging-in-Publication Data:

Water governance for sustainable development / edited by Sylvain Perret, Stefano Farolfi and Rashid Hassan.
p. cm.
"The idea of the present book originated during the international workshop on Water Resource Management for Local Development: Governance, Institutions and Policies (WRM2004, Loskop Dam, South Africa, 8-11 November 2004)"–P. .
ISBN-13: 978-1-84407-319-1 (hardback)
ISBN-10: 1-84407-319-X (hardback)
1. Sustainable development–Africa. 2. Water-supply–Africa–Management. 3. Water quality management–Africa. 4. Water resources development–Africa. I. Perret, S. (Sylvain) II. Farolfi, Stefano. III. Hassan, Rashid M.
HC800.Z9E594 2005
333.910096–dc22

2005036769

At Earthscan we strive to minimize our environmental impacts and carbon footprint through reducing waste, recycling and offsetting our CO_2 emissions, including those created through publication of this book.

Contents

List of Boxes, Figures and Tables

Boxes

Figures

Tables

List of Acronyms and Abbreviations

ABH	Agence de Bassin Hydraulique
ABM	agent-based model
AFEID	International Commission on Irrigation and Drainage
ANC	African National Congress
ARD	Agricultural and Rural Department (World Bank)
ARDA	Agricultural and Rural Development Association
ARWR	annual renewable water resources
BNHR	basic human need reserve
BOT	build–operate–transfer
CCVD	Communauté de Communes du Val de Drôme
CEAG	Comisión Estatal del Agua de Guanajuato
CMA	catchment management agency
CMS	crop management style
CODEVASF	Companhia de Desenvolvimento dos Vales do São Francisco e do Parnaíba (development company of the São Francisco river valley)
CSIR	Council for Scientific and Industrial Research
CV	contingent valuation
CVM	contingent valuation methodology
DD/NF	Dingleydale/New Forest
DFID	Department for International Development (UK)
DOA	Department of Agriculture (UK)
DPLG	Department of Provincial and Local Government (South Africa)
DWAF	Department of Water Affairs and Forestry (South Africa)
DWSS	Domestic water supply system
ECA	Economic Commission for Africa
EO	extension officer
ETa	actual evapotranspiration
ETo	reference evapotranspiration
ETp	potential evapotranspiration
FBW	free basic water
FDI	Foreign direct investment
GGP	gross geographic product
GNU	Government of National Unity
GWC	Ga-Mashishi Water Committee
GWP	Global Water Partnership
HYV	high yielding variety
I&D	irrigation and drainage

ICID	International Commission on Irrigation and Drainage
ICT	information and communication tool
IDA	institutional decomposition and analysis
IDSP	I&D service provider
IFAD	International Fund for Agricultural Development
IFRI	International Food Research Institute
IMT	irrigation management transfer
IWMI	International Water Management Institute
IWRM	integrated water resources management
JICA	Japan International Cooperation Agency
KKT	Karush–Kuhn–Tucker
LDR	linear division rule
LDWA	Lebowa Department of Water Affairs
LNW	Lepelle Northern Water
LPDA	Limpopo Department of Agriculture
LWUA	Lebalelo Water Users Association
MAFS	Ministry of Agriculture and Food Security
MC	management committee
MCM	million cubic metres
MCP	mixed complementarity problem
MDA	Municipal Demarcation Act
MENA	Middle East and North Africa
MSC	Mkoji Sub-Catchment
MWC	Moeding Water Committee
NEPAD	New Partnership for Africa's Development
NGO	non-governmental organization
NIMBY	not in my back yard
NIMP	National irrigation Master Plan
NPDALE	Northern Province Department of Agriculture, Land and Environment
NWA	National Water Act
NWRS	National Water Resource Strategy
O&M	operation and maintenance
OBA	output-based aid
OLS	ordinary least squares
OMM	operation, maintenance and management
ORMVAT	Office Régional de Mise en Valeur Agricole du Tadla
PIM	participatory irrigation management
PITC	policy induced transfer costs
PPP	public–private partnership
PSD	public service delegation
PTO	permission to occupy
PUWS	potentially utilizable water resources
RDP	reconstruction and development programme
RESIS	Revitalization of Smallholder Irrigation Schemes
RPG	role-playing game
SA	South Africa

SALGA	South African Local Government Association
SDM	Sekhukhune District Municipality
SIS	smallholder irrigation scheme
SIWI	Stockholm International Water Institute
SMC	scheme management committee
Smile	Sustainable Management of Irrigated Land and Environments
SSB	Steelpoort Sub-basin
TA	traditional authority
TD	total diversions
TLC	Transitional Local Council
URT	United Republic of Tanzania
WB	water bailiff
WC	water committee
WMA	water management area
WRC	Water Research Commission
WSA	Water Services Act (South Africa)
WSS	water and sanitation sector
WTA	willingness to accept
WTP	willingness to pay
WUA	water users association
WWAP	World Water Assessment Programme

Acknowledgements

The idea of the present book originated during the international workshop on *Water Resource Management for Local Development: Governance, Institutions and Policies* (WRM2004, Loskop Dam, South Africa, 8–11 November 2004). The workshop gathered 90 delegates from 17 different countries. About 50 papers were presented, and case studies from South Africa, Zimbabwe, Tanzania, Nigeria, Mali, Burkina-Faso, Cameroon, Morocco, Senegal and Egypt were discussed. The core idea of the workshop was to create an opportunity for exchange, discussion, and knowledge- and experience-sharing between research teams, and policy and development agents. Full papers and further information on the workshop can be drawn from the website: http://wrm2004.cirad.fr

The event received financial and institutional support from the Department of Water Affairs and Forestry of South Africa (DWAF), the Water Research Commission (WRC), the Embassy of France in South Africa, the Joint Research Unit on Water Management at Cemagref-Cirad-Ird (PCSI, UMR G-Eau), and the French section of the International Commission on Irrigation and Drainage (AFEID). We gratefully acknowledge these organizations for their generous support.

A special note of thanks is due to the following members of the organizations mentioned above for their sustained efforts in making sure the event would be a success: Eiman Karar, Derek Weston, Francois Van der Merwe and Eustathia Bofilatos at DWAF, Gerhard Backeberg and Kevin Petersen at the WRC, Samuel Elmaleh at the French Embassy, Patrice Garin and Jean-Yves Jamin at UMR G-Eau, and Henri Tardieu and Alain Vidal at AFEID.

Considering the success of the workshop, and the quality of the material presented and debated therein, it was decided to develop this book. Papers have been pre-selected and peer-reviewed. Alongside my fellow scientific editors, Dr Stefano Farolfi, and Prof Rashid Hassan, a number of international experts participated in the editing process. The following individuals must be acknowledged and thanked for their efforts, as members of the editorial committee:

Dr Martine Antona (Cirad, France); Dr Gerhard Backeberg (Water Research Commission, South Africa); Dr Felicity Chancellor (Aquademos, United Kingdom); Dr Jean-Yves Jamin (Cirad, France); Dr Damien Jourdain (Cirad, France); Mrs Eiman Karar (Department of Water Affairs and Forestry, South Africa); Dr Kevin Pietersen (Water Research Commission, South Africa); Dr Thierry Rieu (Engref, France); Prof Kate Rowntree (Rhodes University, South Africa); Dr R. Maria Saleth (International Water Management Institute, Sri Lanka); Dr Geert van Vliet (Cirad, France).

A special word of gratitude goes to Cerkia Grant, who has been a very efficient assistant to the editorial process, with the utmost judgement.

The present book is co-published by Cirad and Earthscan. I wish to express my gratitude to Rob West at Earthscan for the quality of his communication and guidance. At Cirad, the excellent support provided by Christine Rawski and Martine Séguier-Guis is noted with thanks.

The book benefited from financial support from the Embassy of France in South Africa, from Cirad and from the Joint Research Unit on Water Management at Cemagref-Cirad-Ird (UMR G-Eau). At the French Embassy, Samuel Elmaleh must be gratefully acknowledged for his constant support. At Cirad and at the UMR G-Eau respectively, Pierre-Marie Bosc and Patrice Garin managed to gather additional funds for publication. I want to sincerely thank them for that.

The University of Pretoria, and more especially Prof Johann Kirsten in his capacity as Head of the Department of Agricultural Economics, Extension and Rural Development, have been most supportive and helpful, hosting the three scientific editors, and providing administrative support to the whole process.

Ultimately, all contributing authors are to be credited for sustaining their efforts throughout the demanding editorial process.

Dr Sylvain Perret
Scientific editor
Organizer of the workshop WRM2004

Foreword

The blue planet that is our fragile and precious home has vast water resources, but only a fraction of these resources are of freshwater. It is that small volume of freshwater that must meet the needs of billions of people, the rich variety of animals and the plethora of plants that make this planet so remarkable. From the dry heat of the deserts to the melting expanse of the polar ice-caps, water traces a common thread through all living organisms: water is the key ingredient of life. Yet freshwater resources are distributed unevenly across the surface of the earth, and the geological scars of old riverbeds, now dry, provide the evidence that where there is water now, there may not always be water.

In the short history of humankind, many civilizations have been built around water – using waterways for transport and trade, using water for agriculture, using water for cultural practices that bind the soul of a society together. Where there has been a shortage of water, human ingeniousness has often transported water across long distances, from one river basin to another. Yet continuing challenges loom large in the relationship between people and water. Lack of infrastructure in developing countries means water cannot be harnessed in the interests of the poor. In developing and developed countries alike, pollution renders our limited water increasingly unfit for use, even lethal. Every year millions die from water borne diseases. Moreover, global climate change threatens altered rainfall patterns and increasingly frequent extreme weather events.

Around the world, billions of people live in poverty, eking out a desperate daily existence in the face of famine, war and an all-pervading lack of resources. The poorest of the poor are often women and women-headed households. The scourge of AIDS has exacerbated poverty and vulnerability in many parts of the world, increasing, horrifyingly, the number of children-headed households. There are many definitions of the condition of poverty, some qualitative, some quantitative. As water managers, instinctively we understand that anyone without access to a reliable source of good quality water is poor. The poor are forced to drink from contaminated rivers, to share their springs with livestock, to watch their meagre crops wilt and die when the rains fail.

Access to safe water for drinking, cooking and washing, access to water for growing crops and watering livestock, and access to water for small businesses could profoundly change the lives of many of the poor. The flow of water would carry to them the potential for healthy lives, for development, for filling hungry bellies. The flow of clean water would save millions of mothers and fathers from watching their children die of diarrhoea. The flow of water would carry hope. Access to water is not, in itself, sufficient to eradicate poverty, but it is a necessary condition of the bigger process of sustainable development.

Africa is particularly vulnerable to these challenges. Much of the continent is vulnerable to droughts. Much of the continent is dry. Most of the continent is underdeveloped. Most of the population lives in poverty. Within this context, many water managers are performing remarkable feats at local, regional and national levels, finding creative and innovative ways to manage water in the interests of sustainable development. Some of these feats have borne remarkable success, while others have failed. We learn through both success and failure. The most important step is to learn and to keep trying.

Over the past ten years, South Africa has profoundly reformed its water governance system. The process resulted in new policy, a remarkable National Water Act (Act 36 of 1998), which enshrined integrated water resources management in a developing country context, and a reformed set of water management institutions to implement the policy and the legislation. These institutions are now grappling with the significant challenges of implementing the new policy and, in particular, ensuring that the implementation meets the challenges of addressing poverty, and historical race and gender disadvantage. In the reform process, South Africa drew deeply from the experiences of other countries around the world, both developing and developed. We did not always agree with the directions taken in other countries, and we were cautious to ensure that any lessons learned were suitable to our own context. Nonetheless, we are very conscious that we stood on the shoulders of the giants that went before us, and that we have benefited from the minds and experiences that shaped water governance around the world.

Having benefited so deeply from international expertise in shaping our water governance framework in South Africa, it was an honour to host the international workshop on *Water Resource Management for Local Development: Governance, Institutions and Policies*, from which this book has been derived.

It is often said that knowledge is power. But access to water too is access to power. It can only be hoped that the publication of this book, and the sharing of the experiences and knowledge contained in this book, will contribute to the empowerment of people, through access to knowledge, and ultimately, through access to water and the benefits that water brings.

Mrs Barbara Schreiner
Senior Executive Manager: Policy and Regulation
Department of Water Affairs and Forestry, South Africa

Introduction
New Paradigms, Policies and Governance in the Water Sector

Dr Sylvain Perret

The 20th century has seen a dramatic global restructuring of the natural hydrology towards increased water resource abstraction and storage. A supply-driven approach has been implemented throughout Africa and indeed worldwide, with massive infrastructural development, resulting in increased storage capacity, irrigation infrastructures, development of distribution and wastewater networks, improved services, quality technology, etc. For decades, such infrastructure and water resources have been centrally controlled and managed, with mostly quantitative concerns. Such processes aimed at ensuring better protection against extreme events and at meeting industrialization and urbanization demands, population growth and resulting increased agricultural production needs, and domestic water supply and sanitation needs.

Although resource development and mobilization remain crucial and feasible here and there, opportunities for further massive development seem unlikely in many countries, owing to financial issues. Moreover, it is unlikely that further significant increase in abstraction of water from nature at reasonable costs is plausible without severe environmental or social disturbance in most countries.

Unlike today's relatively stagnant supply, demand has quickly evolved, along with rapid urbanization, diversification of uses and users, raising environmental and health-related concerns, and issues regarding trans-boundary basins. These elements have motivated in-depth reforms and new policies in the water sector, which have been taking place in recent decades in many emerging and developing countries, in Africa and elsewhere.

As the resource became scarcer, users more diverse and uneven, and quality issues more acute, an alternative approach succeeded previously quantitative phases of utilization and development; resource allocation, demand and quality management, and renewed governance form the core principles of this new phase. As Barbara Schreiner reminds us in her foreword, in many places worldwide sustainable development now inescapably must rhyme with sound water management.

As elements of a renewed 'water governance', key principles and policy options have emerged:

- decentralization and the development of new forms of local governance;
- participation and the quest for greater equity;
- liberalization and the need for financial viability and economic soundness;
- overall State/public withdrawal in technical and financial terms and the need for new private–public partnerships; and
- sustainability, and especially the need for meeting environmental needs and concerns.

Such reforms have been implemented through the following:

- Integrated Water Resource Management (IWRM) policies at the river basin level.
- Irrigation Management Transfer (IMT) policies and the privatization of individual schemes.
- The emergence of new decentralized institutions, at both river basin (basin committees or Catchment Management Agencies) and local (Water Users' Associations) levels.
- The development of alternative environmental, social, economic and policy frameworks and tools (eg registration and licensing of users, water-rights markets, incentives and subsidization, ecological and human reserves, 'free basic water' principles, cost recovery and charging principles, and local business plans for common resource utilization).

Although these processes and changes have been initiated relatively recently in Africa, the continent is no exception to the above listed elements. A large number of countries now acknowledge the need for reform, indeed are already implementing reform, albeit under specific and constraining circumstances.

In Africa, 70 per cent of the poor live in rural areas, while there is an urban bias in development and resource allocation strategies. Population growth calls for increased food production. As a consequence, agricultural production will require nearly 20 per cent more water within the next 25 years, even accounting for increased production efficiency. Just 7 per cent of Africa's arable area is under irrigation compared with 33 per cent in Asia.

The 2005 report from the Commission for Africa notes that agricultural performance remains a driver of poverty trends. Irrigated agriculture is closely linked with rural poverty alleviation and rural food security. The Commission report, in line with latest New Partnership for Africa's Development (NEPAD), World Bank and Food and Agriculture Organization (FAO) thoughts, insists on the need to increase spending on physical irrigation infrastructure and to extend the area under irrigation.

Numerous experts acknowledge that need but highlight the parallel need to take into account other challenges posed by irrigation and agricultural development in Africa: water institutions, economics, policies and governance in a context of increasing resource scarcity, increasing competition between diverse users, rapid urbanization processes, marginalization of rural and peri-urban areas, and pervasive poverty, food insecurity and inequality. Moreover, agriculture in Africa already extracts 85 per cent of all water resources (although there are significant regional variations).

Such a dilemma seems currently universal. The management of and access to water resources are identified as key aspects of poverty reduction, agriculture and food security, and sustainable development in developing, transitional and developed countries worldwide.

In certain African countries – South Africa, Zimbabwe and Tanzania, among others – the State has been instrumental in reforming the water sector, in providing a conducive and enabling policy environment. Yet good 'water governance', so eagerly sought, has, in practice, proven difficult to achieve. The objectives underlying policies and reforms at the national level often appear mutually exclusive, ranging from social equity to economic efficiency and from environmental conservation to rural and urban development concerns. Decision-makers and operators often find it difficult to strike a balance by identifying proper implementation and allocation options combining all these objectives. As a result, implementation programmes sometimes experience problems, delays, and unexpected or undesirable outcomes, especially affecting the poorest and marginalized members of African societies.

This book examines that situation and the recent changes in governance, institutions, economics and policies as pertaining to water from a global point of view and a cross-country perspective, with special emphasis on African and Southern African case studies.

The book is divided into three main sections. Part I includes two keynote papers which offer a global perspective on two major issues of water governance in the 21st century: institutions and economic viability. In chapter 1, Saleth presents a generalized framework for understanding, explaining and evaluating water institutions and their change process. Beyond concepts and analysis, and drawing on recent developments in the literature, the paper further provides some major implications for the practical and policy dimensions of water institutional reforms. In chapter 2, Vidal et al report on a global study of public–private partnerships (PPPs) in irrigation and drainage. Their paper first identifies theoretical and conceptual elements of economic and financial viability in irrigation and drainage schemes and then draws data and analysis from 21 examples of existing or projected PPPs worldwide. The paper compares both theory and empirical findings on partnerships in terms of demand, offer, content, contracts and risks.

Part II includes six papers on governance, institutions and policies, at both the local level (communities and irrigation schemes) and the regional/basin level. Papers refer to case studies and examples from Mali and South Africa. In chapter 3, Döckel investigates the possibility of trade in water use entitlements in South Africa. In chapter 4, Uiterweer et al investigate the implementation of new water policies in domestic water supply in rural South Africa and the implications thereof. Veldwisch (chapter 5) and Letsoalo and Van Averbeke (chapter 6), in two different case studies in South Africa (Limpopo), and Vandersypen et al (chapter 7) in Mali (Office du Niger), investigate the emergence and functioning of local, indigenous institutions on irrigation water sharing and the interplay with processes of irrigation management transfer. Chancellor (chapter 8) investigates the establishment of Water Users Associations in South Africa and makes the case for stronger links between institutional and economic actors in water-related institution-building processes.

Part III includes eight papers on practical tools and approaches that have been used to address institutional, economic and policy-related issues regarding water governance at both the local level and the regional level, from a cross-sector perspective. Papers refer to case studies and examples from Tanzania, Morocco, Mexico, Brazil, France and South Africa. In chapter 9, Kadigi et al present and discuss two analytical frameworks that support a better understanding of conflicts over water at regional level (in Tanzania). In chapter 10, Banda et al propose and use a contingent valuation method to identify the determinants of quality and quantity values of water for domestic uses in South Africa (Steelpoort basin). In chapter 11, Kasambala et al apply the Podium simulation model in Tanzania, investigating the relationships between food security and water trends at different levels. In chapter 12, Rieu and Petitguyot integrate and model the interplay between groundwater and surface water, towards an improved management of dam release for irrigation water supply in a scheme in Morocco. Jourdain uses tools inspired from game theory in chapter 13 to evaluate the impact of institutional changes with small-scale groundwater irrigation systems in Mexico. In chapter 14, Perret proposes an action-research methodology including participatory diagnosis and prospective analysis for local empowerment and institutional development in a smallholder irrigation scheme in South Africa. In chapter 15, Chohin-Kuper et al develop a social learning approach in irrigation management in Brazil, through a role-playing game, and discuss the issues pertaining to the use of such methodology. Finally, in chapter16, Barreteau et al present a specific association of a role-playing game and an agent-based model in France, aimed at local empowerment of final users in collective decision-making.

The contributors to this volume form a diverse set of professionals, from different nationalities, disciplines and backgrounds. But all strive to contribute to a better understanding of situations in water governance, and to provide ideas for further analysis, policy development and integrated water resource management.

We hope that readers will find enough merit, ideas and challenge in the following chapters, and that an increasing number of academics and students in various disciplines will address the theoretical and empirical dimensions of water governance and water management, as two of the major global issues of this new century.

Part I

1
Understanding Water Institutions: Structure, Environment and Change Process

R. Maria Saleth

Introduction

As the limits of the physical and technical approaches to water resource management are becoming more and more transparent, policy attention is shifting increasingly towards institutional reforms. In fact, the institutional arrangements governing the water sector have experienced remarkable changes in many countries around the world, especially during the past decade or so. These changes, which are more due to purposive reform programmes than to any natural process of institutional evolution, can be observed both at the macro and national levels (eg enactment of water laws, declaration of water policies and organizational reforms) and at the micro and sub-sectoral levels (eg irrigation management transfer, informal water markets and privatization of urban supply). These changes and their implications for water resources management are well documented with varying coverage, details and perspectives (eg Easter et al, 1998; Savedoff and Spiller, 1999; Dinar, 2000; Saleth and Dinar, 2000; Gopalakrishnan et al, 2005).

Despite the critical importance of institutional reforms for enhancing the development impact of water resources management at different levels, considerable divergence persists as to the way water institutions are approached and evaluated. This obviously leads to a serious dissipation of research attention and distortion in reform policies. In the process, the critical linkages between formal and informal institutions as well as the structural and spatial linkages among local, regional and national institutions are often ignored and such institutions are treated as if they were operating independently in separate spheres and being governed by different sets of factors. The reform programmes developed from such a limited and segmented understanding of water institutions often fail to have the expected impact on water resources allocation, use and management because of their inability to exploit the tactical and strategic aspects inherent in water institutions and the course of their change process. Based on the recent work of Saleth and Dinar (2004), this chapter tries to show how a better understanding of water institutions and the course of their change process can lead to reform strategies that can advance water institutional change at different levels and contexts with minimum economic and political transaction costs.

As to its specific objectives, this chapter (a) describes the nature and features of water institutions together with their practical and methodological implications; (b) conceptualizes the internal structure and external environment of water institutions; (c) reviews the relevance of existing theories and presents alternative approaches to explaining and describing the process of water institutional changes; (d) indicates how reform design and implementation principles can promote institutional reforms by exploiting endogenous institutional features and exogenous political economy contexts; and (e) concludes with key implications for the theory and practice of water institutional reforms. As regards organization, from here on this chapter is organized, more or less, in line with the objectives listed above. Although essentially theoretical and analytical in scope, it has major implications for practical policy, especially in promoting institutional reforms needed to underpin various strategies for water resources management adopted at different levels and scales.

Water Institutions: Nature and Features

Following the general definition of institutions (Commons, 1934; North, 1990a; Ostrom, 1990), water institutions can be defined as rules that define action situations, delineate action sets, provide incentives and determine outcomes both in individual and collective decision setting in the context of water development, allocation, use and management. For analytical convenience, these rules can be broadly categorized as legal rules, policy rules and organizational rules (Saleth and Dinar, 2004).[1] Water institutions also share the same features as characterize all other institutions. First, institutions are subjective in origin and operation but objective in manifestation and impact (Hodgson, 1998).[2] Second, they are path-dependent in the sense that their present status and future direction are dependent on their earlier course and past history (North, 1990a). Third, the stability and durability properties of institutions do not preclude their malleability and diversity (Adelman et al, 1992; Hodgson, 1998). Fourth, since institutions comprise a number of functionally linked components, they are hierarchic and nested both structurally (North, 1990a; Ostrom, 1990) and spatially (Boyer and Hollingsworth, 1997). Finally, institutions are embedded and complementary not only with each other but also with their environment as defined by the cultural, social, economic and political milieu (North, 1990a).

Implications for Institutional Change

The subjective nature of institutions underlines the central role that perceptional convergence among stakeholders plays in prompting institutional change. Indeed, the perceptional convergence that can occur through the interaction of subjective and objective factors represents the origin of the demand for institutional change. Path dependency, taken together with the relative durability and stability properties of institutions, makes institutional change essentially gradual, continuous and incremental (North, 1990a). The hierarchic, nested and complementary features of institutions suggest that structural and functional linkages among institutional components are quite pervasive. In view of these institutional linkages, a change in one institutional component can facilitate both sequential and concurrent changes in other institutional components. This suggests the scope for scale economies and increasing returns in institutional change (North, 1990a). Since institutions are embedded within the general environment characterized by a configuration of social, cultural, economic and political factors, a change in one or more of these factors can also trigger institutional change. Thus, institutional change can occur due to changes both in endogenous factors (structural features within institutions) and in exogenous factors (spillover effects from the institutional environment).

Implications for Institutional Evaluation

The implications of institutional features for institutional change are well known, but, their analytical and methodological implications are either less familiar or not yet recognized. Let us focus here on two of these implications. First, the technical features of institutions are nothing but different forms of institutional linkages. For instance path dependency relates to institutional linkages in a temporal context. Similarly, the hierarchic, nested and embedded features of institutions actually characterize institutional linkages in a structural and functional sense. Although the transaction cost implications of these institutional linkages are well recognized in the literature, the role of these linkages has not been formally incorporated as part of transaction cost theory. Second, the close resemblance between the institutional system and the ecosystem allows the development of the *institutional ecology principle* (Saleth and Dinar, 2004). While this principle seems trivial, it is powerful enough to provide the conceptual basis for developing the institutional decomposition and analysis (IDA) framework.

The IDA framework is a flexible tool for analytically decomposing water institutions at various levels and contexts. For simplicity, institutional decomposition can proceed in three stages. First, following North (1990a), water institutions can be decomposed into water institutional structure (governance structure) and water institutional environment (governance framework). Second, the institutional structure can be decomposed into its legal, policy and organizational components. Finally, each of these three *institutional components* can be decomposed to highlight their underlying *institutional aspects*.[3] As the IDA framework enables the evaluation of the structural and functional linkages within and between water institutional structure and its environment, it provides valuable insights into the internal mechanics and external dynamics of water institutions. While we will later show the tactical and strategic implications of these insights for designing and implementing practical reform policies, let us turn now to describe the internal structure and the external environment of water institutions.

Water Institutions: Structure and Environment

To show how endogenous features and exogenous influences affect the performance impact and change process of water institutions, it is necessary to conceptualize their internal structure and external context. Water institutional structure includes the structurally nested and embedded legal, policy and organizational rules governing various facets of water resource management. Water institutional environment, on the other hand, characterizes the overall social, economic, political and resource context within which the water institutional structure evolves and interacts with the water sector. Under certain simplifying conditions, it is possible to provide a visual representation of both the structure and the environment of water institutions.

Water Institutional Structure

While it is possible to provide a complete description of the water institutional structure for any given national and regional context, for expositional convenience, it can be characterized in terms of some of the key legal, policy and organizational components that receive major attention in national and global debates (Saleth and Dinar, 2004). Figure 1.1 depicts such a simplified representation of water institutional structure. While Figure 1.1 is largely self-explanatory, a few

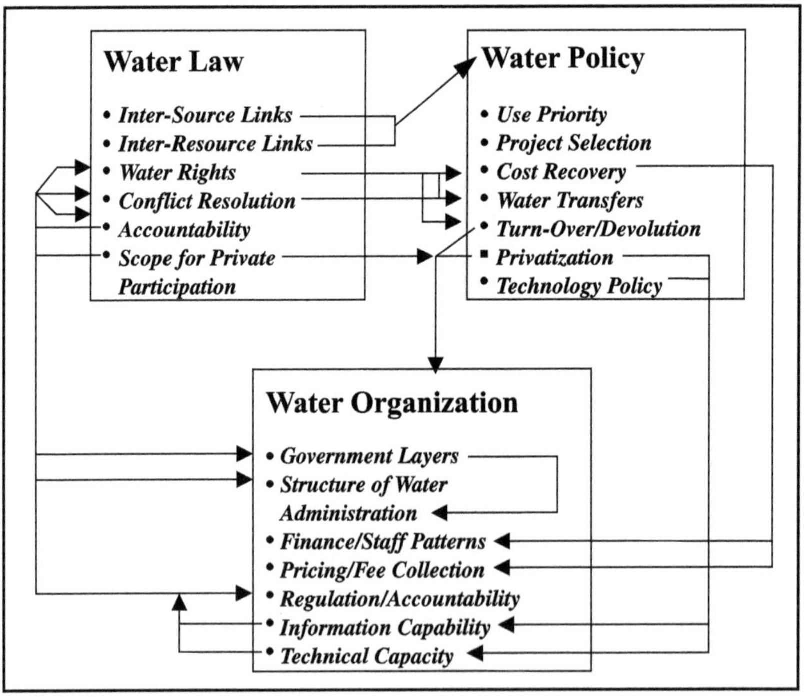

Source: Saleth and Dinar (2004)

points require attention. The overall performance of water institutions and their ultimate impact on water sector performance depends not only on the capabilities of their individual institutional components and aspects but also on the strength of structural and functional linkages among them.

The arrows in Figure 1.1 indicate an illustrative set of linkages possible both within and across the three institutional components. For instance, the legal aspects of how water sources and their relationship with land and environmental resources are treated within water law have linkages with policy aspects such as priority setting for water uses and project-selection criteria. Thus a water law that does not differentiate water by its source but recognizes the ecological linkages between water and other resources is more likely to encourage a water policy that assigns a higher priority to environmental imperatives and hydrological interconnectivity in project selection. Such a legal–policy linkage is also conducive for promoting integrated water resource management. Notably the legal aspect of water rights has multiple linkages with many other legal, policy and administrative aspects. Similarly, the policy aspects pertaining to user participation, management decentralization and private sector participation have strong linkages in terms of the ability to tap user support, fresh skills and private funds while, at the same time, contributing to devolution and debureaucratization.

Water Institutional Environment

Since institution–performance interaction in the context of water resources occurs within an environment characterized by the interactive role of many factors outside the strict confines of either water institutions or the water sector, institutional linkages and their performance implications are also subject to exogenous and contextual influences. The water institutional environment depicted in Figure 1.2, although not fully comprehensive, can enable us to conceptualize some of the pathways of these influences. Figure 1.2 has two analytical segments. The first captures the interaction between water institutions and sector performance; the other captures the general environment within which such interaction occurs. From the perspective of institutional change, the two segments represent, in fact, the two main sources from which the actual pressures for reform originate and are ultimately reflected in various media. Obviously, the first segment represents endogenous sources of institutional changes conveyed largely through economic, hydrological and institutional media whereas the second represents the exogenous sources of change conveyed mostly through social and political media.

Figure 1.2 also provides a context for contrasting the narrow approach to institutional change focused exclusively on institution–performance interaction with a broader approach focused not only on such interaction and its larger institutional context but also on the strategic role of the internal dynamics within the water institutional structure itself. As a result, the broader approach can deal with institutional change both from institutional and political economy perspectives. In this approach, therefore, the context of institution–performance interaction is as important as the internal mechanics of such interaction because of the conditioning effects of exogenous factors. In many instances, the context can even better explain why similarly structured water institutions can be responsible

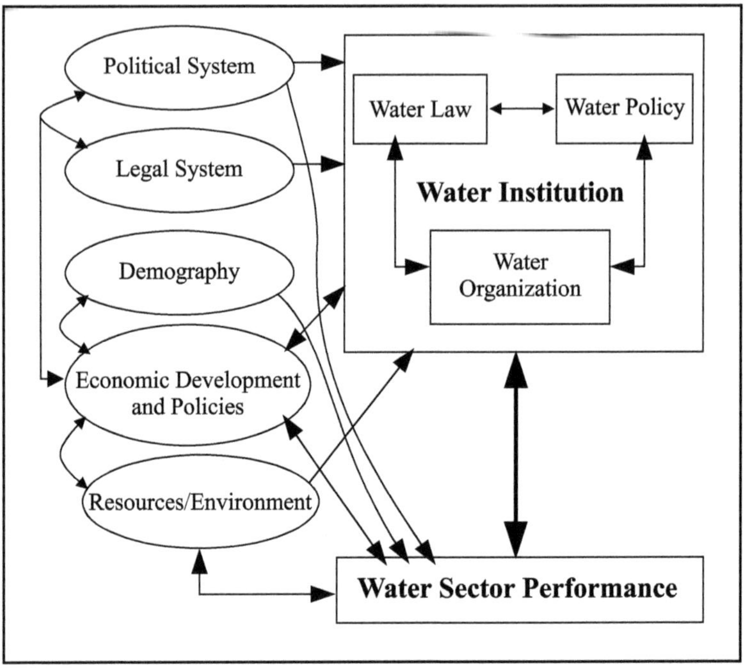

Source: Saleth and Dinar (2004)

Figure 1.2 *Water institutional environment: a partial representation*

for different water sector performance. In fact, recent international case studies show that exogenous factors (eg macro-economic crisis, political reforms, trade policies, environmental problems, international agreements and donor agencies) have played a dominant role in providing impetus for water institutional changes (Saleth and Dinar, 2000). At the same time, there are also many cases where institutional design and implementation principles based on various forms of institutional linkages (eg institutional prioritization, sequencing, packaging, and scale and timing aspects) have also been used to advance water sector reform processes.

Explaining Water Institutional Change: Role of Theories

From the conceptualization of water institutional structure and its environment as well as the underlying broader institutional and political economy approach, it is fairly clear what set of factors can explain water institutional change in different contexts. However, it is still necessary to understand how and in what forms these factors influence the process of institutional change. For this purpose, we can use some of the theories of institutional change available in the institutional economics literature. This exercise can also allow us to see which theories of institutional change are more appropriate to provide a more realistic explanation of the process of institutional change in the particular context of the water sector.

General Theories of Institutional Change

There are a wide variety of theories to explain institutional change from different perspectives and contexts. An exhaustive review of these theories, especially from the perspective of water institutional change, is provided by Saleth and Dinar (2004). For present purposes, we consider only a few major theories that can be used to explain the process of water institutional change. Broadly speaking, the theories of institutional change can be grouped into three distinct categories: evolutionary theories explaining the emergence of social conventions, market-based theories emphasizing institutional selection through competition and bargaining theories explaining institutions in terms of asymmetries of power (Knight, 1995). While the evolutionary theories explain institutional change and survival in terms of social, cultural, and economic factors, the market-based theories (eg public choice theory and transaction cost theory) consider institutional change as endogenous to the economic process itself. North (1990a and 1990b) has generalized these theories to also account for exogenous and non-economic factors such as subjective perception and ideology. As the bargaining theories focus explicitly on the distributional consequences of institutions, they rely heavily on the role of political and social bargaining as a mechanism of institutional change (Knight and Sened, 1995; Levi, 1990).

Two additional but still related categories of theories can also be considered. These pertain to the theories based on 'intentional institutional design' and those based on 'induced institutional innovation'. The former theories adopt a contractarian approach to explain institutional change as a product of free and voluntary exchange in the political market (Buchanan and Tullock, 1962). In contrast, the theories of induced institutional innovation place considerable emphasis on exogenous factors, especially the interactive effects of resource endowments, cultural conditions and technological developments (Ruttan and Hayami, 1984). Interestingly, the theory of institutions based on collective action proposed by Ostrom (1990) falls in between these two categories of theories, as intentionally designed institutions are seen as co-evolving with their social, economic and resource environment. The theory of voluntary and rational self-building of institutions proposed by Boyer and Hollingsworth (1997) also comes closer to this middle ground because organizational innovations are considered to originate from the rational calculus of individuals and firms as derived from a constantly changing economic environment.

As to their relevance for explaining water institutional change, the theories noted above are useful for explaining some, but not all, aspects of institutional change. For instance, the intentional design approach is useful to describe the drafting of water law, design of water policy, and the creation of or change in water organizations, but it cannot explain why and how these changes are brought about. While the transaction costs approach can explain these aspects, it cannot explain the distributional consequences of the change process and how these consequences are handled through the process of political bargaining. And bargaining-based logic clearly applies only to macro and formal institutions. Similarly, market-based theories are useful to explain some aspects of water institutions, but they cannot be generalized, while the transaction cost theory, as generalized by North (1990a and 1990b) and with a few additional adjustments made by Saleth and Dinar (2004), can provide a more appropriate framework to bring together various endogenous and exogenous factors affecting water institutional change. However, while this theory is useful to explain how various factors affect institutional change, analytically it is not capable of explaining the dynamics of the change process. For this, we also need other theories such as those based on subjective perception, demand–supply factors, economic and political markets, bargaining, intentional design, and organization and bureaucracy, which play varying roles in different stages of the process of institutional change.

Institutional Transaction Cost Theory

Although the transaction cost approach was originally developed purely in an economic context (Coase, 1937; Williamson, 1975), it was subsequently amended by North (1990a and 1990b) to allow for the role of the real costs associated with many non-economic and non-market aspects. While this extended framework is useful to capture the individual and interactive effects of both the economic and non-economic factors within a common analytical context, it still excludes the transaction cost implications of the endogenous institutional features. And while these implications are well recognized in the literature (eg North, 1990a; Ostrom, 1990), they have not been formally incorporated into the transaction cost framework. Saleth and Dinar (2004) not only incorporated the role of institutional linkages in their generalized institutional transaction cost approach but also recognized their strategic role as a basis for developing institutional design and implementation principles.

Let us now interpret the institutional transaction cost framework in the context of water institutional change. As already discussed, the factors influencing water institutions can be grouped into endogenous factors that are internal to the water sector and water institutions, and exogenous factors that are outside the strict confines of both the water sector and its institutions. The endogenous factors relating to the water sector include water scarcity, water conflicts, financial and physical deterioration, service levels, and water-related ecological effects (eg waterlogging and salinity). Those relating to water institution include institutional linkages and path dependency. The exogenous factors include economic development, demographic growth, technical progress, economic and political reforms, environmental crises, international commitments, donor roles, and natural disasters such as floods and droughts. Since the exogenous and endogenous factors are interrelated and their relative impacts differ by context, it is difficult either to isolate their individual roles or to generalize the direction of their effects. Nevertheless, their effects can be tracked within the institutional transaction cost framework by conceptualizing them as part of either the transaction costs or the opportunity costs of institutional change.

For water institutions, transaction costs cover both the real and monetary costs of altering the regulatory, monitoring and enforcement mechanisms related to water development, allocation and management. These costs are subject to scale economy effects from institutional linkages as well as the negative (or, occasionally, positive) effects from path dependency constraints. Similarly, opportunity costs cover both the real and economic value of opportunities foregone, which, in fact, represents the net social costs of the status quo. The opportunity and transactions costs of institutional changes are not static but change continuously due to the effects of factors both endogenous and exogenous to the water sector and water institutions.[4] As reforms initiated in the early stages brighten the prospects for downstream reforms, there are intricate linkages between the transaction costs of earlier reforms and those of subsequent reforms.[5] And while it is possible to estimate these costs and benefits *ex-ante* and *ex-post* using both quantitative and qualitative information,[6] the framework is equally valuable as an analytical tool to understand the individual and joint effects of various factors affecting the process of water institutional change (Saleth and Dinar, 2004).

A Stage-based Perspective of the Change Process

The transaction cost theory captures the role of various possible factors influencing water institution change, but it can neither depict the dynamics of the change process nor show how the same factors play diverse roles during different stages in the process of institutional change. As the theory assumes a social planner making the transaction cost calculation for a society as a whole, it cannot recognize the potential for divergence in the reckoning of the transaction costs across individuals and groups, who make the calculation on different aspects and at different stages of the change process.[7] Thus there is a need for giving explicit attention to the scope for convergence as well as how the conflicts from divergence are resolved through political bargaining and implementation adjustments. For a better and more realistic description of the process of institutional change, therefore, we need to use different theories in a complementary way. It is in this spirit that Saleth and Dinar (2004) proposed a stage-based approach as a general framework for linking different theories to provide a simple but complete description of the process of water institutional change.

Stages of Institutional Change

From a stage-based perspective of the process of water institutional change, four stages can be identified along with their underlying factors, processes and applicable theories. The states are: (a) mind change, ie the perceptional convergence among stakeholders at various levels; (b) political articulation, ie the crystallization of the demand for institutional change; (c) actual institutional change, ie the implementation of the reform programme; and (d) ultimate impact of institutional changes, ie the perceptible flow of benefits. These stages progress as a circular process which is subject to constant subjective and objective feedbacks, learning, participation and adaptations. The process is influenced both by subjective factors (eg ideology, bias and ignorance) and by objective factors (eg relative prices, technological change, and other economic, physical and political factors). It is also affected by other factors operating both at the individual and macro levels, such as instrumental evaluation, political lobbying and bargaining, information flow and learning externalities, organizational power and politics, behavioural changes and performance expectations, and public participation and influence. Notably, the circular process is not free from the influence of existing institutions partly due to their technical features, such as path dependency, and partly due to their effects on the worldview of the main actors. Figure 1.3 depicts the four-stage-based process of institutional change in a stylized form.

Despite a fair amount of simplification as to the specifics and time dimension of the reform process, Figure 1.3 is able to highlight the roles that some of the key factors play during the process of institutional change. Of the four stages, the first stage involving mind change assumes a critical significance. The mind change signifies a change in the mental construct of individuals and it gathers power when there is a critical mass of convergence as to the need, extent and direction of institutional changes.[8] This mind change and perceptional convergence occur among individuals not only as a result of their adaptive and instrumental evaluation of subjective and objective factors around them but also from the information feedbacks and learning experience they gain from existing institutions and ongoing changes. Since perceptional convergence means an implicit demand for institutional change, political entrepreneurs with an eye on electoral payoffs articulate such demand in the political spheres with their campaigns and lobbying.[9] This leads to the second stage of political debate and articulation.

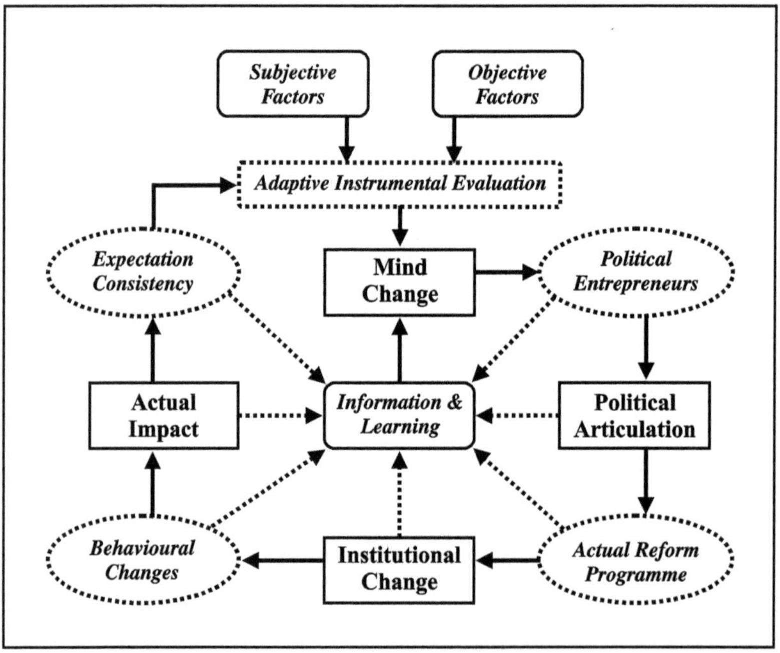

Source: Saleth and Dinar (2004)

Figure 1.3 *A stage-based conception of the process of change*

Political agreement with regard to the need for change, however, does not necessarily mean agreement as to the details and formats of change, in view of the potential for divergence in the economic and political transaction cost calculation of different social and political groups. This can lead to intense debate, bargaining and even counter-campaigns. Thus, a reform programme emerging at the end of the political process is an outcome of the relative bargaining strength of political and interest groups.[10] From the perspective of the supply and demand of institutional change, the second stage is critical as it is there that the implicit demand for change is explicitly articulated and a politically and technically consistent reform programme is developed to take the process to its third stage of reform implementation. There is considerable scope for slippage between reform implementation and actual change in institutions, especially in a democratic system. Often implementation proceeds with changes that are mostly ceremonial and procedural in nature (eg policy declarations, legislative enactments, and renaming or merging of organizations) and ends up with enthusiasm in policy-making circles but without any substantive change. However, although such a false impression may keep the demand for reform dormant for a while, even these procedural changes can also have a facilitative role both in realigning political groups and in creating a pro-reform atmosphere conducive for undertaking real and substantive changes (eg legal reform, devolution and privatization, price revisions, creation of water rights, and changes in water allocation principles).

The implementation stage is crucial as it represents the actual supply process of institutional change. In this stage, financial, organizational and bureaucratic aspects play a major role. In cases where the reform is likely to dilute the power of existing organizations, there will inevitably be some resistance. Similarly, power struggles among water-related organizations can also compromise implementation. However, political and resource commitments at the top and continued oversights and pressures from below can ensure a consistency between a reform programme and its practical translation and take the process to its fourth stage, where the ultimate economic impacts of institutional change are realized.

Even with substantive changes in institutions, its impact on economic performance is not immediate but has a very long gestation period. The direct outcome of institutional change is actually a process of behavioural changes and their ultimate outcome depends on the extent to which these behavioural changes improve the process of actual usage and exchange. The material outcome of this process is, therefore, not immediate but goes far beyond the programme period. As a result, subjective and instrumental evaluation is also important even during this fourth stage.[11] These subjective evaluations constantly feed into the process of mind change, along with the objective factors and learning experience, and get internalized within the circular process of institutional change. In fact, as can be seen in Figure 1.3, the subjective instrumental evaluation of information and learning experience plays a central role during the entire process of institutional change.

Complementary Roles of Different Theories

Having described the stage-based perspective institutional change, we can go on to show how different theories can explain the reform process in different stages of the process. Since the literature on institutional economics offers no theory to explain the process in the first stage, Saleth and Dinar (2004) proposed a subjective theory of institutional change by linking the roles that 'subjective model' of individuals (North, 1990a), 'collective attitude' (Bromley, 1989), and 'adaptive instrumental evaluation' (Tool, 1977; Bromley, 1985; Livingston, 1993) play during the process of institutional change.[12] The basic idea of this theory is that perceptional convergence among individuals or stakeholders, as induced by their adaptive instrumental evaluation of subjective and objective factors, including information and learning, represents the main source of institutional change.[13] The process in the second stage can be explained in terms of bargaining theories including the political economy theory of rent-seeking and interest group politics.

The theory of intentional institution design and contractarian approaches can be used to explain the process of reform formulation. Similarly, the process in the reform implementation stage can be explained in terms of organizational and bureaucracy theories. It is also clear how and by whom transaction costs are calculated during the reform process. The calculation is performed *ex-ante*, both by individuals and groups, in evaluating different components of the reform. It influences the convergence in the perception of stakeholders, lobbying decisions of political entrepreneurs, choice and priority of reform components by the state, and the tactics and strategies of implementation by the bureaucracy. From a general perspective, therefore, the stage-based perspective also indicates the way various theories can be used, in a complementary way, to explain the dynamics in different stages of the process of water institutional change. In this sense, the stage-based approach enables us to link various theories of institutional change within a logical framework.

Water Institutional Reforms: Tactics and Strategies

The stage-based perspective of institutional change sheds light on the role, configuration and relative significance of various subjective, economic, political and organizational factors involved in different stages of the change process. While these factors lead to changes in different stages of the process, the change process is not entirely evolutionary: deliberate and purposive policies can substantially alter or reinforce the course of institutional change. As we have seen, these policy options and reform strategies are implicit not only in the institutional features and their implications for institutional change and transaction costs but also in the mechanics of the stage-based process of institutional change. Let us briefly outline some of them.

Design and Implementation Principles

Saleth and Dinar (2004) looked at the options and strategies for promoting institutional change in terms of the reform design and implementation principles developed from the way institutions and their transaction costs are influenced by internal institutional structure and external institutional environment. For instance, the presence of various forms of structural and sequential linkages among institutional components provides the scope for developing institutional design principles such as institutional prioritization, sequencing and packaging. These principles have the ability to promote institutional reforms with minimum transaction costs and political opposition but maximum success and impact. Thus institutional prioritization enables us to target reform efforts and investments on those components having a greater probability of success and immediate performance returns. Similarly, there is also vast scope for minimizing transaction costs with alternative sequencing and packaging of reform components. For instance, the sequential linkages among institutional aspects (eg user organization, cost recovery, water rights, accountability and conflict resolution) can be used to enhance the prospects for upstream institutional changes by exploiting scale economy benefits and path-dependency properties. Reform packaging (eg linking price revisions with system improvements or quantity assurance; packaging reforms within an investment programme) also have similar effects.

Besides the institutional design principles, there are also implementation principles indicating how and when to initiate and deepen reform efforts. Obviously, these principles are based on the strategic roles played by exogenous factors. For instance, the political economy contexts provided by changes in the overall institutional environment (eg macro economic crisis, political reforms, droughts/floods or international/bilateral agreements) can be exploited with appropriate timing and scale of institutional reform programmes. Thus economic crises and natural disasters provide a favourable context for initiating even radical programmes with the least political opposition. Similarly, when water sector reform forms part of a larger political or economic reform, its overall economic and political transaction costs are likely to be lower due to synergy effects and scale economy benefits. Similarly, when water reforms are undertaken to cover more sectors and regions, such scale economy effects on the political and financial transaction costs can be substantial.[14] Donor pressures and international water-related or general agreements can also be used to promote reforms that are otherwise difficult politically.[15] Interestingly, besides their economic and technical implications, institutional sequencing and packaging also have strategic roles in relaxing political economy constraints. For instance, institutional sequencing (eg undertaking the politically easier reforms first) can bypass political opposition by concurrently creating a pro-reform climate and strengthening pro-reform groups. Similarly, institutional packaging (eg combining reform options favouring different groups) can also help in building pro-reform political coalitions (see While, 1990; Haggard and Webb, 1994). Within the context of the stage-based process of institutional change, these roles are critical for sustaining the change process, especially during the second and third stages dominated by political factors.

Learning, Research and Institutional Supply

Besides the design and implementation aspects, the reform process can also be influenced by altering the general reform climate. The pivotal status of learning and information in Figure 1.3 clearly suggests the important roles that education, information and knowledge can play during the entire process of institutional change. Research can affect not only the demand side of institutional change by providing more information and knowledge products but also its supply side by providing institutional options and implementation strategies based on country-specific and international reform experiences. Unfortunately, the supply side of institutional change is a less studied aspect in institutional economics and the role of research is an underestimated dimension in actual reform policies (Saleth and Dinar, 2004). Nevertheless, the role of institutional supply is becoming more and more important as purposive changes in institutions are critical in order to shortcut the long process of natural evolution and to shape the direction of the ongoing process of institutional change.

From the perspective of institutional supply, the state also plays a major role as a source of institutional change. In fact, current institutional economics assigns a key role to the state in creating and enforcing stable systems of property rights; it is these institutionalizing functions that enable the state to reduce the transaction cost per unit of exchange (North, 1990a). Such a scale economy effect makes the state a more efficient mechanism than other governance arrangements for lowering the overall transaction costs of both market and non-market institutions (Eggertsson, 1996).[16] In addition to these maintenance roles, the state also has a more active role as the supplier of institutional change (Alston, 1996).[17] Moreover, given the coercive and persuasive powers of modern states, they can also play mediating and facilitative roles, particularly in avoiding dead ends possible in markets and negotiations (North, 1981 and 1990a). Unfortunately, researchers and donors often ignore this fundamental institutional role of the state and take an angular position of viewing the state in terms of conflicting with market and private initiatives. Donor and funding agencies, national and international research and technical organizations, and multilateral and bilateral economic and political agreements can also affect the supply side of institutional changes both in general and in water contexts (White, 1990; Haggard and Webb, 1994; Saleth and Dinar, 2004).

Conclusion

Utilizing the theories and principles of institutional economics and the recent work of Saleth and Dinar (2004), this chapter has presented an alternative but theoretically consistent way of understanding, explaining and evaluating water institutions and the course of their change process. Although its focus is on the macro and formal segments of water institutions, the framework based on IDA, a transaction cost approach and a stage-based perspective of change can be generalized and applied to basin and local contexts. Despite its analytical and theoretical orientation, the chapter does have major implications for the practical and policy dimensions of water institutional reforms. Specifically, the endogenous and exogenous features of institutions are used to derive institutional design and implementation principles which have considerable tactical and strategic advantages not only in exploiting institutional linkages and synergies but also in countering the political economy constraints for change. Similarly, the stage-based perspective of institutional change is used to demonstrate not only the variations in the configuration of factors operating in different stages of the change process but also the centrality of stakeholders' learning and information during the entire process of institutional change. This demonstration clearly underlines the roles of education, research and institutional supply as specific strategies for setting the general reform climate promoting institutional change.

Although the standard explanation for the absence or lack of reform is linked with transaction costs, political economy constraints and stickiness of institutions, this chapter shows how these constraints can be overcome through careful reform strategies and public policies. Indeed international review of recent reforms clearly suggests that many countries have, in fact, advanced their water sector reforms through a clever exploitation of the opportunities and contexts provided by the very factors that are usually considered to constrain the reform process. Although the demand-side role of education and the supply-side role of research are now increasingly recognized, they are not yet getting the policy attention they deserve because of the mistaken view that their effects are slow, remote and marginal. Considering the fact that ambiguity in the understanding and divergence in the interpretations of institutions often constitute the initial but critical stumbling blocks for institutional reforms in many contexts, the institutional roles of education and research can be immediate, substantial and indispensable. The way water institutions and their change process are conceptualized here can be seen as a starting point for developing institutional learning and evaluation tools to facilitate a better and more consensual understanding of institutions among both the public and policy-makers. Public policy in this respect is as least as important as the design and technical aspects of water sector reform policies.

Notes

1 These categories, in fact, correspond to the three rule groups identified by Ostrom (1990) in the context of local resource management institutions: constitutional-choice rules, collective-choice rules and operational rules. Laws are seen as the outcome of constitutional choice and policies the result of collective choice through political process, while organizations are the arms for implementing, enforcing and monitoring laws and policies.

2 The subjective nature of institutions is evident as they are treated as 'belief system' (Veblen, 1919), 'behavioural habits' (Commons, 1934), 'mental construct' or the 'subjective model' of individuals (North, 1990a), or 'artifacts' that think and act through human media (Douglas, 1986; Ostrom, 1990).

3 While decomposition can proceed further to the point of primordial rules and conventions forming the building blocks of institutional aspects, it stops at the third stage, as most of the theoretically and policy-relevant issues can be addressed by the IDA framework based on this three-stage decomposition.

4 For instance, as water scarcity becomes acute, the real and economic costs of inappropriate water institutions tend to rise. Similarly, these costs are relatively high in the early stages of

reform but tend to decline as institutional maturity through stronger institutional linkages facilitates further changes.

5 For instance, with the establishment of transferable water rights, the prospects for other institutional aspects such as conflict resolution, water markets and cost recovery become brighter in view of the linkages that the transaction costs of the latter aspects have with those of water rights. Scale economies are also possible in transaction costs as the costs of effecting water institutional changes will be lower when water sector reform forms part of a larger country-wide economic and political reform programme.

6 For instance, the performance implications of institutional linkages can be quantified in terms of impact transmission coefficients (Saleth and Dinar, 2004) and the political economy constraints can be captured in terms of political risks (Dinar, Balakrishnan and Wambia, 2004). Such a reckoning of these costs and benefits need not cause any distortion or dilute the neoclassical rigor of the transaction cost theory.

7 For instance, a water user considers an institutional change in terms of its socio-economic benefits and costs whereas a political leader considers the same in terms of electoral prospects or rent-seeking possibilities. In both cases, the calculation can also vary between people and over time.

8 Perceptional convergence also has implications for both the overall costs of and the ultimate gains from institutional transactions. The magnitude of this effect, however, depends on the extent that changes in subjective perception lead to actual changes in attitudes and behaviour.

9 The issue of whether such initiatives – considered as public goods – will be taken by political entrepreneurs depends not on any *ex-post* benefit-cost analysis but on their *ex-ante* perception of a tangible political benefit to themselves or to their political parties (Knight and Sened, 1995).

10 But the bargaining process is not free from the financial implications of reform, changing economic and resource realities, or the technical and financial constraints imposed by factors such as path dependency, donor conditionalities and international agreements.

11 Thus perceived behavioural changes can be used to evaluate the effectiveness of institutional changes and the perceived gap between expected and actual performance can be used to evaluate the impact of institutional changes.

12 The process of 'mental accounting', in which people organize the outcomes of transactions and evaluate them relative to a 'reference point' (Kahneman and Tversky, 1984), can be identified as the mechanism that is used by individuals for adjusting their subjective evaluation. The reference point can be either the instrumental values or status quo outcomes, or both.

13 While perception can diverge due to the effects of ideology, bias and ignorance, it can also converge due to cultural influences and the persuasive powers of the state or other moral authorities (Bates, 1994) and the powerful effects of interaction, information, experience and learning (North, 1990a).

14 South Africa provides an interesting case for the scale economy benefits of both undertaking a nationwide water reform and its packaging with the overall economic and political reform programme. The relative transaction costs of water reforms have been reduced due to wider coverage and strategic links with the national reconstruction programmes.

15 Donor pressures can cut both ways as they are used as much to promote reforms (eg India) as to derail the reform proposals (eg Sri Lanka). Instances for the way international agreement can affect the reform prospects include the role of the Water Framework Directive in the European Union and that of the subsidy reduction requirements of general trade agreements under the provisions of the World Trade Organization.

16 While it is good to develop and supply institutions that strengthen voluntary and market-based exchanges, it is not necessary to leave institutional supply entirely to the market. This is because society also needs non-market and purposively designed institutions (Boyer and Hollingsworth, 1997).

17 For instance, economic development in Western Europe is historically connected with the state's role in establishing and enforcing property rights, weights, measures, legal institutions, banking institutions and capital markets (North, 1990a).

References

Adelman, I., Morris, C. T., Fetini, H. and Golan-Hardy, E.(1992) 'Institutional change, economic development and the environment', *Ambio,* vol 21 no 1, pp106–110

Alston, L. J. (1996) 'Empirical work in institutional economics: An overview' in Alston, L. J., Eggertsson, T. and North, D. C. (eds) *Empirical Studies in Institutional Change*, Cambridge University Press, Cambridge, MA

Bates, R. H. (1994) 'Social dilemmas and rational individuals: An essay on the new individualism', unpublished manuscript

Boyer, R. and Hollingsworth, J. R. (1997) 'From national embeddedness to spatial and institutional nestedness' in Hollingsworth, J. R. and Boyer, R. (eds) *Contemporary Capitalism: The Embeddedness of Institutions*, Cambridge University Press, Cambridge, UK

Bromley, D. W. (1985) 'Resources and economic development', *Journal of Economic Issues*, vol 19, September, pp779–96

Bromley, D. W. (1989) 'Institutional change and economic efficiency', *Journal of Economic Issues*, vol 23, no 3, pp735–759

Buchanan, J. M. and Tullock, G. (1962) *The Calculus of Consent,* The University of Michigan Press, Ann Arbor

Coase, R. H. (1937) 'The nature of the firm', *Economica*, vol 4, no 2

Commons, J. R. (1934) *Institutional Economics*, Macmillan, New York

Dinar, A. (ed) (2000) *The Political Economy of Water Pricing Reforms*, Oxford University Press, New York

Dinar, A., Balakrishnan, T. and Wambia, J. (2004) 'Politics of institutional reforms in the water and drainage sector of Pakistan', *Environment and Development Economics*, volume 9, pp409–445

Douglas, M. (1986) *How Institutions Think?*, Syracuse University Press, New York

Easter, K. W., Dinar, A. and Rosegrant, M. (eds) (1998) *Markets for Water: Potential and Performance*, Kluwer Academic Press, Boston

Eggertsson, T. (1996a) 'A note on the economics of institutions' in Alston, L. J., Eggertsson, T. and North, D. C. (eds) *Empirical Studies in Institutional Change*, Cambridge University Press, Cambridge, MA

Gopalakrishnan, C., Biswas, A. K. and Tortajada, C. (eds) (2005) *Water Resources Management: Structure, Evolution and Performance of Water Institutions*, Springer-Verlag, New York

Haggard, S. and Webb, S. B. (1994) 'Introduction' in Haggard, S. and Webb, S. B. (eds) *Voting for Reform: Democracy, Political Liberalization, and Economic Adjustment,* World Bank, Washington, DC

Hodgson, G. M. (1998) 'The approach of institutional economics' *Journal of Economic Literature*, vol 36, no 1, pp166–192

Kahneman, D. and Tversky, A. (1984) 'Choices, values and frames', *American Psychologist*, vol 39, no 4, pp341–350

Knight, J. (1995) 'Models, interpretations, and theories: Constructing explanations of institutional emergence and change' in Knight, J. and Sened, I. (eds) *Explaining Social Institutions,* The University of Michigan Press, Ann Arbor

Knight, J. and Sened, I. (1995) 'Introduction' in Knight, J. and Sened, I. (eds) *Explaining Social Institutions,* The University of Michigan Press, Ann Arbor

Levi, M. (1990) 'A logic of institutional change' in Cook, K. S. and Levi, M. (eds) *The Limits of Rationality*, The University of Chicago Press, Chicago

Livingston, M. L. (1993) 'Normative and positive aspects of institutional economics: The implications for water policy', *Water Resources Research*, vol 29, no 4, pp815–821

North, D. C. (1981) *Structure and Change in Economic History*, Norton, New York

North, D. C. (1990a) *Institutions, Institutional Change, and Economic Performance*, Cambridge University Press, Cambridge, MA

North, D. C. (1990b) 'A transaction cost theory of politics', *Journal of Theoretical Politics*, vol 2, no 4

Ostrom, E. (1990) *Governing the Commons: The Evolution of Institutions for Collective Action*, Cambridge University Press, Cambridge, UK

Ruttan, V. W. and Hayami, Y. (1984) 'Toward a theory of induced institutional innovation', *Journal of Development Studies*, vol 20, July, pp203–223
Saleth, R. M. and Dinar, A. (2000) 'Institutional changes in global water sector: Trends, patterns and implications', *Water Policy*, vol 2, no 3
Saleth, R. M. and Dinar, A. (2004) *The Institutional Economics of Water: A Cross-Country Analysis of Institutions and Performance*, Edward Elgar, Cheltenham, UK
Savedoff, W. and Spiller, P. (eds) (1999) *Spilled Water: Institutional Commitment in the Provision of Water Services*, Inter-American Development Bank, Washington, DC
Tool, M. R. (1977) 'A social value theory in neo-institutional economics', *Journal of Economic Issues*, vol 11, December, pp823–849
Veblen, T. B. (1919) *The Place of Science in Modern Civilization and Other Essays*, Huebsch, New York
White, L. G. (1990) *Implementing Policy Reforms in LDCs: A Strategy for Designing and Effecting Change*, Lynne Rienner Publishers, London
Williamson, O. E. (1975) *Markets and Hierarchies, Analysis and Antitrust Implications: A Study in the Economics of Internal Organization*, Free Press, New York

2

Public–Private Partnership in Irrigation and Drainage: The Need for a Professional Third Party Between Farmers and Government

Alain Vidal, Bernard Préfol, Henri Tardieu, Sara Fernandez, Jacques Plantey and Salah Darghouth

Introduction

The irrigation and drainage (I&D) sector plays a vital role in the food supply as well as world economics. However, after almost a century of growth, it now faces a variety of issues, the most controversial being poor water efficiency, heavy public contribution, lack of asset maintenance and, often, socio-economic inequity. Drawing lessons on institutional and financial tools from the water and sanitation sector (WSS), the introduction of the private sector in the mostly public context of I&D is thought by many to be a possible remedy.

Wishing to spearhead and promote this idea within the I&D community, the Agricultural and Rural Department (ARD) of the World Bank decided to produce a background paper on public–private partnerships (PPPs) in irrigation and drainage, the main aspects and results of which are presented in this chapter.

A brief review of the I&D sector leads to the following problem statement: I&D has become, and will probably be more and more in the future, the main provider of the world's food needs, despite the wide range of improvement of rainfed agriculture; new investments are then needed in the I&D sector to allow irrigated agriculture to play its role in food security and poverty alleviation.

Simultaneously a major effort must be carried out to develop sound practices in the operation and maintenance of I&D systems and to relieve public budgets of these recurring charges. It is a well known fact that the heavy I&D investments are deteriorating all over the world, some of them being in such poor condition that they do not allow normal operation for production purposes; this lack of maintenance is attributed to lack of funding, which itself is a consequence of insufficient water charges, faulty fee collection and general bad management.

Numerous attempts have been made to try to find a way out of the vicious circle that less public funds and lack of asset maintenance tend to establish; farmer involvement through participatory irrigation management (PIM) and subsequent irrigation management transfer (IMT) have shown promising results.

However, in most situations, such attempts are proving insufficient, which is why the idea of trying to involve the private sector through PPPs has lately been growing, helped both by its good record in the WSS and by encouraging 'pilot' experiences in the I&D sector itself.

Analytical Framework

For a better understanding of PPP opportunities and risks in the I&D sector, an analytical framework has been designed addressing the diversity of I&D systems, their main components, their main functions and the actors that are or should be involved in their development and evolution.

I&D systems are very diverse throughout the world, depending on criteria such as scale, water resource and management; however, each world region can be characterized by a dominant type, the global dominant type being the large, collective and publicly managed system estimated to represent 50 per cent of the I&D total and by far the bulk of I&D problems.

I&D systems can be broken down into components (water mobilization, water conveyance, and water distribution and their drainage equivalents), each having its specific problems. Even more essential for analysis purposes is breaking them down into functions belonging to four main categories: (a) investment functions (investment decision, investment financing, project design and project implementation); (b) the governance functions of regulation and control (water allocation, water police, maintenance auditing and price regulation); (c) operation, maintenance and management (OMM) functions (management of water allocation, system operation, system maintenance and system management, including customer relation management); and (d) the agricultural production function, whose aim is to achieve optimal use of irrigation water.

Finally, all I&D systems can be located and moved about in the 4-box analytical support diagram which has been used in this study (see Figure 2.1) according to the public or private characteristics of their management and the level of commercial risk involved. The two dotted lines in Figure 2.1 represent indicative 'thresholds' on both variables, the vertical one particularly separating public accounting systems from private accounting ones.

Lessons from PPP in the Water and Sanitation Sector

In the experience of the WSS, PPP arrangements are, by definition, contracts between a public client and a private supplier, operator or service provider. The form of PPP contracts in the WSS are very diverse, but can be divided into two

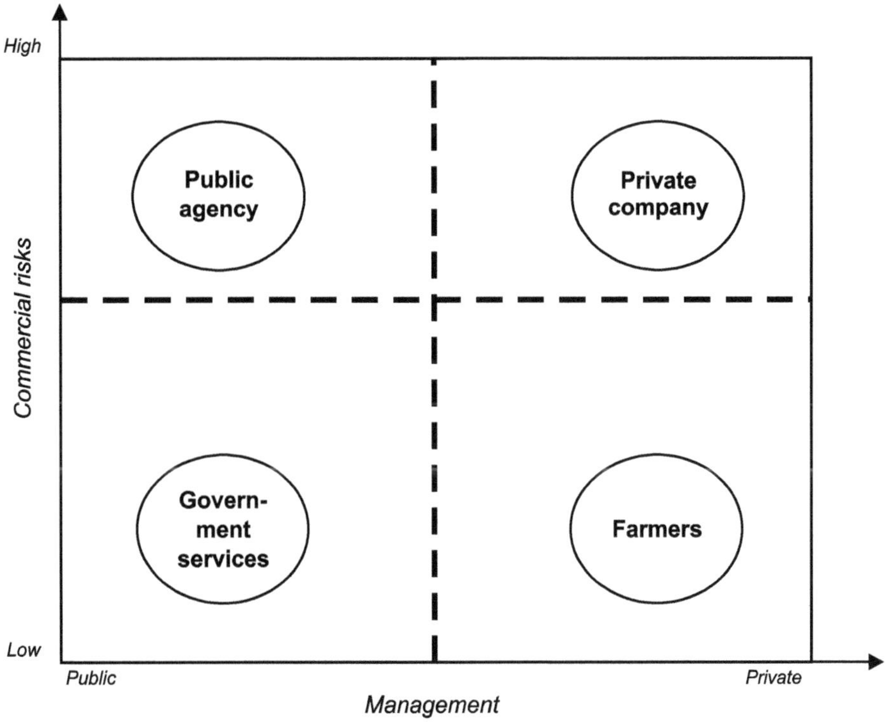

Figure 2.1 *A 4-box analytical support diagram representing the commercial risk for the operator, depending on public vs. private management*

major categories according to whether payment of the service depends on operational results or not:

1. If the private service provider is paid, usually a fixed amount, by the public client, the PPP contract is a mere public contract, which can be either partial (service contract) or comprehensive (management contract).
2. If the private service provider is paid according to the operation results, ie usually by the end user, the PPP contract is a Public Service Delegation (PSD); under this heading come the five arrangements known as lease, *affermage*, concession, build–operate–transfer (BOT) and divestiture.

The crux of this distinction is really how the commercial risk, allocated through the project agreement (Merna and Smith, 1996), of recovering the service fee is allocated between public client and private operator: in a public contract, the private operator bills the public client and gets paid, at least theoretically, regardless of the collection of the service fees, thus leaving the bulk of the commercial risk with the public client; in a PSD the private operator bills the end users and faces the major risk of having to collect service fees directly from a large number of end users.

Together, the variety of contracts and the commercial risk involved constitute one major lesson to draw from PPPs in the WSS. A second lesson is that the early successes have developed into a series of shortcomings to which the I&D sector should pay great attention. Among these, given the high capital intensity of water sector projects (Hallmans and Stenberg, 1999), is the financial issue: the low sector rate of return and the foreign exchange risk of investing in strong currencies and being paid back in weak ones (World Water Council, 2003). Socio-political risk also threatens PPP development: in order to succeed in the new water service, which generally implies a strong rise in OMM expenditure kept dangerously low hitherto through slack service management and poor asset maintenance, the private contractor must frequently choose between two unpopular measures, ie slashing the usually overstaffed former public agency when the staff is contractually transferred and/or increasing the water price, which is often a heavier burden to politicians. The private sector also needs to rely on a consistent regulatory framework for allowing large and immobile investments (Gray, 2001). PPP projects are viable only if a robust, long-term revenue stream, over the period of the concession, can be established (Grimsley and Lewis, 2002).

Few examples from the I&D sector confirm the need for strong government regulation for the inception and development of profit-oriented irrigation services (Solanes, 2002). Indeed, in most developing countries a major issue is low recovery rates for any water charges set due to lack of strong collection mechanisms or sanctions (Bosworth et al, 2002).

Nevertheless, the relevance of PPPs in the WSS for PPPs in I&D must not be overlooked as it pinpoints one major advantage and one major drawback. The main advantage for governments, particularly in developing countries, is the fact that a PPP water service becomes autonomous and can embark on long-term management improvement. The PPP's first positive impact is the introduction of improved management, based on corporate culture, not necessarily fresh funds, which are difficult to obtain in view of the length of the investment pay-back period. It also contributes to distinguishing governance from management – conflicts of interest or monopoly risks are taken care of by negotiated contracts; thanks to productivity gains, the reallocation of recurrent public funds from OMM to investment is made possible.

But the PPP issue is not one-sided: the end of subsidies entails increased water charges, which justifies the call for high value added crops rarely seen in large public irrigation systems where staple crops still dominate (Requena, 2002). Other drawbacks include a bad public image (particularly true in the Middle East and North Africa (MENA) region, where water is seen as God's free gift and not liable to pricing), limited enthusiasm from civil servants and the fact that the cost of private capital is structurally higher than that of public funds.

Emerging Experience of PPP in Irrigation and Drainage

As already discussed, while the WSS was experimenting with new financial and legal arrangements, some evolution was also taking place in I&D. If the 1950s were devoted to large development programmes, the following decades showed some concern and effort to improve the performance of the sector, with the focus switching from on-farm improvement to farmer involvement through PIM, IMT and quality service-oriented organizations. The PPP should really be viewed as a logical successor to these successive or simultaneous efforts, particularly PIM, without which little can be achieved in the way of performance enhancement (Malano and Van Hofwegen, 1999).

At the dawn of the 21st century, therefore, the PPP in I&D is already a fact, albeit a modest one, with implications for the different existing non-governmental organizational models for managing irrigation systems (Minami, 2002), each set of basic functions being treated differently: investment functions remain largely public, and so do – as they should – governance functions; OMM functions show a slight tendency toward private sector participation; while the agricultural production function remains – as it should – entirely private. Investment or OMM seem to be the two domains of potential PPPs.

The best way to check these theoretical conclusions is to test them in a number of case studies, and this has been done using 15 existing and 6 projected PPPs in I&D, analysed and modelled in terms of demand, offer, content, contracts and risks. The 21 case studies are located in Figure 2.2 and their characteristics are summarized in Table 2.1. The following were the main findings:

- PPP demand is mostly a government – department, agency or local government – initiative aimed at reducing recurrent public subsidies to I&D system operations.
- On the offer side, service providers are more reactive than pro-active, especially if the project includes private capital participation in I&D investment.

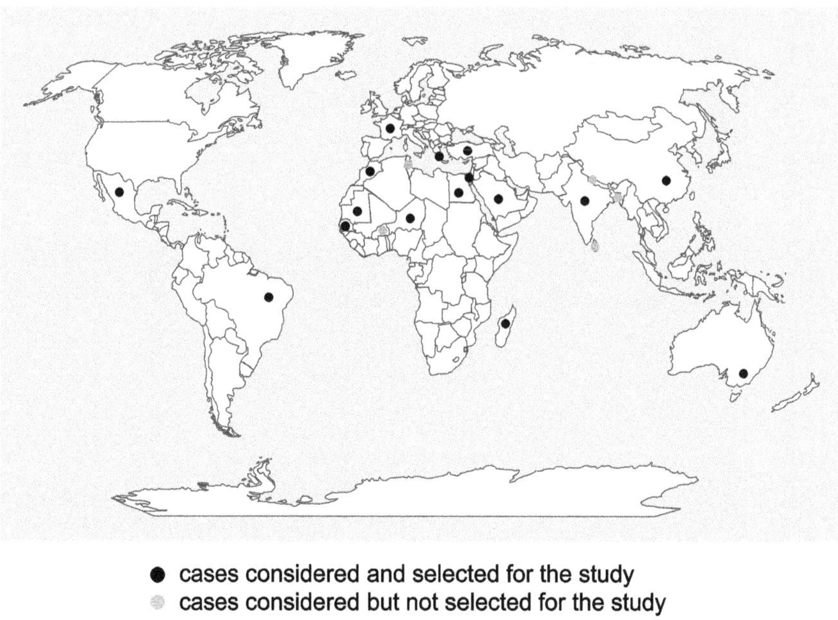

Source: chapter authors

Figure 2.2 *Location map of case studies*

Table 2.1 *Main features of case studies*

Case	*Country*[1]	*Case name*	*Progress*	*I&D function*	*Service provider*	*Client*	*Results for client*	*Major risks for supplier*
1	France 600mm	CACG[2]/WUA (23,000 ha)	Since 1972	Maintenance	*Subcontractor, local:* CACG	*Public IDSP*[3], *WUA*[4]	*Obtained:* improved reliability of water supply	*Commercial:* low (heterogeneous spare parts)
2	Senegal 200mm	SAED (40,000 ha)	Early stages	Maintenance	*Subcontractor, local:* SAED (Autonomous Maint. Division, DAM)	*Public IDSP, WUA*	*Expected:* improved reliability of water supply	*Commercial:* moderate (competition from local businesses)
3	Madagascar 1000mm	Alaotra (4,000 ha)	Starting	OMM[5]	*Subcontractor, local branch of int. company:* BRL Madagascar	*Private IDSP, WUA*	*Expected:* improved water service, improved rice yields	*Country/political:* high (no tariff protection for local rice)
4	Mauritania 200mm	Nakhlet (27 ha)	Early stages	OMM, Design	*Subcontractor, local:* to be identified	*Public IDSP, WUA:* Village 'cooperative'	*Expected:* improved water productivity	*Commercial:* moderate, related to small size
5	Brazil	Juazeiro (4,300 ha)	Early stages	OMM	*PSD*[6]: Local manpower and businesses turned into Irrigation Districts	Farmers	*Obtained:* improved water distribution, improved recovery rate	*Commercial:* high (confusion between O and M incomes)
6	Niger	Toula (350 ha)	Since 2000	OMM	*PSD to private IDSP:* local service provider (SENAGRHY)	Farmers	*Obtained:* reduced pumping costs	*Commercial:* high (water service price too high)
7	Albania	Peqin Kavaje (10,000 ha)	Early stages	OMM (water supply)	*PSD to Federation of WUAs* becoming actual IDSP	Farmers	*Obtained:* improved collection rate, improved water allocation,	*Technical/Financial:* the sustainability cost is not recovered

Case	Country[1]	Case name	Progress	I&D function	Service provider	Client	Results for client	Major risks for supplier
8	Mexico 400–1200 mm	Sonora (3,300,000 ha)	Since 1991	OMM and Modernization	*PSD to empowered WUAs* becoming actual IDSP (Irrigation districts)	Farmers	*Obtained*: improved performance of I&D, less public funds to irrigation	*Financial*: new water fees to be introduced to reduce risk *Commercial risk* for farmers
9	China 1300–2000 mm	Tieshan (25,800 ha)	Early stages	OMM, Design	*PSD to empowered WUAs* becoming actual private IDSP	Farmers	*Obtained:* better maintenance, less water use conflicts, control on maintenance funds	*Financial*: secured by water service provided to other (urban) clients *Technical*: no volumetric measurement
10	Jordan	Adasiyeh (400 ha)	Not started yet	OMM	*Expected PSD to private IDSP:* to be identified	Farmers	*Expected:* improved water distribution	*Commercial:* high recovery risk *Water demand:* conflicts
11	Morocco 200–800mm	ORMVAs[7] (375,000 ha)	Reflection stage	OMM	*Expected PSD:* to either ORMVA subsidiaries, or SEMs or fully private companies	Farmers	*Expected:* improved water distribution, more water available	*Political*: changing staff status
12	India	Eastern UP (2.4 to 3.2 m ha)	Since mid-1980s	Investment and financing, design and construction	*Take-over by private IDSP:* diesel-pump dealers replacing govt.	Farmers (plus govt subsidies for investing in tube wells)	*Obtained:* access to underground water for 800,000 poor farmers	*Financial:* very low in a (seemingly) 'win–win–win' solution
13	Turkey	GAP (100,000 ha)	Since mid-1990s	Investment and financing (IAS possibly later)	Local private irrigation equipment dealer substituting government	Farmers	*Obtained:* access to credit and irrigation equipment	*Financial:* medium to high, depending on cotton (subsidized) price

Table 2.1 *Main features of case studies*

Case	*Country*[1]	*Case name*	*Progress*	*I&D function*	*Service provider*	*Client*	*Results for client*	*Major risks for supplier*
14	Senegal 200mm	CSS (12,000 ha)	Since 1970	All functions, incl. agricultural production	Concession contract to private local business (Senegal Sugar Co, CSS)	Government. and local sugar consumers!	*Obtained:* 8,000 jobs supporting Rich.Toll's entire pop. (100,000)	No risks: monopoly enjoying free water plus automatic tariff barrier
15	Egypt 100mm	Dina Farm (4,400 ha)	Since 1987	All functions, incl. agricultural production	None	None	*Obtained:* A sustainable vegetable farming system but subsidized?	Low risk
16	Saudi Arabia 0mm	Business farms (2 m ha)	Since 1980	All functions, incl. agricultural production	None	None	*Obtained:* An increasing wheat farming system	*Political:* non renewable resource; *Financial:* stop price guarantee
17	France 600mm	SCP (80,000 ha)	Since 1960	Partnership and solidarity bet-ween water uses	Concession contract (water supply) granted to SAR[8] (SCP)	Farmers	*Obtained:* Better service quality with a reduced tariff	*Political:* Change in the socio- economic context against solidarity
18	France 600mm	CACG Neste (60,000 ha)	Since 1960	Water allocation in the Neste System	Concession contract (water resources) granted to SAR (CACG)	Farmers	*Obtained:* reduced fail rate in rivers, improved equity between farmers	*Political:* bureaucracy *Financial:* service must be coupled with OMM
19	Australia	Murray	Since 1994	Investment and OMM	PSD to a corporatized state-owned authority	Farmers	*Obtained*: Increased economic and technical efficiency	*Financial*: low turnover (1%) compared to asset
20	Egypt 0mm	Toshka (230,000 ha)	Starting	Investment and OMM	Proposed concession/ PSD contract	Farmers, Agrobusiness	*Expected:* new investment with private funding.	*Financial*: lack of private offers

Case	*Country*[1]	*Case name*	*Progress*	*I&D function*	*Service provider*	*Client*	*Results for client*	*Major risks for supplier*
21	Morocco 200mm	Guerdane (10,000 ha)	Bidding in progress	All functions (except farming)	Proposed concession/ PSD contract through ongoing call for tenders	Farmers	*Expected:* new investment with private funding, reliable water supply	*Financial risk:* high water charges (for stockholders) *Water demand:* aquifer

Notes: 1 Average annual rainfall was indicated when available
2 Compagnie d'Aménagement des Côteaux de Gascogne
3 Irrigation and Drainage Service Provider
4 Water Users Association
5 Operation, Maintenance and Management
6 Public Service Delegation
7 Offices Régionaux de Mise en Valeur Agricole: Regional Agencies for Agricultural Development
8 Société d'Aménagement Régional: Regional Development Company

- While in two-thirds of the case studies contracts concern private participation in one or more of the investment functions, 90 per cent of the contracts concern private participation in one or more of the OMM functions.
- In terms of contracts, service (and management) contracts account for 13 per cent of the total, while PSD accounts for 81 per cent of the contracts, nearly always (8/10) including some OMM functions as well.
- Finally, in terms of risks, while investment PPPs are understandably more sensitive to country risks (primarily devaluation), PSD arrangements are just as understandably more averse to commercial risks (fee collection from farmers) than service or management contracts; as regards specific water resource risks, these obviously need special allocation arrangements with the public sector.

The emerging picture of PPP in I&D is therefore that instead of being 'reduced' to either a concession/BOT type of arrangement for new projects, or the empowerment of Water Users Associations (WUAs) through IMT of existing systems, PPP in I&D is in fact much more about raising the level of professionalism in the systems considered: the important point is not so much to find an 'absolutely private' partner but rather a professional 'third party' between farmers and government, whether it be public (eg a reformed and financially autonomous government agency) or private (eg a private irrigation and drainage service provider (IDSP) looking for business or a WUA turning into a private corporation). In a more specifically development perspective, performance differences observed in irrigation systems managed (a) by farmers, (b) by private companies and (c) by parastatal agencies in Sudan also appeared to be mainly due to lack of professional capacities to deal with services (Samad and Dingle, 1995).

Conclusions and Recommendations

Four conclusions can be drawn from the above:

1 The vicious circle (insufficient maintenance–downgrading infrastructure–poor performance–low cost recovery) in I&D systems calls for more professionalism throughout all the I&D functions.
2 More professionalism is not necessarily brought in only by the private sector: whether public or private, I&D systems need a professional third party between farmers and government, especially one with a private-type accounting system.
3 Among all I&D functions, OMM functions are the most crucial to improve and should be the primary 'target' for new PPP projects, even if, in some cases, PPP on investment could be reasonably envisaged.
4 This 'professionalization' process can be conducted through service or management contracts, but preferably through PSD, which guarantees commitment, efficiency and overall coherence.

Three recommendations can be drawn from the above conclusions. The first recommendation is for PPP projects to focus on *raising the professional level of I&D*

systems, ie specific skills, experienced staff, appropriate technical means and financial reactivity, in a manner adapted to each I&D function addressed:

- At the investment level, in addition to the recommended farmer participation in project design, the future professional third party's point of view is paramount because of its future responsibility in the daily OMM of the investment: issues such as analysis of client needs, realism in modernization, and equipment standardization and maintainability should be central to any I&D investment.
- Governance functions must remain under government control, but more professionalism must also be achieved through capacity building within the public entity in charge of regulation and control.
- As for OMM functions, failures of many government agencies and difficulties met by WUAs after transfer on the one hand, and success stories on the other, are now well researched and documented: the required professionalism consists of cost-effectiveness, pro-activity and a strong sense of negotiation.
- Finally, in the agricultural production function the professionals are the farmers themselves, provided governments are convinced that they should let the farmers do the farming. This of course is also an aspect of professionalization – through the acquisition of skills and know-how by farmers, where the public sector might well still be involved through training and extension, for example.

The second recommendation is that PPP projects should also concentrate on *addressing risks properly*. Although numerous risk-mitigating tools exist (different types of guarantees, tariffs indexation and resets, transition periods, etc.), PPP risks remain significantly higher in the I&D sector than in the WSS, due to the following I&D specifics:

- The strong political and social issues relating to water, food and agricultural production maintain high country risks.
- Commercial risks – especially the non-recovery risk – remain high in countries and regions where water is considered a free gift from God and where the political wording maintains a confusion between water itself and water service (which should be paid for).
- Water specific risks are high in all countries where water is scarce and where agriculture competes with other uses that strongly affect standards of living.

To help PPP projects raise the professional level of their actors and address risks properly, the third recommendation is for the *development banks and donors to get involved in the emergence and development of PPPs in I&D*. The potential is high, as significant efforts are needed to achieve high standards of professionalism and reduce risks. The range of instruments available is broad, from technical assistance and policy advice, adjustment loans/credits and standard investment approaches, to new products such as output-based aid (OBA) and guarantees. Achieving sustained progress often takes a number of years, and banks and donors' involvement will be needed in the long term, sizing and phasing technical assistance in line with progress achieved, including, in particular, capacity building for both local public and private operators.

References

Bosworth, B., Cornish, G., Perry, C. and Van Steenbergen, F. (2002) 'Water charging in irrigated agriculture: Lessons from the literature', Report OD 145, HR Wallingford consultants for ODI, London

Gray, P. (2001) 'Private participation in infrastructure: A review of the evidence', World Bank working paper, Washington, DC

Grimsley, D. and Lewis, M. K. (2002) 'Evaluating the risks of public–private partnerships for infrastructure projects', *International Journal of Project Management*, vol 20, pp107–118

Hallmans, B. and Stenberg, C. (1999) 'Introduction to BOOT', *Desalination* no 123, Elsevier, pp109–114

Malano, H. and Van Hofwegen, P. (1999) 'Management of irrigation and drainage systems: A service approach', IHE Monograph 3, A. A. Balkema,. Rotterdam, The Netherlands and Brookfield, VT

Merna, A. and Smith, N. J. (eds) (1996) *Projects Procured by Privately Financed Concession Contract*, vols 1– 2, Asia Law and Practice Books, Hong Kong

Minami, I. (2002) 'Irrigation management entities in bank-financed projects: An overview', chapter 2 in *Institutional Reform for Irrigation and Drainage: Proceedings of a World Bank Workshop,* World Bank technical paper no 524, pp35–56

Requena, S. (2002) 'PPP in water for irrigation in MENA', World Bank draft framework paper for discussion, Washington, DC

Samad, M. and Dingle, M. A. (1995) 'Privatization and turnover of irrigation schemes in Sudan: A case study of the White Nile Pump Schemes', IWMI draft final report, Colombo, Sri Lanka

Solanes, M. (2002) 'Provision of profit-oriented irrigation services: Institutional issues', chapter 4 in *Institutional Reform for Irrigation and Drainage, Proceedings of a World Bank Workshop*, World Bank technical paper no 524, pp77–80

World Water Council (2003) 'Financing water for all', report of the World Panel on Financing Water Infrastructure chaired by M. Camdessus; report written by Winpenny, J., 3rd World Water Forum, Global Water Partnership

Part II

3

The Possibility of Trade in Water Use Entitlements in South Africa under the National Water Act of 1998

J. A. Max Döckel

Introduction

The realization that water resources are getting scarcer and that the development of new sources is only possible at great cost has changed the way in which the management of water resources is viewed (Saleth and Dinar, 1999), the emphasis shifting towards the more efficient use of the existing stock of water. Water should accordingly be considered an economic asset. Such an approach implies that the old centralized administration accompanied by the bureaucratic allocation of water must be replaced by decentralized procedures in which users can participate directly and where provision is made for the use of economic instruments such as the trade in water use entitlements.

It is against the background of this paradigm shift that a reform process in water management in South Africa took place during the 1990s (Backeberg, 1997; Department of Water Affairs and Forestry, 1997). Ultimately, a water management framework was agreed on and incorporated in the National Water Act (NWA) of 1998 (Republic of South Africa, 1998). Since then steady progress has been made in implementing the provisions of the NWA.

This chapter is based on a research project commissioned by the Water Research Commission (Conningarth, 2004) that investigated whether a formal water market could function within the framework provided by the NWA. The merits and demerits of a formal water market in permanent and temporary water use entitlements are not discussed. These issues have received full attention in the literature by authors such as Perry et al (Perry et al, 1997) and do not need to be repeated here. The issue of whether trade leads to greater efficiency was also addressed in a recent study in two areas in South Africa, namely the Crocodile

River and the Lower Orange River regions (Nieuwoudt, 2004; Gillit and Nieuwoudt, 2004). The emphasis of this chapter will be on:

- whether conditions were created that provide for a water market to function;
- the underlying principles, as well as the specific provisions of the NWA regarding water markets; and
- the way in which the trade in water use entitlements is dealt with at the practical level.

The discussion must be seen against the backdrop of the limited trade possibilities that existed under the old legislation and the absence of infrastructure necessary to trade. The challenge is to provide the necessary infrastructure to accommodate a market-based approach to water use management. Policy induced control measures on the market are needed to protect parties not directly involved in the trade and these must be designed to keep transaction costs as low as possible.

Legislative Framework Prior to 1998

The trade in water use rights was rare under the legislative framework applicable to water use prior to the 1998 NWA. The lack of trade was due to the complicated way in which the rights to water use were dealt with. For instance, rights might be based on the status of a river, which could be classified as either a private stream or a public stream. It was sometimes difficult to determine whether a specific river was a private stream, a public stream or a source of a public stream. In many cases the law spelled out the rights to the use of water but the stipulations were not definable in terms of units of measurement, reliability or priority of right.

The possibility of trade in water use rights was also dependent on whether or not the use occurred in a designated water control area. A government water control area could be declared if control of the abstraction, utilization, supply or distribution of public water were deemed to be in the public interest.

In areas not declared government water control areas all owners of riparian land shared proportionally in the normal flow of a river for irrigation and urban purposes. The upper owner had a preferential right to surplus water for irrigation and urban purposes. In such areas trading could take place directly between owners through agreements or indirectly by way of an application to the Water Court. The Court granted rights and determined the compensation to be paid to those whose rights were reduced. However, due to high transaction costs and legal uncertainty about the rights, trade was limited in such areas. In addition, there was no effective institutional framework to enforce rights and persons could abstract water to which they were not entitled by official authorization or by way of trade arrangements.

In a government water control area where no government waterworks were involved the rights to the use and control of water in public streams were controlled by the Minister of Water Affairs and Forestry. A person could use water only on the basis of a provisional right, permission or allocation. Although the law allowed for trading, it was not the policy of the State to allow the trading of permits. In many areas water rights were only provisional and as such not tradable. If a person needed water an application was made to the Department of Water Affairs and Forestry (DWAF) and, if supply were available, permission was given for the use for a specified period only.

In a water control area where government waterworks were involved, water rights could be allocated on a temporary basis or, for irrigation, on a permanent basis. The water could be supplied and distributed by canals or by way of releases into a river from which each person could extract his or her allotted share of the water. The provision of water for irrigation for each piece of land was determined on the basis of an agreed formula. No consideration was paid for this entitlement but irrigators had to pay a user fee for the provision of the water. The list reflecting the area that could be irrigated by each individual was reviewed annually.

In such cases the law allowed trading of rights to use water, subject to the approval of the Minister. Prior to 1994 trade in water use entitlements was a rare occurrence; there are various reasons for this such as high transactions costs, water rights linked to land ownership and a preference for judicial and bureaucratic allocations of water rights (Armitage and Nieuwoudt, 1999).

The Possibility of Trade in Water Use Under the NWA

The NWA does not refer to a water market specifically but creates the conditions that can accommodate such a market. The point of departure for the NWA is that water belongs to all and cannot be owned by an individual.

- An individual can only obtain the right to use water, while the State has the responsibility to allocate such rights in the public interest. The right to use water is not determined by land; it is allocated to a specific person and the right can be transferred to another person. The riparian principle, applicable under the previous legislation, no longer applies; in its place there is a framework to achieve the best possible use of water.
- A licence is issued to all water users specifying what their specific entitlement entails, ie the quantity, time of use and return flows.
- Water use entitlements are no longer permanent but are for a 5-year cycle and renewable up to a maximum period of 40 years.
- The NWA does not prohibit the creation of a market in the trade of entitlements to the use of water should it become desirable in certain areas (Conningarth, 2004).
- The state must ensure that water is used efficiently based on social and economic considerations and that the intended transfer is sustainable over the long run. The authorities also acknowledge that trade can involve third parties not directly involved in the intended trade but whose interests must be protected.

In order to attain these objectives trade will be subject to varying degrees of control. The extent to which these considerations are present in a specific transaction will be directly related to whether the intended trade is within a single user sector or between sectors, within or between the area of jurisdiction of a water services provider or within water management areas.

The consideration by the authorities of an application to trade may be on a case-by-case basis initially, but once clarity is obtained regarding the parameters that trading must adhere to it must be guided by a set of regulations. Such regulations will be in line with countries or regions where a formal water market has already been functioning for a considerable time, such as Western Australia (Banyard and Kwaymullina, 2000).

Institutional Requirements Needed for a Water Market and the Extent to Which These are Provided by the NWA

A market in water use entitlements is not likely to start spontaneously (MacDonnell, 1998). The institutions needed for a water market to function must therefore be established and supported by the authorities and accepted by the respective trading parties. A primary requirement here is the existence of a water use right that is both measurable in precisely defined terms and easily measured (Simpson, 1994). Various other conditions can be added to this basic requirement, each emphasising a different perspective. For instance, Simpson (ibid) adds conditions such as that water must be scarce, trade in water must have societal acceptance, good administrative and regulatory structure is present, water must be mobile via a delivery infrastructure, and the initial allocation of water use must be fair and equitable.

Easter and Hearne (1995) emphasise that water rights need to be established which are recorded, tradable, enforceable, separate from land and supported by mechanisms that can resolve conflicts, third party impacts and water quality issues. The water market in Chile has been in operation since 1981 and the experience gained over this period provides an excellent case study to investigate the role of water rights, institutions and markets (Schleyer and Rosegrant, 1996). A more recent study by the World Bank (Marino and Kemper, 1999) investigated institutional frameworks in Brazil, Spain and Colorado. They reduce the institutional requirements to four categories: a clear description of the rights to the use of water, the contracting mechanisms, the availability of and access to information, and the establishment of an administering and enforcement agency.

This classification will be used to assess whether the provisions of the NWA and the subsequent implementation steps provide the necessary institutional framework.

Rights to the Use of Water

The primary requirement for a market to function is that the characteristics of the product to be traded should be known. This requirement is especially relevant in the case of the use of water since the amount and the source of the water, as well as the other defining attributes, must be known and free from dispute. Water rights should therefore be definable in terms of a unit of measurement such as volume or rate of flow for a set period, the reliability of such supply and the priority of the right. These conditions are specified in section 29 of the NWA and will be specified in the use of a water licence.

All significant water uses will eventually be licensed. Licences can be issued on an ad hoc basis by individual application or by a compulsory procedure whereby all water use is reconsidered in a specific catchment area. The compulsory process involves, among other things, determining the volume of water needed for the reserve requirements, international obligations, future needs and existing lawful use. In this way claims on the water source and supply are balanced. The sequence in which compulsory licensing will take place is based on the degree of water scarcity experienced in the respective catchment areas.

The process of licensing all water use entitlements is, however, a formidable task. The speed at which it can occur depends on the financial as well as the human resources available for this purpose. According to the draft National Water Resource Strategy (NWRS) a schedule for compulsory licensing has been drawn up with the last of the catchment areas scheduled for completion only by 2021 (DWAF, 2002).

Contracting Mechanisms

Until the proposed licensing (individual or compulsory) is implemented, water use can be continued in the same way as prior to the introduction of the NWA, provided that such use was lawful.

The trade of entitlements can only take place if the use is licensed. Trade can take place according to the provisions of section 25 of the NWA. Section 25(1) provides for the temporary transfer of water use. An applicant can use some, or all, of the entitlement for a different purpose or on another property in the same vicinity for the same or similar purpose. The possibility of trade on a permanent basis is created by section 25(2). This provides the opportunity for a person holding an entitlement to sell all or part thereof. The buyer must then apply for a licence to use the water, and the sale becomes effective only once the application is granted and the licence issued.

The compulsory licensing process will take a long time to complete. In the interim, procedures must be created to allow for the reallocation of water use as needs change. One way in which such reallocation can occur is through trade in entitlements. For this purpose an ad hoc procedure has been created so that an application to trade can be considered on an individual basis before compulsory licensing has taken place. Since licensing will only be phased in slowly this procedure will be the norm for many years.

Directives have been issued to guide the process of both temporary and permanent transfer of the use of water (DWAF, 2001). A description of the steps involved to get approval for an application to trade is discussed later. Ideally, general rules specifying the conditions under which trade is allowed should be known. However, complete clarity on this matter is not evident and applications are considered on a case-by-case basis (Adv. Uys, personal communication).

Availability of and Access to Information

Information is a prerequisite for the proper functioning of any market. Although this is an obvious requirement it is difficult to comply with during the transition period as the water management system moves from a system where trade occurred only rarely to one where it is possible to trade but where all the structures and procedures are not yet in place.

In terms of the NWA the Minister must, as soon as reasonably practicable, and under the terms of section 139, establish national information systems regarding water resources. The information system may include, among other tools, a register of water use authorizations. The Minister may, under the terms of section 141, require in writing that any person must, within a reasonable given time or on a regular basis, provide the Department with data, information and documents for:

- purposes of any national monitoring network or information system; or
- management and protection of water resources.

The information contained in these systems must in turn be made available by the Minister to interested parties.

The fulfilment of the above formal requirement is in the process of implementation; provision of information regarding the trading process itself is still incomplete. Transactions that have taken place have been registered with the DWAF, but potential participants in the market are still not fully informed of the possibilities of trade and the process involved. The authorities are also experiencing difficulty in dealing with the new situation. The uncertainty created by the new system leads to a process of learning by doing, as authorities both at the central and local levels move towards implementing the new water management system.

An Administering and Enforcing Agency

A three level structure has been proposed by the NWA consisting of the national, catchment and local levels.

At the national level the DWAF will provide the national policy and regulatory framework within which other institutions will manage water resources. The DWAF will progressively withdraw from the direct involvement in water development, financing, operation and maintenance of water resource infrastructure. These functions will be transferred to other water management institutions according to the NWRS (DWAF, 2002).

At the regional or catchment level, catchment management agencies (CMAs) are provided for. These institutions will deal with water management at the regional level. Their functions include, inter alia, the development of a catchment management strategy, the coordination of activities of both water users and water management institutions within a water management area and the promotion of community participation in all aspects regarding water management in the area. It is at this level that licences will be issued and the administration of the trade will reside.

At the local level there is provision for water users associations (WUAs), whose activities are restricted to this level. These are cooperative associations of individual water users and will undertake water-related activities for the common good.

In addition, provision is made for various other bodies such as advisory committees, forums, international water management institutions and institutions for infrastructure development. The NWA provides for an appeal process to the Water Tribunal against decisions made by the various authorities regarding, among other issues, the trade in water use entitlements.

Non-compliance with the provisions of the NWA is a criminal offence. The Act includes a list of offences and associated penalties (NWA, chapter 16) and gives the courts and water management institutions certain powers associated with prosecutions such as the power to remove the cause of a stream flow reduction.

The eventual role of each of the structures to facilitate trade has not yet been determined and will most probably receive attention only as the structures become operational.

Progress Towards the Establishment of the Proposed (Decentralized) Infrastructure to Deal with Trade

It is impossible to accurately describe the progress made in establishing the structures discussed above because it is an ongoing process; the discussion here is therefore very selective. Most of the implementation plans are outlined in the NWRS (DWAF, 2004). Other information on the progress has been obtained by personal communication with various DWAF officials.

Nineteen CMAs are provided for in the NWRS. A programme for the establishment of these CMAs has been prepared. Five CMAs require priority, namely Inkomati, Olifant/Doring, Breede, Crocodile (West) and Marico and Mvoti to Umsimkulu. Two to three years have been allowed for establishing the CMAs and a further five years for establishing the executive structure of each CMA in order for it to become fully functional.

WUAs have already been established; in many cases the irrigation boards that functioned under the previous legislation have been transformed into WUAs. Further agencies will be established as the need arises. The Water Tribunal has been established and is functioning.

The difficult issue of getting agreement on water rights to water sources that form the border between South Africa and neighbouring countries is still to be dealt with. A schedule has been drawn up for discussions on international water use agreements.

The DWAF has started to verify some of the registered water uses, mainly in the Upper Vaal River catchment area and the Mhlatuzi catchment area. A project to compile a guideline that can be used to determine the lawfulness and extent of existing water use is in progress and is due for completion in 2004.

The DWAF has started to verify various tools to allocate water in the case of the compulsory licensing process, such as a tool kit for water allocation, which includes social factors to be considered when existing entitlements are curtailed as part of the licensing process. The first pilot project to test the compulsory licensing process will be undertaken in the Mhlatuze River catchment area. The lessons learnt from this process will be valuable in smoothing the future ongoing compulsory licensing process.

At present, not all the structures provided for in the NWA have yet been established. For instance, only one of the CMAs has been established (and is not yet operational), while only some of the WUAs are functioning. One of the functions of the CMAs will be to deal with applications for trade entitlements, but until the envisaged structures are in place, the DWAF will act as the responsible authority in dealing with applications for trade in water use. Similarly, until such time as compulsory licensing has progressed meaningfully, the DWAF will administer the ad hoc licensing process provided for to deal with trade applications that are needed to effect the reallocation of water according to changed priorities.

Issues that Complicate the Licensing Process

In an idealized situation where one starts out with the process of compulsory licensing in a catchment area, the following calculation steps are involved for defining the net allocable water:

- total resource availability;
- minus water for reserve requirements;
- minus water for international obligations;
- minus future water needs and inter basin transfers;
- minus water needed for strategic considerations; and
- net resource availability.

This net availability must provide for the existing lawful uses. In the fortunate situation where there is surplus water available in a catchment area, new applications can be invited or the water use could be auctioned or distributed by another mutually agreed process. If there is a deficit, reductions in entitlements will have to be introduced and reflected in the licences issued.

In cases where compulsory licensing has not taken place (and at present this applies to all cases), the luxury of such a clean slate approach is not available. Where trade is to take place it is based on an ad hoc procedure, and in such cases, ie where an ad hoc application for a licence is made, the necessary parameters specified in the formula above – such as the final reserve requirements and international obligations – may not be in place. Preliminary criteria can be used to determine how much water is available in terms of the entitlement and therefore available for trade, but so doing creates uncertainty about the terms specified in the licence. In the absence of the macro criteria, or even the preliminary parameters, uncertainty is created which can be responsible for delays in approving an application.

The common concern in the case of compulsory and ad hoc licensing is the evaluation of the existing lawful use. Before the application can be considered the validity of the existing use must be determined. In the case of an application for a proposed trade a whole range of inconsistencies in specific water use are possible, ie declarations of existing use may be an overstatement or an understatement of the original terms. The terms of the original application to register the use may not have been complied with, or the use patterns may have changed over time and bear no resemblance to the original use. The verification is even more problematic in cases where someone exercised a riparian right and is abstracting what they regard as their fair share of the normal flow in accordance with the previous legislation but does not have documentation to verify their use or entitlement. A large proportion of irrigation uses fall within this category.

The verification of existing lawful use can be categorized according to the degree of complexity. Conningarth (2004) identifies six categories here. The case of a transfer within an irrigation sector that falls under a government water scheme or schemes belonging to a WUA is the least complicated and is classified as category 1. The most complicated cases involve trade between different sectors where water is not supplied by any schemes, water users drawing water using their own waterworks, and are classified as category 6.

This classification is built mainly around the issues of whether the water use in question is secure in its attributes, how the water use is undertaken, whether trade occurs within or between sectors and whether the level of assurance of supply differs. Implicit in this classification are also issues pertaining to third party rights, the requirements of the aquatic environment, which it is proposed tend to become more complicated as fundamental attributes become more uncertain, whether transfers are inter-sectoral and variation in the level of assurance of supply.

The transaction cost of executing a trade will therefore also increase (in terms of time involved and other application costs) from category 1 to category 6. Based purely on this argument, one would expect trade in water use entitlements to be more common in the lower categories. This observation is borne out by a recent study which records that trade in the Lower Orange River, which is a government water scheme, is much less troublesome than that on the Crocodile, where this is not the case (Nieuwoudt, 2004). The associated cost of trading in the case of the Lower Orange River region was also fairly low.

Table 3.1 *Degree of complexity for evaluating existing lawful water use*

Category	*Intended transfer*	*Delivery mechanism*	*Fundamental attributes*	*Level of assurance*
1	Intra-sectoral	Govt or WUA water scheme	Secure	Same
2	Inter-sectoral	Govt or WUA water scheme	Secure	Different
3	Intra-sectoral	Owners' waterworks	Secure	Same
4	Inter-sectoral	Owners' waterworks	Secure	Different
5	Intra-sectoral	Owners' waterworks	Not secure	Same
6	Inter-sectoral	Owners' waterworks	Not secure	Different

Source: adapted from DWAF (2002)

Procedure Followed in an Application to Trade

The procedure followed in dealing with an application to trade permanent water use entitlement from buyers and the authorities is outlined in Figure 3.1. Once all the structures are in place the procedure will be handled at three levels as follows:

- The DWAF will set the national criteria such as the Reserve requirements, international obligations etc. based on the NWRS.
- The CMAs deal with aspects directly relating to the application, such as verifying the existing lawful use and the impact on the environment based on the Catchment Management Strategy. They are also the body that approves an application.
- Officials available to the specific CMA will carry out the assessment. This might involve officials at the central or regional level.

At present the DWAF handles the entire process, in which the complexity of the approval process is clearly implied in Figure 3.1.

The parameters that must be set by the DWAF are formidable, but this is a one-off occurrence. The establishment of existing lawful use is similar. Once these elements are in place in a catchment area the evaluation of a trade application will be less onerous. Each of the factors listed under the assessment process are requirements specified in section 27 of the NWA and are evaluated based on a methodology that ensures its consistent application (Conningarth, 2004). These considerations are developed to protect third parties, the environment and even the future development potential of a region from the effects of the intended trade. One of the considerations that is unique to the local circumstances is the equity issue, which is of considerable importance. The imbalance due to past racial and gender discrimination as far as ownership of water use entitlements is concerned needs to be addressed. If excess water is available for allocation previously disadvantaged persons should get priority. If surplus water is not available ways will have to be devised to provide for access. Capacity creating programmes to assist new entrants in successfully taking advantage of their acquired entitlements should accompany redistribution efforts.

If all the factors listed in Section 27 are relevant to a specific application the assessment process can be formidable. Many of the considerations may, however, not be applicable in a specific case and the assessment process can then be dealt with fairly quickly. The way in which the licensing process is dealt with by the authorities and the complexity of a particular application has a direct bearing on the transaction cost, both in terms of the time involved and supporting documentation that may be needed.

The application for a licence by a buyer must be accompanied by the following supporting documents: an agro-economic report that indicates whether the proposed trade is reasonable; a report from Land Claims Commissioner specifying the status of land claims in respect of the affected land; a WUA recommendation of proposed transfer based on practical management implications of the transfer and the state of payments of WUA accounts; a confirmation that all bondholders

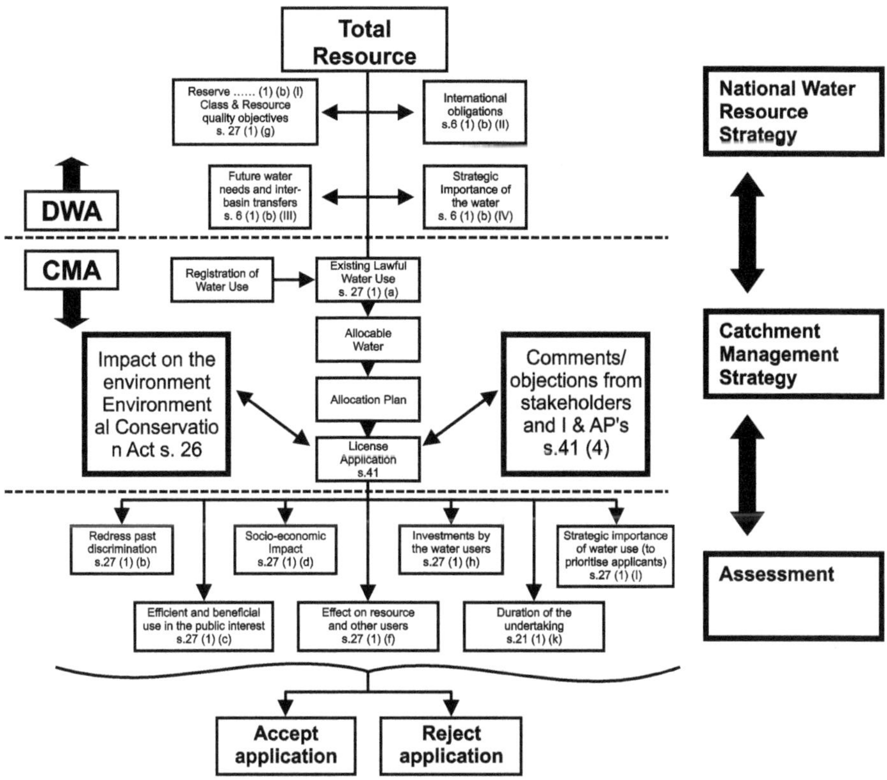

Source: adapted from DWAF (2002)

Figure 3.1 *Steps involved in assessing a licence application*

in the area have no objection; and supplementary documents, such as authorization in terms of section 22 of the Environmental Conservation Act (1989), proof of ownership etc.

The Regional Director makes a recommendation to the Head Office of the DWAF and specifically to the Chief Director: Water Use, who, after satisfying himself that all the provisions of the Act have been addressed, approves or refuses the trade with conditions that are considered appropriate.

Evidence of an Emerging Water Market

The NWA has been in force for a relatively short period and not all the institutions and criteria defined therein are yet in place. In the following discussion, therefore, it should be borne in mind that a critical analysis of the functioning of a water market is premature because of its short history and the lack of a fully developed infrastructure.

Table 3.2 *Trade since NWA entered into force*

Year	*Number of transactions*
1999	323
2000	272
2001	203
2002	223
2003 (to end of Aug)	90

Source: Personal communication with Mr Barkhuisen, DWAF

Number of Transactions

Initially, when this study was commissioned in 2002, there was doubt as to whether transactions were taking place to the extent that one could speak of a water market. Subsequently, it became clear that a substantial number of transactions have taken place since the NWA came into force. The number of permanent transfers of water use entitlements since the NWA came into force is presented in Table 3.2 above.

Based on discussions with officials at the DWAF it would appear that a substantial number of the transactions are of a one-off nature and occurred after the declaration in section 33 of the NWA came into effect. This allowed that in particular cases, unused entitlements within a Government Water Scheme or an area under the control of a WUA could be recognized as an existing lawful water use, provided that these unused entitlements were utilized within three years. Holders of these entitlements, who could not develop their properties, sold their water use entitlements rather than run the risk of losing them. The original three-year limit imposed on unused entitlements has since been withdrawn. The number of transactions that are recorded in the first year may nonetheless be artificially high due to this factor.

The largest number of transactions from 1999 to 2003 occurred in the Lower Sundays River, Lower Orange River, Middle Orange River, Orange River (Namaqualand) and Orange River (Van der Kloof).

It would appear that the number of transactions has stabilized to in the region of 200 per year. The number of transactions for 2003 represents only part of the year up to August.

Trade and the Efficiency of Water Use

The purpose of this study was to determine whether the institutional requirements for trade have been provided by the NWA. The behaviour of the traders was not specifically addressed and will form part of a follow-up study commissioned by the Water Research Commission (WRC, 2005). Some information about traders' behaviour can, however, be obtained from previous studies on water transfers in the Lower Orange River, the Nkwaleni Valley and the Crocodile River. The results confirm that transactions were motivated by the expected economic factors.

An investigation into water market transfers in the Lower Orange River (Armitage and Nieuwoudt, 1999) showed that water use entitlements were transferred to farmers with the highest return per unit of water applied. Buyers were farmers who produced table grapes and those who owned high potential arable land away from the river, without water entitlements. It is concluded that the market promoted the more efficient use of water. Buyers of water use entitlements use only slightly more water conservation technology than sellers. The reason for this difference is that the opportunity cost faced by the buyers is higher because the cost of the transaction must be added. Trade involved only unused water use entitlements while water saved (through adoption of conservation practices) was retained, possibly to attain a higher level of assurance.

The Lower Orange River was also investigated by Gillit and Nieuwoudt (2004). Results indicated that farmers held more water entitlements than their actual irrigated area. Sellers had, on average, about 22 per cent more hectares of water entitlements than actual area planted, while buyers had 41 per cent more water use entitlements. Few temporary transactions took place because farmers need long-term security of water for perennial crops. Excess water entitlements are held usually for future development and not necessarily for insurance against drought. Few water shortages have occurred during the past ten years, which respondents attributed to the stabilizing influence of the Van der Kloof Dam. A principal component of variables indicated that the purchase of water entitlements was positively associated with table grape production, advanced irrigation technology, income per cubic meter of water applied and negatively associated with area planted to field crops and other grapes.

Policy risk and risk aversion also appear to be important in explaining future investment in irrigation farming in the Lower Orange River. Results show that farmers who view their water entitlement as less secure expect to invest less. The implication is that better information about the practical implications of the NWA and specifically water licences will increase efficiency.

Studies on the Crocodile River, where water supply is irregular, show that water is transferred from farmers with high-risk crops to farmers with lower risk crops such as sugar cane. In the Lower Crocodile crops in the purchasing area have lower production risk (sugar cane) or lower financial risk and better cash flow (bananas and sugar cane). Other attributes of buyers are that they irrigate larger areas and probably have a deficit allocation (use more water than their quota), which motivates them to legitimize their current use.

In a previous study in the Crocodile River catchment area a wide range of trade prices was observed (Bate et al, 1999). This was attributed to asymmetric information between a few large buyers (4 accounted for 90 per cent of trade by volume) and many small sellers. This view is strengthened by the fact that the price paid by a large buyer differed for each of the small sellers. Water renting was also found to be more common than permanent trade in this area. The survey indicated 23 permanent trades (563 ha) and 46 temporary trades (2140 ha). Short-term leases are likely to be used on sugar cane as it is a shorter-term crop and production can be changed more quickly. Sugar cane production expanded in spite of lower relative returns per ha of land. The lower risk in the sugar industry appears to outweigh the greater profitability of other products further up the gorge. A negative externality resulting from trade is that river flow is reduced, causing increased concentration of industrial sewage and farming effluent. Several farmers sought extra water as assurance against drought, so their water use entitlement was not fully used.

A further study shows that trade will not necessarily occur in all situations. Armitage and Nieuwoudt (1999) looked at the situation in the Nkwaleni Valley in northern KwaZulu-Natal and found that no transfers of water use occurred despite the scarcity of water in the area. No willing sellers of water rights existed. Transaction costs appear larger than benefits from trading. Farmers generally retained surplus rights as security against drought because of unreliable river flow. In addition, general crop profitability in this area was similar for potential buyers and sellers since they grew the same crops.

Challenges to be Faced in Trade in Water Use Entitlements

A preliminary evaluation of issues encountered in water trade identifies three areas of difficulty. First is the newness of the system and the various teething problems encountered. Once all the criteria to approve an application for a licence to use water are in place permanent trade in water use entitlements will become more of a routine matter than at present. Second, at present there is an inevitable scarceness of knowledge and expertise to deal with the required administrative procedures. And third are the complications caused by the need to protect society, water sources, the environment and other users

Unintended Effects of Trade in Water Use Entitlements

The dormant entitlement issue

Provision is made for unused water use entitlements to be considered an existing lawful use under specific circumstances. One of the requirements is that the entitlement must be activated within a certain period or can be lost. Owners of these rights who did not have the means to use their dormant entitlements decided to sell rather than risk losing them. The buyers of these entitlements exercised their rights and, in catchment areas already under stress, this increased supply problems.

Concentration

Economic efficiency that requires that a scarce resource should be used where its marginal productivity is highest can, under certain circumstances, have negative social effects. A case in point is the pressure on small farmers to sell to large undertakings that can afford to pay high prices for water. Large owners have an incentive, especially in stressed areas, to buy more entitlements to ensure a higher degree of water supply assurance. Another example is that of the irrigation farmers on the Orange River who buy out entitlements from the Sundays River area. The water can be used more productively by, for instance, the table grape growers along the Orange River. This more efficient use of water does not take into account the threat to the economic development in the Eastern Cape that will result if this transfer is allowed.

Socio-economic and distribution impact of trade

High prices for water use entitlements provide an incentive for emerging farmers to sell their water use entitlements. The underlying reason for selling entitlements may be that other factors prevent the farming enterprises from becoming viable. The selling of water entitlements is an easy way to generate cash. In such cases strategic interventions may be necessary; it may be a case that inter-departmental cooperation between the Department of Agriculture and the DWAF is needed to ensure that small-scale agriculture can become viable. Section 27 does provide a mechanism for considering social issues when an application to trade is considered.

Administrative Issues

Present situation

There is at present a perception that the time it takes to deal with an application has increased. In some instances this is due to a lack of administrative capacity or availability of the necessary hydrological, legal and economic expertise relating to an application at the regional offices and therefore a tendency to refer matters to head office.

Slower decision-making at the central level may be due to the fact that the so-called national criteria are not yet fully in place. In addition it must be acknowledged that the time it takes to complete an application is directly related to the complexity of a specific case, as indicated by the hierarchical structure discussed earlier.

The approval process will also be slowed down as the scarcity of a specific water source increases. In such cases the third party effects will be much more prominent and require a more careful evaluation of the impact of the proposed trade. One of the ways to judge the effectiveness of the administrative procedure to deal with an application to trade is the time it takes to complete the process.

It must be acknowledged that to compare the present situation with experience elsewhere would be somewhat unfair given the initial problems associated with the licensing process. Nonetheless, given this situation it appears that the

Table 3.3 *Time taken to approve an application to trade*

Country or region	*Permanent trade (time to approve)*	*Temporary trade (time to approve)*
New South Wales*	6 to 12 months	Up to 7 weeks
Chile***	Up to 2 years	Up to 7 weeks
Colorado**	2 to 3 months	Up to 7 weeks
South Africa:		
Lower Orange**	Up to 2 months	Up to 7 weeks
Nkomati***	2 to 8 months	Up to 7 weeks
Crocodile**	No trade	Up to 7 weeks

Sources: * Ministry of Agriculture and Forestry, 2002
** Nieuwoudt, 2000
*** Conningarth, 2004

South African authorities are dealing reasonably effectively with applications to trade.

The difference in response time in the different regions in South Africa can in part be attributed to the degree of difficulty in assessing applications. The Lower Orange is a government water scheme where the water entitlements are secure and correspond to a category 1 case as described above.

The impression obtained so far is that notwithstanding the relative slowness, the present procedure for dealing with applications is satisfactory. It can also reasonably be expected that as the national criteria fall in place and the assessment procedures become institutionalized, the handling of applications will become a more routine matter.

Decentralized administrative procedures

A decentralized administrative structure for dealing with applications for a licence as proposed in the NWA and outlined in Figure 3.1 has many advantages, not least of which is the opportunity it allows for local participation and consideration of region specific issues.

A decentralized administrative structure can, however, be more expensive and very demanding due to its demand for available expertise, which includes technical and management skills which may not yet be readily available in a developing country such as South Africa.

Eventually there will be 18 CMAs, each with its own administrative structure and technical support staff to deal with regional issues including applications to trade water use entitlements. Doubt can be expressed, however, as to whether a decentralized system will, in the foreseeable future, be capable of providing the expertise needed to approve applications for trade.

Although the broad functions of the CMAs are outlined in the NWA the details are still to be worked out. One such issue is whether the governing board of the CMAs will consist of members serving on a voluntary part-time basis with the emphasis on representativeness, or whether it will be formed on the basis of expertise on water management issues. There is uncertainty as to whether the CMAs will be able to draw the required expertise if the members are not full time or at least well remunerated. The adequacy of funding for each of the 18 CMA infrastructures is also not clear since it is funded by abstraction fees and waste management fees levied in the catchment area. Waste management fees have received little attention thus far, although they could be the dominant source of funding in urban and industrialized areas.

It may be that experience gained as the decentralized process unfolds may cause a rethink of the role of CMAs with regard to trade in water use entitlements. The positive role that the CMA creates by forming a closer link with the local community must be weighed up against the probability of weaker regional administrative capacity.

It might be the case that the positive elements of the CMAs could be maintained if their functions are adjusted somewhat towards that of an integrating body that must coordinate the various environmental and planning concerns in the catchment area and serve as an information facilitating body.

Table 3.4 *Administrative clusters*

Cluster	*Headquarters*	*Area covered*
South	King Williams Town	Eastern Cape & Western Cape
East	Durban	Natal and Mpumalanga
Limpopo	Polokwane	Limpopo, Olifants and Polokwane
Central	Pretoria	Orange River, Northern Cape, Gauteng and North West

Source: adapted from DWAF (2002)

At present the lack of administrative capacity in the regions has been realized and, in response to this fact, the regional offices have been partly restructured into four clusters. Each cluster will provide the technical expertise to the area it is designated to serve.

Most probably this will be an interim arrangement whose future will depend on how the view regarding the role of CMAs unfolds.

The success of the cluster system will also depend on the ability of the authorities to create adequate capacity that will allow them to function efficiently.

Alternative model for an administrative structure to deal with water trade

A different administrative structure, such as a dedicated privatized division that can pool the necessary skills needed to deal with trade in water use entitlements, can also be considered. Whether such a structure should be established and what form one might take depends on whether the number of transactions warrants it and the extent to which the long-term vision of the water management structure unfolds.

It must also be admitted that at present the issue of trade in water use entitlements is probably one of the smaller issues in the changeover from the old to the new approach to water management.

Whether one would proceed on the way towards decentralizing the decision-making process on water trade applications or keep the more centralized model that is used at present depends on the availability of technical and managerial expertise and whether a decentralized system will turn out to be very costly. Should it prove in the end that a centralized system is considered more prudent, various alternative administrative arrangements (such as a dedicated privatized division) that will make it operate more efficiently can be considered.

Impediments to Trade

At the moment the number of transactions in the permanent trade of entitlements is relatively small. This could be because there is little need for trade or it might be due to other circumstances. It would be useful to investigate specifically from the point of view of potential buyers and sellers what they perceive the impeding factors to be.

This aspect was not investigated in the South African case but has received attention internationally. Some of the common impediments are restrictions on trade, spatially or between different user groups, physical constraints and policy induced transfer costs (PITCs). An important impediment identified by Brown

(1982) involves the level of proficiency. Although this research is somewhat old it is probably still very relevant to the case of present day South Africa. Under proficiency factors the existence of experienced market intermediaries and consultants, ease of administrative procedures and flow of market information are listed. Similar limitations have been experienced in Australia (Bjornland, 2002).

A recent survey (Nieuwoudt, 2004) reported on the perception of farmers regarding the NWA. The result is that although a majority of farmers feel positive about the NWA there is a lot of uncertainty regarding it.

In answering the question of whether conditions have improved since the introduction of the NWA the majority of farmers indicated that they were uncertain. When farmers were asked to rate whether the NWA has made trade in water use easier more than 60 per cent said they were uncertain. The majority of farmers agreed that the NWA had increased the protection of the environment.

Similarly, there is uncertainty regarding the issue of whether the five year review period of entitlements affected investment decisions and the related issue of whether the five yearly review of licences led to uncertainty.

Part of the reason for this tentative view on the NWA could be a lack of information. There is in fact a lot of information available from the DWAF, but much of the information may be too technical and cumbersome to a non-technical reader.

Policy rules indicating the situations where trade is allowed are not always clear or understood. Examples of such implicit rules are that trade is only allowed if the transaction involves water use from the same source and also a reluctance to allow transfer of water use entitlements from agricultural use to other uses.

Temporary Transfers of Water Use (the Rental Market)

Temporary trade is much less regulated than permanent trade. Section 25(1) of the NWA allows for the temporary transfer of water use. The power to approve such transactions has been transferred to the regions (and now to clusters). It needs to be noted, however, that even in the case of temporary transfers the verification of the lawfulness of the water use must be determined.

Temporary transfers frequently take place, but there is uncertainty as to the extent a formal procedure is followed. In the past the irrigation boards handled this function, but the WUAs have not been delegated to deal with temporary transfers.

In Australia, temporary transfers of water use have become a major marketing activity. The popularity of this market has led to the establishment of large temporary water market exchanges. This popularity seems to depend on the need to consider the certainty factor. Temporary transfers are much more common in the case of short-term cash crops. In the case of long-term agricultural enterprises such as fruit and wine farming, the need for certainty of water supplies may count against short-term transfers.

Because rental transactions (temporary trade) have been decentralized, no statistics from the DWAF are available on the magnitude of such trade. Preliminary discussions with regional officials indicate that this kind of trade is fairly active, but it is not always clear whether all such transactions follow the official channels.

It is a potential market that might be stimulated by extending the existing infrastructure.

Conclusion

South Africa embarked on a new water management structure that commenced with the promulgation of the NWA. It provided for, among other things, the possibility of trade in water use entitlements. The implementation of the institutional requirements for a market is, however, more complicated and time consuming. It requires the establishment of institutions, the development of rules of operation, the provision of the necessary human resources and financial resources, and the acceptance of potential market participants.

The starting point of the implementation process is the clear definition of the product to be traded. This is especially important in cases where there is transition from a system where such entitlements were not defined clearly and where, as in some cases, over-allocation of available water occurred. The process of systematically balancing various demands, such as the reserve requirements and existing lawful use, with the available water in a catchment area is done by a compulsory licensing process. It will take a considerable amount of time and effort to complete this process. In the meantime the ad hoc process of issuing licences that is required in order to deal with applications to trade will be the norm.

The establishment of the necessary infrastructure to deal with applications to trade in a decentralized structure is also not fully developed. Such process necessitates a central approval process. Applications have to be dealt with on a case-by-case basis because of the newness of the system and the fact that regulations and rules to cover all the possible complexities are not fully developed.

Notwithstanding these complications trade is taking place and, based on the available evidence, the time it takes to approve an application compares favourably with that in other countries.

References

Armitage, R. M. and Nieuwoudt, W. L. (1999) 'Establishing tradable water rights: Case studies of two irrigation districts in South Africa', *Water SA*, vol 25, no 3, pp301–304

Backeberg, G. R. (1997) 'Water institutions, markets and decentralised resource management: Prospects for innovative policy reforms in irrigated agriculture', *Agrekon*, vol 36, no 4, pp350–384

Banyard, R. and Kwaymullina, A. (2000) 'Tradable water rights implementation in Western Australia', *Environmental and Planning Law Journal*, vol 27, no 4, pp315–341

Bate, R., Tren, R. and Mooney, L. (1999) 'An econometric and institutional economic analysis of water use in the Crocodile River Catchment, Mpumalanga Province, South Africa', report no 855/1/99, WRC, Pretoria, South Africa

Bjornland, H. (2002) 'What impedes water markets?', *4th Australasian Water Law and Policy Conference*, Sydney

Brown, L., McDonald, B., Tysseling, J. and Dumars, C. (1982) 'Water realloacation, market proficiency and conflicting social values' in G. D. Whetherford, L. Brown, H. Ingram and D. Man, *Water and Agriculture in the West US: Conservation, Reallocation and Markets*, Studies in Water Policy and Management, no 2, Westview Press, Boulder, CO

Colby, B. G. (1990) 'Transaction Costs and Efficiency in Western Water Allocation', *American Journal of Agricultural Economics*, vol 72, no 5, pp1184–1192

Conningarth Economists (2004) 'The facilitation of trade in water use entitlements by the National Water Act in South Africa', Water Research Commission Report No K5/1297, Pretoria

Department of Water Affairs and Forestry (1997) *White Paper on National Water Policy for South Africa*, Government Printers, Pretoria

Department of Water Affairs and Forestry (2001) 'Temporary water use and surrendering of water use entitlements', unpublished departmental directive, Pretoria

Department of Water Affairs and Forestry (2002) 'National water resource strategy', draft paper, Pretoria, www.dwaf.gov.za

Easter, K. W. and Hearne, R. (1995) 'Water markets and decentralized water resource management: International problems and opportunities', *Water Resources Bulletin*, vol 31, no 1, pp9–20

Gillit, C. G. and Nieuwoudt, W. L. (2004) 'Private water markets in the Lower Orange river catchment of South Africa', unpublished paper, University of KwaZulu-Natal, Pietermaritzburg, South Africa

MacDonnell, L. J. (1998) 'Marketing water rights', *Forum for Applied Research and Public Policy*, vol 13, p52

Marino, M. and Kemper, K. E. (1999) 'Institutional frameworks in successful water markets: Brazil, Spain and Colorado', technical paper no 427, World Bank, Washington, DC

Ministry of Agriculture and Forestry (2002) 'Economic efficiency of water allocation', Ministry of Agriculture and Forestry, New Zealand, Appendix B, www.maf.govt.nz

Nieuwoudt, W. L. (2000) 'Water market institutions in Colorado with possible lessons for South Africa', *Water S A*, vol 26, no 1, pp27–34

Nieuwoudt, W. L. (2004) 'Water marketing in the Crocodile River, South Africa', unpublished report, Water Research Commission, Pretoria

Perkins, J. C. (2000) 'The assessment of water use authorisations and license applications', unpublished memorandum, Department of Water Affairs and Forestry, Pretoria

Perry, C. J., Rock, M. and Seckler, D. (1997) 'Water as an Economic Good: A Solution or a Problem?', Research Report no 14, National Irrigation Management Institute, Colombo, Sri Lanka

Republic of South Africa (1998) *National Water Act, Act no 36*, Pretoria, South Africa, www.dwaf.gov.sza

Saleth, R. M. and Dinar, A. (1999) 'Water challenge and institutional responses: A cross-country Perspective', policy research working paper no 2045, World Bank, Washington, DC

Schleyer, R. G. and Rosegrant, M. W. (1996) 'Chilean water policy: The role of water rights, institutions and markets', *Water Resources Development*, vol.12, no 1, pp33–48

Simpson, L. D. (1994) 'Are water markets a viable option?' *Finance and Development*, vol 31, no 2, pp30–33

Water Research Commission (2005) 'Towards the establishment of water market institutions for effective and efficient water allocation', WRC Project number K5/1569, Pretoria

4

Redressing Inequities through Domestic Water Supply: A 'Poor' Example from Sekhukhune, South Africa

Nynke C. Post Uiterweer, Margreet Z. Zwarteveen, Gert Jan Veldwisch and Barbara M. C. van Koppen

Introduction

The democratic election of a new government in South Africa in 1994 was a huge turning point not only for the country as a whole but also in terms of water distribution and services legislation. The new Constitution (1996) was followed by a Water Services Act (WSA, in 1997) and a National Water Act (NWA, in 1998). These statutes were aimed at the redressing of past inequalities based on race and gender and made poverty eradication a major guiding principle of the new nation. The application of these principles to water makes South Africa one of the few countries in which water is seen as a fundamental tool for achieving social justice and pro-poor economic growth (Van Koppen et al, 2003). The tool has two important axes: redistribution and decentralization (or democratization):

1 Redistribution is about redressing the inequitable situation that was inherited from the Apartheid era: while only about 45 per cent of the black population had piped water in 1994, about 100 per cent of the other groups could access it. Inequities in water are one symptom (and cause) of wider social inequalities: in 1994 South Africa's Gini-coefficient, which measures the degree of inequality, stood at 0.61, close to the figure for Brazil, which then had the highest level of inequality of the world (Klasen, 1997; Corder, 1997). Since the abolishment of Apartheid in 1994, however, all inhabitants have in theory been equal and all remnants of poverty and oppression in the past are to be removed. First on the agenda of the Reconstruction and Development Programme (RDP) of the first post-apartheid government was meeting basic needs, which are: 'jobs, land, housing, water, electricity, telecommunication,

transport, a clean and healthy environment, nutrition, health care and social welfare' (ANC, 1994, p7 in Corder, 1997). Free basic water for all, and a more equal distribution of water for productive means (eg agriculture, mining and industry) are seen as important instruments to achieve the new goals and the Department of Water Affairs and Forestry (DWAF) has put several mechanisms in place to try and get the relevant instruments working. Basic water needs were set at 25 litres of clean water per capita per day, from a tap no further than 200 metres from the homestead. In early 2000, just after the re-election of the ANC, the Free Basic Water (FBW) policy was launched. This policy is aimed at providing all South Africans with those water services that are regarded as basic under the RDP for free. From the standard per person, the DWAF moved to a more workable standard of 6000 litres per household per month. After this, it launched a framework around the FBW policy aimed at 'stepping up the ladder' (DWAF, 2003), improving water and sanitation services for poor households step-by-step. One part of that framework looks at cost recovery mechanisms to finance the free basic water for the poor.

2 Decentralization is about increasing and improving the participation of all citizens in the formulation and implementation of water policies. In principle, the management of water resources is regarded as a national task, with the national department, DWAF, as its main executor. DWAF Regional Offices also have a lot of ruling power in terms of water resources. For domestic water supplies, though, provincial and local government have a strong say. Most important in the terms of this chapter is that the realization of the domestic water policy targets as stipulated in the WSA largely depends on the functioning of the newly created local government bodies. In most cases, municipalities are responsible for implementing the FBW policy and for part of the cost recovery, assisted by the DWAF, the Department of Provincial and Local Government (DPLG), the National Treasury and South African Local Government Association (SALGA). Part of the financing of free basic services (including water) comes from the so-called 'Equitable Share', a tool to redistribute tax revenues both vertically (from national to local level) and horizontally (at provincial and sometimes local levels). Next to that, and starting in 2005, all DWAF personnel working on domestic water supply delivery at local levels are supposed to be transferred to the municipalities. The DWAF will then only keep the responsibility for monitoring water quality and for integrated water resources management. At the community level water distribution is to be organized by water committees. These are to be democratic and equitable, fairly representing the water users. In most communities, water committees did already exist before the WSA was accepted but functioned as one of the committees representing the traditional authority. According to the democratic guidelines the WSA stipulated, the existing committees were often not 'equitably representative' nor did they function in a democratic way.

These two policy axes are inter-related and inter-dependent: successful redistribution depends on successful democratization. This chapter aims to question this interdependence and argues that making access to water dependent on the performance of local institutions only works when conditions are favourable for such institutions to emerge and flourish. The case that is presented to support this argument, Sekhukhune District Municipality (SDM) in the Limpopo Province, is one that is strikingly poor compared to other areas in terms of reaching policy targets. Whereas in Gauteng (South Africa's richest province and business centre) almost 100 per cent of the target was reached, the overall figure for Limpopo is 44 per cent and for Sekhukhune a very minimal 0.5 per cent (DWAF, 2004).

The collection of data and information for the research on which this chapter is based was done at two levels: the policy or government level and the level of local water systems. For the first level, primarily aimed at identifying and understanding policies regarding local water supply services, use was made of the literature and interviews were held with some 30 key informants, policy-makers and implementers at different levels. For the second level, two villages in Sekhukhune (Ga-Mashishi and Moeding) were selected and field research was done in these villages. Practically all the people involved in domestic water supply in the area were interviewed. An additional and central element of the methodology was the direct observation and inspection of the water service infrastructure and management. Pumps, basins, pipelines, valves and taps were traced and assessed in terms of technical performance, and sketches of system designs and layouts were made and used to discuss the functioning of the system with people concerned. Meetings of water committees were attended. Further, interviews were conducted in a representative sample of some 40 user households in each village to assess water conditions.

After this introduction, the chapter continues with a description and analysis of the processes of water policy formulation and implementation at the level of the district municipality. The next section presents water distribution and management in the two studied villages, Ga-Mashishi and Moeding. The chapter ends with a discussion of the conditions that need to be better fulfilled to guarantee that the targets of South African's water policies are met.

Sekhukhune District Municipality

On leaving Burgersfort, a typical modern but dirty mining town, in a westward direction, one enters a completely different world. Sekhukhune is dry, red, hot and full. To drive the one asphalt road from Burgersfort to Lebowakgomo is also a dangerous undertaking as minibuses, chickens, children, water-carrying women and goats cross it unexpectedly, and all the time. There are houses and huts and people, everywhere. Some 95 per cent of the people earn less than US$100 a month (South African Census, 2001), this in the part of South Africa where most platinum and other minerals are found. This Sekhukhune is our research area.

The SDM (Figure 4.1) covers South East Limpopo Province and part of Mpumalanga and is part of the Olifants river basin, where seasonal and spatial water shortages often occur (Stimie et al, 2001). Mean annual rainfall in the basin is 631mm. The negative water balance manifests itself especially during the dry, hot season when extra water is needed for irrigation, electricity generation and domestic use. Due to the expected population increase and the improvement of living standards in the former homeland areas, it is estimated that the total urban water demands in 2010 will be twice those of 1995 (Van Veelen et al, 2001).

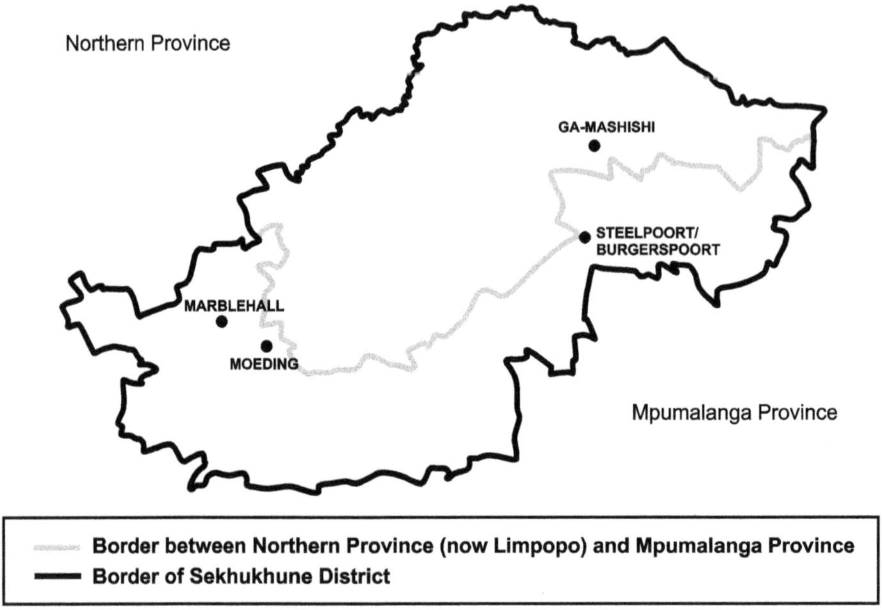

Source: Municipal Demarcation Board (MDB)

Figure 4.1 *Sekhukhune cross-boundary district municipality, showing location of towns involved in case study*

The overall condition of local infrastructure is at as low a level as the water. Communication, roads and bus connections all leave much to be desired, tremendously complicating effective lines of decision-making. The municipalities, NGOs, Lepelle Northern Water water-board and the DWAF have started to try and bridge the huge backlog in infrastructural development in the area, but progress is slow and there is still a long way to go.

Before 1994, Sekhukhune was part of the Lebowa homeland. Presently, it consists of five local municipalities and is home to 1 million people (MDB, 2003). Whereas Sekhukhuneland was one of the most powerful kingdoms in southern Africa in the 19th century, the area currently is one of the poorest in South Africa and is no longer even well known. Disputes about the status of the chief, and especially his succession, are frequent and widespread. These disputes can be seen as a result of the 'anthropological engineering' by the Apartheid regime that resulted in a multiplication of the number of chiefs in Sekhukhune from 9 to 50 in the 1970s (Oomen, 1999). Nowadays the democratically elected government officially rules, but chiefs still play an important role, often leading to tensions.

Several studies, and in particular the Northern Province Economic Development Strategy, have revealed that the lack of regular water supply is one of the major constraints hampering development in the region (Klvechuk and Jenkins, 2003). More than 64 per cent of the district population is not economically active, and only just over 13 per cent is fully employed. The regional per capita gross geographic product (GGP) in 2001 was ZAR1000 (approximately US$143), ie one-third of the average GGP in Limpopo Province and about one-sixth of the national GGP. Over 95 per cent of the population is black (MDB, 2003). HIV/AIDS is a major, but largely ignored problem.

The legacy of the apartheid era is still omnipresent: the owners and managers of mines, townhouses and commercial farms are predominantly white, while townships and the marginal dry rural areas are inhabited by blacks. The southern part of Limpopo Province (including Sekhukhune) is the worst off.

Infrastructure

Limpopo Province is among the poorest provinces of South Africa and Sekhukhune is its poorest district. The general progress in water service provision mentioned by government has not reached Sekhukhune as yet. Over 40 per cent of the villages do not even have basic domestic water supply systems or a borehole (DWAF, 2004) and rely for their water on private water vendors, on rivers or wells, or on more fortunate relatives or friends in a neighbouring village. Yet the existence of reliable infrastructure to capture and transport water from its source to the places where it is used is one important condition enabling implementation of the FBW policy.

Infrastructure development is a constant topic of discussion in the Olifants river basin. As water is scarce, and as there are a lot of stakeholders interested in getting more water for either domestic (villages, municipalities) or economic (mines, industry) reasons, dam and distribution infrastructure development is seen as one of the priorities in water resource management. At the time of study (March 2004) stakeholder participation meetings were being held by the DWAF to discuss different implementation plans regarding dams and other infrastructure. The present situation and today's discussions are mainly on the dams (developments) in the Olifants river.

At all meetings, the importance of the dams for enabling the increase of basic water provision is extensively touched on, but other factors also play a role. For industry and mine development more water is needed (personal communications, Maartens (Technical Manager), LWUA, 2003; Spockswood (consultant), 2004) and the commercial agricultural sector has launched a long-term lobby to increase the height of the Arabie or Flag Boshielo dam. These parties are often much better informed than any municipality in the region, and they are also the ones to take the initiative for instigating the participation meetings, together with the DWAF. Often consultants initiate dam or other infrastructure development in the region. In Sekhukhune, the Local and District Municipalities are overloaded with work and understaffed. Sometimes, consultancy firms from Polokwane examine the municipalities' Water Services Development Plan (often co-written by consultancies) and propose a building plan directly to the regional DWAF office, thereby bypassing the municipalities completely. When asked, one consultant easily presented two cases in which his firm did this in developing domestic water systems in the Greater Marble Hall and Greater Groblersdal municipalities (personal communications, Spockswood (consultant), 2004; Greyling (consultant), 2004). When a consultancy firm is granted permission (by the DWAF) to start construction of infrastructure, a participatory approach is still requested and always performed, and in this phase municipalities often play a role. However, although the required participation of all involved parties takes place, the real influence communities and even local government have on decision-making in infrastructure development is questionable. It is still the same rich and powerful actors as before that play the most important role. Communities and local government do not play a meaningful role in deciding on dam development. Inclusion is achieved on paper, not in practice. And this problem is exacerbated by the fact that the SDM, its local municipalities and the area in general suffer from an overall lack of means, which makes it difficult and costly to participate in meetings.

Understaffed and with Limited Experience

As mentioned in the introduction, much of the responsibility for water policy implementation lies with local governments. Much therefore depends on the abilities, motivation, skills and knowledge of local government staff, and in Sekhukhune these leave much to be desired. This is largely the result of the overall attempts in South Africa to transform its administration from predominantly white to predominantly black in a period of just ten years. The transformation time from white to black in local governments has been even shorter: only four years. Through democratic elections, affirmative action policies for 'blacks' and redundancy schemes for 'whites', local governments in both the former white areas and the former homelands, like Sekhukhune, are predominantly black. The apartheid homeland education system gave only second-class education to black people; Apartheid labour policies relegated black people to unskilled jobs: the new black personnel stand little chance of performing well given their educational background and experience. Affirmative action requirements by the new government imply that the few well-educated, well skilled black personnel often move away to the better-paying, more-challenging world of the private companies. Above all, in the former homelands, local government is a totally new institution, for which hardly any experience has been developed as yet.

The problem of the lack of skilled and experienced staff in municipalities at the local and district levels is a recognized one, and a proposal for solving it has been launched. From 2005 onwards, in a huge transfer process, all DWAF personnel working on local level water delivery, more experienced in local water supply, are to be transferred to the municipalities. They will also have to work on capacity building in local level government. However, the budgets allocated to local government do not allow financing all DWAF staff, so retrenchments are unavoidable.

As a consequence both the DWAF and local municipalities experience a staff shortage for local water operation and maintenance. In cases where the DWAF hires (white) consultants to provide technical assistance at district level, race-based distrust hampers relationships.

The resulting lack of technical support services implies that in day-to-day reality, communities are more or less left on their own when implementing water policies and managing water. They thus rely on their own capacities, skills and ingenuity to secure the water service levels the national government promised to deliver.

Reorganizations and Current Cooperation

Communication between the SDM and local municipalities is often problematic and full of tensions. During apartheid, white area municipalities were responsible for domestic water supply in urban areas of the former white areas. From 1995 onwards the transitional local councils became responsible nationwide. Often these had just the same area of jurisdiction as the Apartheid era municipalities, extended with a piece of a former homeland. In 1998 the local municipalities took over, again with the inclusion of former black areas. Evidently, domestic water supply in the poor, largely rural black areas was quite different from urban water supply, but in most cases, local municipalities had at least some experience in supplying water. The newly created district municipalities had none, but water service authority was vested in them in 2003. Staff of local municipalities in Sekhukhune are of the opinion that they can do a better job and that the SDM is not sufficiently open to advice. Many are frustrated with the disappointing performance of the SDM.

There are also tensions and potential conflicts between traditional authorities (TAs) and the democratically chosen local government bodies. The chieftaincy of Sekhukhune is famous because of its strong kings and headmen, who played a major role in resisting Boer and English rule. And currently the *Kgoshis* still play an important role, although not all are equally respected (Oomen, 2002). As the TAs have considerable decision-making power over the distribution and use of communal land and other community natural resources, local and district municipalities have to work with them. The municipal ward councillors play a key role here: the success of policy implementation processes at the local level crucially depends on them having good relations with the TA. If this is not the case, the struggles between the two can constantly hamper development in the area at the expense of the community. Local level DWAF officers and water committees in the villages studied usually worked together either with people from the TA or with people from the municipality, but rarely with both.

In all, the local government bodies in Sekhukhune on which the implementation of water policies largely depends demonstrate difficulties with being the democratic, representative and accountable institutions they were envisioned as and which they need to be for effectively managing water.

Two Cases: Ga-Mashishi and Moeding

From the previous section it can be concluded that conditions in the SDM for arranging effective and equitable water services are challenging and difficult. The district is very poor and lacks basic infrastructure in terms of communication and transport and water distribution. The efforts to establish well-functioning local government bodies are seriously hampered both by a lack of trained and educated staff and by a history of strong and influential traditional authorities who are not keen to give up their powers. Overall financial means to develop capacities and infrastructure are few. Water distribution in Ga-Mashishi and Moeding (see Figure 4.1 for their locations) is highly influenced by this complicated context.

Ga-Mashishi

Ga-Mashishi is located along the R37 from Burgersfort to Polokwane. It lies in Ward 9 of the Greater Tubatse Local Municipality, which is one of the five local municipalities of Sekhukhune. Most people in Ga-Mashishi belong to the same chiefdom. The chief, *Kgoshi* Mashishi, lives in the village and his ancestors and their people are said to come from Gamake, in the northern part of contemporary Limpopo. The village has around 1000 households, with an average of some 6.5 persons per household. Most of the households are (temporarily) female- or grandparent-headed as many men left the village to look for paid work, either at the mines or in the cities of Gauteng. Unemployment rates are high, and almost 40 per cent of the households depend entirely on old-age pensions or child allowances for their survival. The 'social facilities' comprise a preschool, two primary schools and one secondary school. There is a mobile clinic that attends to patients once a week.

The Lebowa Department of Water Affairs (LDWA) laid the basis of the present domestic water supply system (DWSS) in 1984/1985. Two boreholes were drilled and equipped with diesel pumps to fill two separate reservoirs, which in turn provided 43 communal and 45 household standpipes. The people living above 'reservoir level' were provided with four hand pumps. The LDWA hired an operator and the Ga-Mashishi Water Committee (GWC) was formed to monitor water use and to facilitate communication between water users and the operator. The diesel was paid for by the LDWA. The water was used for multiple purposes, both domestic and small scale productive use at the household level. In 1987 the Randberg Rotary Club donated a borehole equipped with an electrical pump to the village to backup the existing system and meet the growing demand for water. In 1995, after a chief-led application to the Council for Scientific and Industrial Research (CSIR; a national research institute) and the Mvula Trust (a South African NGO), the latter invested in an extension of the existing water system. The extension consisted of two new boreholes equipped with electric pumps, eight new communal taps and 3500m of extra pipeline. The system was designed to provide 420 households with water. The total water supply capacity of the system is $220m^3/d$ and the reservoir capacity is $240m^3$ (CSIR, 1995).

The village water supply is subdivided into five sections, each with their own branch off the main DWSS pipeline. Each section has its own section water committee. There is also a 6th branch, leading to the houses of the chief and his brother. Water is delivered following a rotational schedule, operated by DWAF employees through opening and closing the valves between the main pipeline and the branches. Some of the section branches also contain valves; those too are handled by the operators. The sections with schools get water in the morning, the other sections get water in the evening. The chief's branch has water day and night. Water use is not restricted. Whenever a tap contains water, users can take as much as they like and for whatever purpose.

The GWC meets every week and is responsible for communication between the community and operators, for noticing and repairing breakdowns in the system, and for fee collection. Members are elected and re-elected during a mass meeting held every year.

Infrastructure at present

The original domestic water supply system in Ga-Mashishi has deteriorated substantially since its extension in 1995. Presently the four hand pumps and one of the electric pumps are out of use due to technical problems. The diesel pumps are also not used – the pumps are in good shape, but no diesel is being bought to run them. Of the whole domestic water supply system, extended in 1995, only two electrical pumps remain in use. The water users that depended on the hand pumps currently use an untapped well and buckets for their drinking water. The electricity used by the electric pumps is not paid for by anybody; Eskom (SA's electricity company) did not install meters and cannot quantify how much electricity is used.

From 1995 onwards, there has been a profusion of illegal connections at private yard level to the DWSS. Officially these connections are forbidden but in practice they are tolerated by the DWAF. Eventually they got formalized by the GWC and now anyone who has the money to do so can make a yard connection to one of the branches of the DWSS. The GWC is largely concerned with the location of the connections (mindful of the fact that people were buried all over the village) and takes little interest in the water distribution aspect. The people that have the financial means invest in yard connections to have a more secure water supply. Some users even manage to connect to the mainline of the system and thus have a constant supply. Cracked pipes and leaking taps are also a common sight.

An important factor influencing the technical performance of the domestic water supply system in Ga-Mashishi is the 'overpopulation' of the system, a combination of population growth and system failure. Originally, the system was supposed to be able to provide the connected households with almost 16,000l per month (which is almost triple the RDP standard of 6000l/household/month). Even with the present 1000 households using the system, it is theoretically possible to meet basic human needs, provided that the system performs well and that there is adequate equality in water distribution within the system. At present, however, with only three electrical pumps working, private yard connections which the system was not designed for, unlimited use and a lot of cracks in the system, total water availability is estimated to be only half of the original capacity (personal observation at tank by GWC head operator). The total water availability thus falls far short of RDP standards.

What water is used for is largely a function of its availability and of the reliability of supply. Only users who have secured a reliable access to water sometimes also use water for small-scale productive uses, such as watering garden crops.

Management

During working hours – ie not in the evenings or during weekends – the system is operated by DWAF employees. However, the GWC has a spare key to the pump houses, valves and the reservoir and, although they are formally not entitled to do so, operate the system during weekends for practical considerations. The chief's branch is connected to a private reservoir that is filled during the day in order to make sure the pipe also carries water at night. Some water users are also reported to have opened and closed the valves themselves. These users claim they are helping the operators, a contention supported by the GWC but denied by the operators themselves. The issue of the operation of the valves is a delicate one and somewhat difficult to fully get to grips with.

There is no rule limiting water use although there is a schedule as to which section has water at what time. Users can take as much water as they want while there is water in their taps. This leads to under-valuation of the water and high losses and causes pressure problems in the whole system and water availability problems in the tail-end part of it.

On paper, the Ga-Mashishi water committee is a democratic organization that is controlled by the people and responds to its constituency in mass meetings that are organized every two months. In reality, the present chairman of the water committee has been in place since the start of the committee in 1984 without at anytime being re-elected, as have some other of its members. In interviews with water users, the GWC was said to be greatly influenced by the demands and needs of committee members and the traditional authorities. Users do not feel that the committee takes responsibility for the performance, or lack thereof, of the water supply system, and existing social hierarchies often play in role in the committee's operational decisions. For example, care is taken that the chief is assured of a constant water supply, a service that he rewards by backing up the committee with his protection and authority.

The relationship between the ward councillor (of the local democratic government) and the TA is difficult. Theoretically they should work together in day-to-day community management and on specific topics like a future mine-funded community development project, but which issues are to be addressed by whom is still subject to discussion and communication between the two is marginal.

Almost all water users, especially those downstream, regularly spend several days or even weeks without water in their private or closest communal taps, while some individuals enjoy good relations with influential water committee members and make use of this for securing their water access. The users that do not get water through the system obviously have to look for other means of getting it. Most often they go to relatives or friends in another section. Some people buy water from a local water vendor or in some cases they even go to the nearby river.

Finances

The lack of maintenance is strongly influenced by the lack of financial means the GWC faces. Before 1994 the LDWA used to pay for everything. Nowadays the operator is still paid by the DWAF, but other running costs are supposed to be covered by the water users. The national government allocates a budget to local municipalities to implement the FBW, but in Sekhukhune municipalities are not using this budget to provide FBW related support and services to communities. Furthermore, the amount municipalities receive from the so-called Equitable Share is often much less than is needed, mainly because of administrative deficiencies. Some municipalities, for lack of information, do not even claim the FBW funds; others use them for other purposes such as roads, schools or first getting all inhabitants connected to working water supply infrastructure (although without it necessarily being for free). For most municipalities implementation of FBW is not the highest priority because most communities do not even have RDP standard levels of water services supply to start with and for municipalities as well as the DWAF this is the first priority. The SDM has been receiving FBW funds since 2001, but it was not possible to trace how these have been spent.

Some municipalities are paying diesel or electricity bills for some communities but consider this a transitional measure for the period in which cost-recovery is still difficult to enforce because of the absence of an accurate system to measure and account for uses. In Sekhukhune, given the current low availability of funds, some kind of cost recovery is indispensable to allow proper maintenance and operation of systems. Apart from simple balancing budgets and putting limits to the amount of public money that can be spent on water supply, the principle of cost recovery is also premised on ideas of ownership. Supposedly, when people have paid more for their 'development' (be it construction of a system or fees for receiving services) they will feel more responsible and take more care of things. Furthermore, paying for water services provides water users with an instrument to hold water providers accountable: if water services are poor, users can, in principle at least, withhold or reduce payments. In this way, payments can increase users' control over water supply systems. The majority of the interviewed people in Ga-Mashishi stated that they would be willing to pay for their water as long as supply was secured and the system improved and well maintained.

Up till now Ga-Mashishi people have not paid (or have paid very little), but neither has this basic amount of water been secured. This is a chicken-or-egg situation that risks becoming a downward spiral: since water provision is erratic, users are not inclined to pay, and since there is no money the system cannot be well maintained and the water committee feels little pressure to manage it equitably (as they are not rewarded for doing so), which is why water provision is erratic and inequitable – and so on.

Moeding

In the surroundings of Marble Hall, about five kilometres as the crow flies from the road from Groblersdal to Leeufontein, lies Moeding. With its just under 200 households, it is quite a small village, only reached by a 10km dirt road winding into the hills. *Induna* (headman) Mahila represents chief *Kgoshi* Matlala in the village while Mr Sedibang is the ward councillor of Moeding and five surrounding villages. There is little direct influence on village life from either the local municipality or Mr Sedibang. The headman takes care of day-to-day matters, and the councillor makes sure to visit the headman once in a while to get updated. *Kgoshi* Matlala was the leader of a group of chiefs who walked out of the participation meetings with the Greater Marble Hall Municipality, as he did not agree with the fact that his function would only be an advisory one. In his area of jurisdiction he still acts as the de facto authority, although he does in most cases cooperate with the local municipality.

Before 1994, people in Moeding used to draw their water directly from the river, buy water from water vendors (at ZAR12 per 200l drum) or dig shallow holes in the banks of the river to get water. Some rainwater collection was also practised. In 1994, the Mvula Trust NGO drilled a borehole and built a reservoir and some communal taps. The DWAF paid for the diesel and operation of the scheme. The DWSS of Moeding was updated very recently (finished in 2003). The infrastructure now consists of a reservoir and pipes leading to 30 communal taps, making sure that water provision meets RDP standards. Lepelle Northern Water (LNW), the water board in Limpopo Province, provides bulk water to the reservoir. The Mvula Trust also paid for the extension of the system, with a contribution from each household.

The Moeding Water Committee (MWC) takes care of the collection and administration of fees, of water distribution by opening and closing the communal taps and of contacts with LNW. Currently, three positions in the water committee are vacant. The external communication of the water committee is handled solely by the chairman. He has no contact with the DWAF but does keep in touch with the ward councillor and the headman of the village.

The water supply system is functioning without any technical defects. Maintenance of taps and distribution pipes within the village is performed by the water committee and paid for by the users.

When not operated, the taps are locked, with only the chairman of the water committee having keys. Officially no water is given or sold at weekends, but in case of emergency, for want of a better word, the chairman of the water committee fills drums on request.

Finances

When the system was initially installed, the users were reluctant to pay for water. After all, in election slogans they had been promised basic water for free. Eventually, when they saw the water committee (backed by the Mvula Trust) would not bend, people started paying. The FBW policy is not implemented in Moeding.

When a household does not pay its monthly fee, it does not get any water. According to the water committee, there are no payment problems, although they admit that sometimes people do not have the means to pay on time. Normally, family and relatives are then asked for either a loan or water. It is not possible to get any credit from the water committee.

All households in Moeding have access to a basic amount of water that is set at 800 litres per week (3600l/month), as long as the monthly fee of ZAR25 is paid. RDP standards (600l/hh/month) are thus by no means being met. People can increase the amount of water available for their households at the relatively high price of ZAR10.4/m^3 (most people in the region (in other schemes) pay ZAR5/m^3).

At the time of research, just before the 2004 national elections, an effort was being made to start implementing the FBW policy in the villages surrounding Moeding. The Greater Marble Hall Local Municipality had asked bulk water supplier and water board LNW to make an inventory of how many people are using LNW-supplied water to calculate the costs of supplying FBW. Furthermore, it has asked LNW to check whether the FBW could be partly funded through currently used cost recovery systems. At the time of research, LNW had just finished making the inventory and no further steps were being taken to implement FBW.

Transparency

Contacts between the traditional authority and the ward councillor in Moeding only started in 2003, and on the face of it their relations seem to be rather amicable. Before they started to meet regularly, one village member functioned as contact person between the two. This person also arranged for the Mvula Trust to construct the domestic water system. The Mvula Trust never directly contacted either the ward councillor or the headman. After construction, and almost unavoidably, the village member acting as intermediary became the chairman of the water committee. After complaints from users of the water supply system about the transparency of the water committee, both headman and ward councillor realized that their mutual contact person might not be as neutral as they thought and decided to stay in contact once a month directly.

While the water committee emphasizes the democratic principles that guide its functioning, users complain that the committee is not sufficiently transparent, especially where finances are concerned. When the committee has to account for its finances in meetings actual figures are not presented. Moreover, many people are of the opinion that they pay too much for the water they use. Although the water committee says it uses the money for maintenance, it has not been able to prove this with a financial statement.

Vacancies in the water committee have remained unfilled since it started functioning. The water committee itself states it is the lack of compensation for committee members that keeps people from joining. Some water users were asked why they do not join the water committee; they stated that the hierarchic structure of the committee and the lack of influence individual members have kept them from doing so. Furthermore, people do not want to stand up to a person who managed to get proper water supply to the village when no authority (local government or traditional authority) was interested in doing so. The success in the matter can be attributed almost solely to this person, who as a result stands above critical scrutiny. Although the water committee officially is a democratic body, where all members work together, in Moeding this is not the case. In the constitution of the water committee, a provision is made for the water committee to vote one of its members out, but the committee (ie its other members) declares itself very content with its chairman and will not use this right. The community cannot vote to change the composition of the water committee.

There is some indication that the complaints of water users are justified. First, all water distribution powers are concentrated in the chairman, who is the only person to have the keys to the taps and who can single-handedly decide what situations to consider as emergencies (and thus justify additional water allocations). Second, there is a very substantial mark up from the price the water committee pays for the water from LNW (3 rand) and the price the users pay (7 or 10 rand). Deducting the money paid to LNW, around 5500 rand (US$690) is available every month for maintenance, or for paying committee members. However, despite maintenance costs not being likely to be very high (since the system is quite new) and the committee members' claiming that they do not receive any compensation, the committee's bank account does not show a reserve. When, in the last interview of this research, we asked for an explanation for this discrepancy, the treasurer of the water committee pointed to the chairman, who was not available for comment.

Moeding water users have no insight into the budget of the committee. The financial accountability of the water committee to the water users in this case is low.

Conclusion on Case Studies

Equitable water distribution in Sekhukhune does not prove to be easy, both in terms of basic water and in terms of inclusiveness in democratic structures. The FBW implementation level is very low, as is the commitment of the SDM to implement it in the near future, and a lot of households lack even basic access to water.

Community managed or committee managed

In the two cases studied, the water committee to a large degree determines the success of redressing inequities at the community level. In both cases, part of the community of water users does not feel well represented by their committee. The water users still do not have much influence on the composition or the performance of their committee. Formal mechanisms of accountability are not functioning. The illegal running of the system during weekends, the implicit permitting of illegal yard connections, the inability to repair breakdowns and the waterless periods in Ga-Mashishi, and the opaque water distribution and finances in Moeding all provide serious grounds for questioning the performance and legitimacy of the water committees.

In the case of Ga-Mashishi, all people interviewed (whether they were part of the water committee, the chiefs kraal or water users) agreed that the authority of the GWC had severely declined after the power turnover in 1994. Formerly the GWC had derived legitimacy from its political support from the Lebowa homeland government and the chief, these having both money and authority. Today the local municipality is the officially supporting agency, but its presence in the village is negligible. Furthermore, the legitimacy of the GWC's authority is negatively affected by its failings in dealing with the maintenance of the system. There is a risk here of a downward spiral, with ever-decreasing legitimacy of the committee leading to ever-decreasing satisfaction and willingness to pay on the part of water users. Already the underperformance and lack of legitimacy of the water committee leads to an inequitable distribution of water. Several users are not even provided with basic need level services.

In Moeding the committee's ability to provide basic water is not in question; what is in question is the way it deals with finances. Users have to pay a relatively high price for water, but they are not kept informed on how these prices are established. They also do not know how the collected money is spent. Executive and legislative powers are concentrated in one person, the chairman of the committee, and people have so far not succeeded in holding him accountable for the financial affairs of the committee.

Cost recovery

The two studies cases differ significantly in terms of cost recovery practices. In Ga-Mashishi very low or no fees are paid – and any payments are made on an irregular (in both place and time) basis – and it is not clear where the collected money is going. In Moeding users paid for part of the construction of the DWSS and also pay the running costs for maintenance. The last case might suggest that the people of Moeding should have a greater sense of ownership and more control over the water distribution in their community. This, however, is only partly true. Moeding water users are willing to pay, since the quality of their water services is high, yet they have little or no influence on how the paid money is used, nor can they influence important water distribution choices. Their dependency on the chairman of the committee prevents them from seriously exercising their rights to an insight into the committee's performance.

Apart from simple balancing budgets and putting limits on the amount of public money that can be spent on water supply, the principle of cost recovery is premised on ideas of ownership. Supposedly, when people have paid more for their 'development' (be it construction of a system or fees for receiving services) they will feel more responsible and take more care of things. Furthermore, paying for water services provides water users with an instrument to hold water providers accountable to them: if water services are poor, users can, in principle at least, withhold or reduce payments. In this way, payments can increase users' control over water supply systems. And in South Africa's water policies, cost recovery plays an important part. In Sekhukhune, given the current low availability of funds, some kind of cost recovery is indispensable to allow proper maintenance and operation of systems.

No functioning help desk

In both cases no external functioning help desk or other external authority to which users could address themselves with complaints about the performance of the water committee exists. Given the described lack of operational and/or financial accountability of the water committees to the water users, this is all the more worrisome. The lack of a help desk explains why there is little pressure on the water committees to improve their performance. The chiefs' kraal is officially not involved in the DWSS but is still seen by a lot of water users as the institution with the most power to change things in the system.

In the case of Moeding, several users are not satisfied with the way the water committee is functioning. In this case, nobody but the water committee is responsible for water distribution, maintenance and fee collection, and there is only one very limited way to appeal – in a mass meeting at which questions are often not appropriately responded to.

At a national level the assumption is made that possible problems will be solved in a representative, democratic manner and that water committees are accountable to the water users. At district or local municipality levels it might be reasonably contended that this democratic ideal is not functioning everywhere, and in both the case studies here presented there is a lack of means and staff to solve the problem. The democratic tools users thus have are quite limited.

Conclusion and Discussion

In the first years of the new South Africa, the legal and policy environment changed dramatically. Expectations about the pace in which living conditions would change were as huge as the optimism among the brand new government. South Africa has made enormous progress in improving infrastructure and water management structures for domestic water supply. Basic water services to the rural poor have improved significantly over the last decade, with more than 60 per cent of the country's population now provided with RDP standard water for free. There are, however, still millions of people that do not have access to RDP standard water and for whom FBW is still far away.

South African national policies are not easily implemented in Sekhukhune as they require a certain technical, socio-economic and political context which, as yet, does not exist there.

Rights or Control

Legal water rights play an important role in the water distribution goals of the new South Africa as water legislation is being completely reformed and is supposed to be or become an important tool in the redistribution of water. In itself, however, having legal rights to water is meaningless: control is essential. Rights only become valuable when people are able to cash them in in practice. Recognition of such rights (both inside and outside the group of right-holders) plays a role, but having adequate technology to extract water and transport it to where it is needed is also essential.

Next to the legal and technical aspects of water control, there are other dimensions that also play a role. Water distribution is not a context-neutral phenomenon. As water is mostly managed by committees, organizations or companies (ie by people) and as technical infrastructure is always manipulative, it is a contested resource and its distribution also (or largely) depends on organizational, socio-economic and political factors (Mollinga, 1997). South Africa's water-related national policies for redressing inequities count on a certain available context. The policies are developed mainly along legal, political and organizational lines, providing for a basic right for everybody and for institutions for equitable water management. Technical and socio-economic factors (and inequities) that influence water distribution in practice do not always receive enough attention in the design of the policies, although they are just as essential.

Participation

Democratization and participation as envisaged by the South African government implies user participation and community participation in higher level, existing institutions. The democratization policies are based on a politically liberal conception of democracy, assuming that in the public sphere everybody is equal and that resource distribution can be negotiated in a socio-economic and political power neutral environment. The risk exists that in this way the socio-economical or cultural differences that can exist between actors participating in the same public democratic process are ignored, because of the assumption that political institutions are autonomous and not influenced by societal context. This liberal view on democracy is especially criticized in the gender inequities literature (Cleaver, 1999; Fierlbeck, 1995), but in essence poses problems for all unequally empowered social groups. Fraser (1997, p78) argues that, 'Insofar as the bracketing of social inequalities in deliberation means proceeding as if they don't exist when they do, this does not foster participatory parity'. Distributional equity (of income, property, etc.) is an important factor shaping social hierarchies and power relations and thus in determining social equity. In a society where resource distribution inequalities and thus social inequalities persist, the existence of meaningful participation of all, including marginalized, stakeholders in democratic bodies would seem to be a somewhat utopian idea.

At least some parity needs to exist, however, and structural relations of dominance and subordinations should be dismantled (Fraser, 1997). It can be questioned whether such power neutral situations exist at all, but what we can say with some certainty is that South Africa does not provide such an environment. Social, economic and political power inequities are large, including at the local level, as a legacy of former politics, and reflect the fact that redistribution of power and resources has not yet gone very far. The ideal of a neutral public sphere is, at least in near-future South Africa, out of reach; it can thus be questioned whether a liberal democracy structure and emancipation expectations arising thereof will contribute to redressing inequities.

Infrastructure Development

Infrastructure development plays an important role in political relations and negotiations. On one side of the equation, municipalities and villages need the infrastructure for water services but do not have the means (either capacity or money) to construct. On the other, the consultancy firms and mines/industry have these means and seek influence in water distribution (mines/industry) or budgets (consultancy firms). The field material of this research shows that by providing the means for infrastructure development, consultancy firms and mines/industry can directly secure their influence on water distribution and on municipal budgets.

Where democratic processes are to take place a 'Cowardian' scenario looms – those who are able to invest are the ones who get control (Coward, 1986).

National Policy, Local Context

The FBW policy was designed at the national level, yet it is the municipal and local context that defines the pace of its implementation. Implementation depends on municipal budgets, the development state of infrastructure and the commitment and ability of the municipalities and local committees. The FBW policy was especially designed to redress inequities in basic water supply and to lift the poorest of the poor to a certain basic services level, even if they are not able to pay. The result of the context-dependence nature of the policy is, however, that in poor rural areas implementation levels are far lower than they are in metropolitan areas, for example. The measurement and cost recovery requirements that are placed with the FBW policy are not so easily reached in those areas where often skills, means and staff are lacking. Consequently, former homeland areas often lacking appropriate infrastructure such as Sekhukhune lag far behind on average service levels. Progress is being made, but what is difficult to accept is that inequities from the past are again influencing development, even for basic services.

As all powers and responsibilities in the water services sector are decentralized, a lot depends on the capabilities and number of staff at the local and district levels. In Sekhukhune, rivalries between traditional authority and local municipalities and between district municipalities and local municipalities have shown to negatively influence development. The SDM faces a severe lack of staff in general and of appropriately trained staff in particular. At local municipality level, the problem is even more acute, and worsened by the lack of communication between local and district levels. In performing water services delivery and in developing new infrastructure, local municipalities and traditional authorities often end up in power struggles. And it is the beneficiaries-to-be, local people, who suffer most from these political delays.

Upward accountability links between water committees, local and district municipalities and the DWAF are weak, largely because of a lack of educated staff and means. In Ga-Mashishi and in Moeding no external functioning help desk or other external authority exists to which users might address themselves with complaints about the performance of their water committee. Given the described lack of operational and/or financial accountability of the water committees to the water users, this is all the more worrying. The non-existence of a help desk explains why there is little pressure on the water committees to improve their performance.

Importantly, this study revealed that the crafting of effective local water institutions is a process that does not just happen by decree. The effective implementation of water policies is premised on the not always legitimate assumption that water users, committees and municipalities have the willingness, skills and means to make decisions and solve problems in a representative, democratic manner and that water committees are accountable to all water users. But the implementation success of reform policies depends on the context they are implemented in. Therefore, policies should be anchored in profound understanding and acknowledgement of the context.

The cases and analyses above show that in some situations practice does not easily follow policy. In the highly unequal context of today's South Africa, it is unthinkable that one overall policy at the national level will have the same consequences in the many very different contexts in which it is to be implemented.

For equity-driven policies like FBW to succeed, it is essential to work on the local context first. Of course, the South African government has realized this and put financial and organizational means in place to try and decrease the differences in local context. However, such means are not proving sufficient. More core attention should be paid to capacity building, accountability mechanisms and equitable infrastructure building at the local level. It is important to realize that these are long-term requisites for equitable development. Investments are needed to change the unequal context the rural former homelands have to deal with. Short-term success stories are not to be expected, however, which possibly explains why such investments do not appear to be politically interesting.

The poor rural regions in the former homelands especially need attention. Without specific focus on getting the local context more equitable and providing realistic opportunities for improvement, national policies will not be as effective as they could and should be. And as a result, historical gaps can easily remain, or even widen.

Acknowledgements

The authors wish to thank Cobie Penning de Vries and Stef Smits for their valuable comments on earlier drafts.

References

African National Congress (1994) *The Reconstruction and Development Programme: A Policy Framework*, Umanyano, Johannesburg

Cleaver, F. (1999) 'Paradoxes of participation: Questioning participatory approaches to development', *Journal of International Development*, vol 11, pp597–612

Corder, C. K. (1997) 'The reconstruction and development programme: Success or failure?', *Social Indicators Research*, vol 41, pp183–201

Coward, E. W. (1986) 'State and locality in Asian irrigation development: The property factor', in Nobe, K. C. and Sampath, R. K. (eds) *Irrigation Management in Developing Countries: Current Issues and Approaches (Studies in Water Policy and Management, No 8)*, Westview Press, Boulder, CO, pp491–508

CSIR (1995) Map of Ga-Mashishi village

Department of Water Affairs and Forestry (2003) 'Strategic framework for water services: Water is Life, Sanitation is Dignity', South Africa.

Department of Water Affairs and Forestry (2004) www.dwaf.gov.sa/, accessed in January 2004

Fierlbeck, K. (1995) 'Getting representation right for women in development: Accountability, consent, and the articulation of women's interests', *IDS Bulletin*, vol 26, pp23–30

Fraser, N. (1997) *Justice Interruptus. Critical Reflections on the'Postsocialist' Condition*, Routledge, New York and London

Klasen, S. (1997) 'Poverty, inequality and deprivation in South Africa: An analysis of the 1993 Saldru survey', *Social Indicators Research*, vol 41, pp51–94

Klvechuk, A. and Jenkins, G. P. (2003) *Evaluation of the Olifants-Sand Water Transfer Scheme in the Northern Province of South Africa*, Cambridge Resources International Inc, Cambridge, MA

Manyaka S. and Odendaal, A. (2004) Communication campaign for FBW; stakeholder issues related to the implementation of FBW, DWAF, South Africa

Mollinga, P. P. (1997) 'Water control in sociotechnical systems: A conceptual framework for interdisciplinary irrigation studies', in P. P. Mollinga and K. Van Straaten (eds) (1999) *Irrigation and Development Reader*, Wageningen University, the Netherlands

Municipal Demarcation Board (2003) 'Assessment of municipal powers and functions', Limpopo Province provincial report

Oomen, B. (1999) 'Hoepakranz: Between sjambok and school governing body – A report on law, administration and the position of traditional leadership in a South African village', unpublished first interim report on PhD research, Wageningen, the Netherlands, available at http://home.mweb.co.za/b./b.oomen/report.htm

Oomen, B. (2002) 'Chiefs! Law, power and culture in contemporary South Africa', PhD thesis, Leiden University, the Netherlands

Republic of South Africa (1996) *Constitution of the Republic of South Africa, Act 108 of 1996*, Constitutional Assembly, Cape Town

Republic of South Africa (1997) *Water Services Act*, Office of the President, Cape Town
Republic of South Africa (1998) *National Water Act*, Office of the President, Cape Town
South African Census (2001) *Population Census 2001*, available at www.statssa.gov.za/publications/populationstats.asp
Stimie, C. M., Richters, E., Thompson H., Perret, S., Matete, M., Abdallah K., Kau, J. and Mulibana, E. (2001) 'Hydro-institutional mapping in the Steelpoort River Basin, South Africa', Working Paper 17, International Water Management Institute, Battaramulla, Sri Lanka
Van Koppen, B., Jha, N. and Merrey, D. J. (2003) 'Redressing racial inequities through the National Water Act in South Africa: The prospect of revisiting old contradictions?', unpublished report, International Water Management Institute, Battaramulla, Sri Lanka
Van Veelen, M., Coleman, T., Thompson, H., Baker, T., Bowler, K., De Lange, M. and Sibuyi, I. (2002) 'Proposal for the establishment of a catchment management agency for the Olifants Water Management Area', draft proposal, Pretoria, South Africa

5

Local Governance Issues after Irrigation Management Transfer: A Case Study from Limpopo Province, South Africa

Gert Jan Veldwisch

Introduction

In the Limpopo Province of South Africa, 114 smallholder irrigation schemes are currently in the process of being rehabilitated and subsequently turned over to local management through a programme called the Revitalization of Smallholder Irrigation Schemes. This became necessary following the withdrawal of subsidies by the National Government in 1994 (Perret, 2002; Hope and Gowing, forthcoming). The history of many smallholder irrigation schemes can be traced back to the 1950s, following the Tomlinson Commission's advice to set up investment projects in rural areas (Hope and Gowing, forthcoming). In the decades that followed the schemes were mostly run by parastatal entities, on huge external budgets, which organized contract farming arrangements with the people working on the land (Shah et al, 2002).

After these subsidies were stopped (at the national level) almost all parastatal management entities collapsed. The provincial government of Limpopo then initiated its revitalization programme, aimed at the establishment of water user associations (WUAs) and the hand-over of both management responsibility and ownership of the schemes. Participative rehabilitation of the infrastructure was seen as a cornerstone of the approach (Northern Province Department of Agriculture, Land and Environment (NPDALE, 1998), while the long-term objectives were financial viability and self-management, which are common objectives in Irrigation management transfer (IMT) programmes (Vermillion, 1997). The Thabina irrigation scheme (1998–2001) was a pilot within the revitalization programme (LPDA, 2002a and 2002b; NPDALE, 2001a and 2001b).

As management of irrigation systems involves a complex set of social and technical aspects (Mollinga, 2003), transfer of management involves the transformation of a complete socio-technical system (or network), with all its intricate relations.

Until now, the understanding of how the programme affected the organizational situation and water sharing practices has been limited, even though implementation research has since been conducted on the scheme (Perret et al, 2003). Thus there is a need for better understanding of both the content of the intervention and its effects on local water management.

Methodology

Field work for this chapter was conducted between December 2003 and May 2004. Some eight different irrigation schemes in the Limpopo Province were visited. Among them were six schemes currently in the process of revitalization, a historically farmer-managed irrigation scheme and the Thabina irrigation scheme. Thabina was studied in more detail as a case of early transformation.

In total six visits were spend researching Thabina, with a total stay in the area of about 25 days. A (mainly) qualitative and 'actor-oriented' approach was adopted. By 'qualitative' it is meant that the focus was on the mechanisms of relations and processes rather than on numbers. By 'actor-oriented' I mean that the actor was put in the centre of research by reconstructing the relations, actions, strategies, etc. of real people (Long, 2001; Chambers, 1997). Understanding the intricate and numerous relations and processes in a local situation requires spending time within the scheme's area and the irrigators' community. The research focused on how various actors related to (1) infrastructure, (2) each other, (3) governance structures and (4) external influences. However, social processes take place through material elements as well; for instance people interact with each other through infrastructure (Latour, 1991; Callon, 1991). Getting a good understanding of the technical functioning and the practical management of infrastructure was therefore also an important aspect of the methodology. The methods used consisted mainly of direct field observations, semi-structured interviews and some simple water measurements. They were structured around two strategic approaches – 'follow the actor' and 'follow the water'.

A key working principle was to talk with as many people as possible who were directly involved in the scheme – farmers, members of the management committee, extension officers, etc. 'Follow the actor' is a method that creates possibilities for informal talks. One advantage of the informal setting was that people were more inclined to say what they really thought. This required creating situations in which such informal talks could be held. 'Hanging around' in the fields and 'becoming part of the furniture' in the office contributed to this. The method basically implied spending at least part of a day with key actors during their activities. The water bailiff, for instance, was joined on his patrol along the canals and farmers were interviewed in the fields where they were weeding, ploughing or irrigating. Such a practical, on-site method helped to crosscheck oral information, to observe practices as well as to build trust relations. Besides non-structured interviews during incidental encounters, semi-structured interviews were held more systematically with some 30 farmers. These were interviews not in questionnaire style but in journalist style and were prepared in advance with clear, pre-formulated objectives and a pre-determined list of subjects to be covered. The advantage of semi-structured interviews over non-structured, completely informal interviews is that the outcomes of the former are to a greater extent comparable with one another.

The second strategic approach, 'follow the water', is an irrigation-specific form of a transect walk; it implies following the water flows in (and outside) the system, looking for sources and destinations. Both the original functions of water management constructions and their modifications were analysed. This was done by viewing them in the perspective of the overall logic of the irrigation scheme and by working out the hydraulic formulas under various (boundary) conditions, their flexibilities and their possibilities of use. This was complemented with some simple flow measurements. The following questions were raised: Why were some outlets bigger than others (while serving the same area)? Why did one stop log disappear and the other not? Who profits the most from the current pump management and who profits least? In this way 'following the water' is a method relating not only to the technical dimension of irrigation schemes but also to the socio-political dimension.

Water Management at Thabina

The Thabina irrigation scheme is situated close to Tzaneen, about 25km along the R36 to Lydenburg. It consists of 234 plots (with a total area of 200ha), which are held under customary land tenure; traditional authority (TA) chief Muhlaba is the landowner, while the plot holders have a permission to occupy (PTO), which is a usufructary right. There are about 160 plot holders, with an average land holding of 1.3ha. The scheme has probably existed since 1948, although some people put the start-up in the late 1950s or early 1960s as the history of many smallholder irrigation systems in South Africa can be traced back to the Tomlinson Commission, which in 1955 proposed small-scale irrigation as a mechanism to support the lives of the rural poor in the homelands (Hope and Gowing, forthcoming). Most probably the existing Thabina Scheme was improved and extended under this programme. From its inception Thabina, like similar schemes, was run with continuous capital input from the central government. Thabina gradually collapsed after the first post-Apartheid government removed these state subsidies in the early 1990s and was one of the three pilot schemes in the provincial revitalization project. Over a period of roughly three years, a group of consultants worked together with the farming community. A Water Users Association (WUA)[1] was established, the main infrastructure was rehabilitated, training on field and system water management was provided and management of the scheme was finally handed over to the WUA.

A recent study (Perret et al, 2003) revealed that only about 60 per cent of the area is currently being cropped. Furthermore, the study showed that there is a high variety of crops as well as many different farming strategies in Thabina. There are some large-scale commercial farmers, a substantial group of small, intensively cropping commercial farmers and a large group of farmers who partly crop for

Table 5.1 *Timeline of relevant events in the Thabina irrigation scheme*

1948	Start of Thabina irrigation scheme on half the area with earthen canals
Late 1950s	Doubling of the area and improvement of the canals
1994	Withdrawal of government subsidies
1994–1996	Gradual collapse of the system
1998–2001	Rehabilitation, establishment of a WUA and transfer of management
Nov 2003–Apr 2004	Field work for this research

Sources: Perret et al (2003); NPDALE (1998); author's own data

home consumption and partly for commercial purposes. At the same time there is also a large group of subsistence farmers. In the summer maize is predominantly cropped, while in winter various vegetable and other cash crops are grown.

Layout of the Scheme

The scheme is fed by the Thabina river, a tributary of the Letaba river. A diversion weir leads the water into the main canal, which has a capacity of about 700m^3/hr (194l/s). There is a storage dam in the river (the Thabina dam), which was initially built for irrigation purposes serving the Thabina irrigation scheme but which is currently only used for domestic purposes. The conveyance system consists of open canals that are lined with concrete. The first part of the canal runs through a residential area and is covered by concrete slabs, which were installed during the recent rehabilitation. The main canal carries a continuous flow of water, 24 hours a day. From 8.00am till 6.00pm the first 12 sub-canals take water. During the rest of the day the water flows into a night storage dam, which supplies the remaining six sub-canals during the following day. The sub-canals are about 200m to 800m long and each supplies up to 20 farmers.

The off-take structures from the main canal into the sub-canals consist of a check structure in the main canal, which is a long crested weir, and between one and five pipes in the side of the main canal just upstream of the check structure. Both from the original design and from a re-design document, it is evident that these outlets are supposed to have a capacity proportional to the area served by them. A sub-canal serving 20ha should therefore have twice the capacity as a sub-canal serving 10ha.

The outlet pipes have an internal diameter of 75mm and are operated with a sort of plug, making an on/off system. When opened each pipe releases a discharge of about 7l/s under a constant head of 24cm (due to the broad crested weir). The number of pipes at an outlet structure ranges between one and five, depending on the area it serves. Due to the on/off system a sub-canal supplies 7, 14, 21, 28 or 35l/s, which is never perfectly proportional to the area served. This lack of proportionality is shown in Figure 5.3. The dashed diagonal line shows a (theoretical) true proportional relation between discharge in the sub-canal and area served, while the short horizontal lines show the theoretical/design relation.

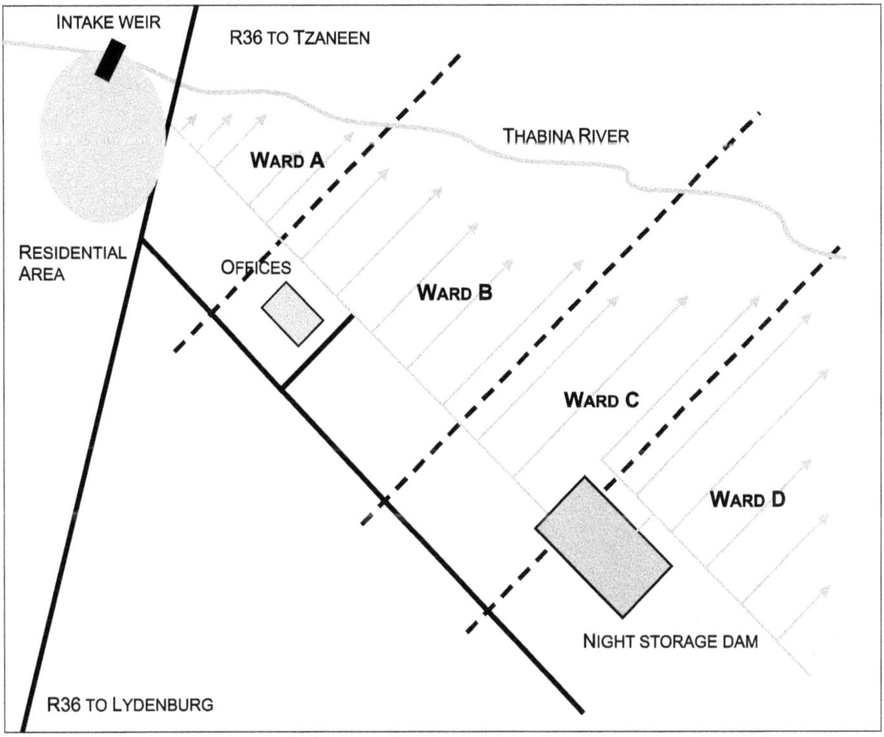

Source: Chapter author

Figure 5.1 *An overview sketch of the Thabina irrigation scheme*

This shows how the use of standard-size pipes makes it impossible to have a truly proportional distribution. The diamond-shaped dots in the figure show the actual distribution. The figure shows that five outlets have an obviously wrong layout; three of them have too few pipes and two have too many. Explanations might respectively be found in expansion of the served area and the adding of extra pipes.

Water from the river is diverted into the main canal at a weir situated a few kilometres upstream. In order to increase supply, five pumps were added to the system during the 1980s. They pump water directly into the main canal at several points along the river, close to the scheme, just downhill. All pumps are collectively owned (ie they belong to the scheme). During the intervention period they were renewed and repaired. As a result of theft and breakdowns since 2001, only one pump is functioning at the time of writing.

The average water availability is far below the requirements for cropping the whole area, especially during the dry winters. The capacity of the main system is big enough, but losses at field level are very high, which causes low overall efficiencies. Furthermore, the actual water flows in the main canal are much lower than the design capacity (during the period of fieldwork it was only one-third of this).

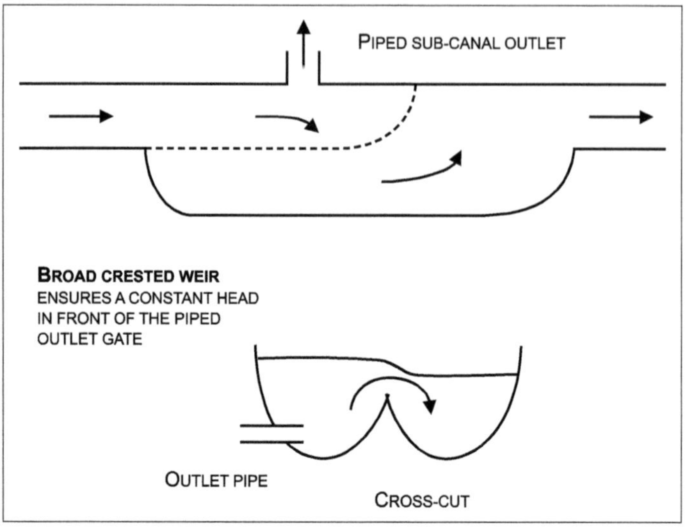

Source: Chapter author

Figure 5.2 *The typical layout of the outlet structure to the sub-canals*

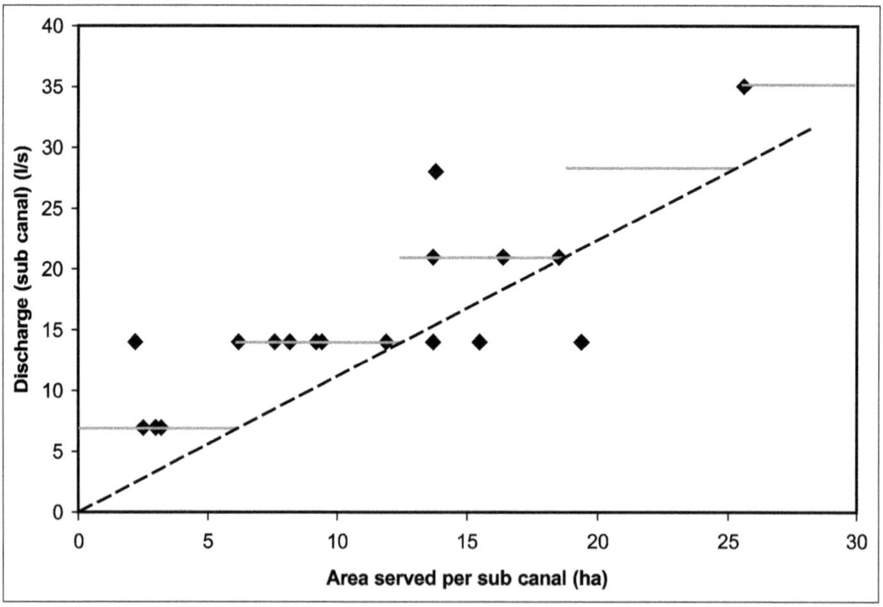

Source: Chapter author, based on field observations and calculations

Figure 5.3 *The relation between the design discharge in a sub-canal and the area served by it*

This is caused by silt and other obstructions in the canal. Regularly cleaning the main system could partly solve this problem (Veldwisch, forthcoming).

Managing the Scheme

During the intervention a representative management committee (MC) was formed. This was in fact a continuation of an existing group of farmers taking a central position in governing the scheme but only formalized during the rehabilitation process. The MC consists of 13 members – 2 representatives of each of the 4 wards and 5 executive members that are directly chosen at a general meeting (see Figure 5.4). The full MC usually meets about once every month, and while meetings of the executive board are exceptional, executive members frequently make contact with each other. Once every quarter a general meeting is held to which all plot holders within the scheme are invited. Usually about 60 people attend these meetings, during which issues such as election of the executive members (once every 5 years), plans for maintenance works, the financial statement and advice on strategic planning of crops to be planted are discussed. Most decisions are, however, made at the MC meetings, which function as a management pivot at the scheme level.

Ordinary farmers, ie not members of the MC, frequently complained about insufficient information sharing. Moreover, many of them did not understand the function of the MC nor the procedures related to it; some did not even know who was representing them on the MC.

The wards (boundaries shown in Figure 5.1) are clusters of three to five subcanals and vary considerably in size and number of people. During periods of intensive cropping and periods of water stress ward meetings are regularly held. At other times six or more months may pass between meetings. Each ward has its own committee consisting of four to six people.

The ward leader plays an important role in formalizing informal arrangements, in dispute resolution and in a sort of fine/repercussion system. In practical

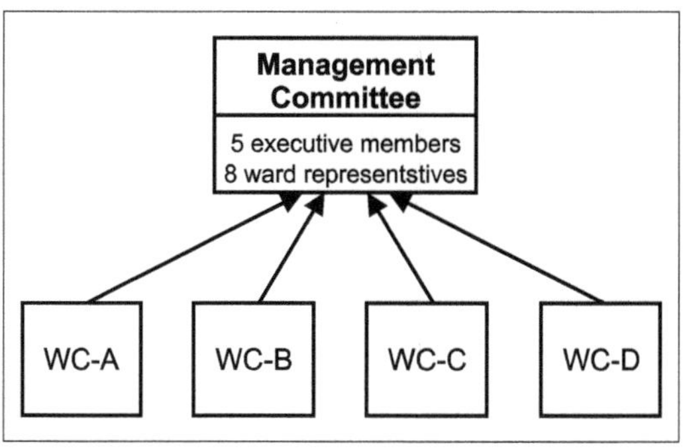

Source: Author's own data
Note: WC = Ward Committee

Figure 5.4 *Organizational structure of the Thabina irrigation scheme*

terms this means that two people who want to exchange their shifts jointly go to their ward leader to inform him[2] about this. In the case of a dispute between two farmers within the ward it is the ward leader's task to intervene. In addition, people come to him if they have complaints relating to the management of the system, and he is usually the person who talks to somebody caught in illegal practices relating to water use. This last is the first step in a very informal repercussion system. People refer to it as 'disciplining', which seems to be restricted to a stern talk. There is not a clear relation between illegal practices and their consequences. In this context, people most often mentioned the 'disciplining' and in some cases depriving somebody of a weekly water shift or assigning somebody the task of cleaning a canal stretch. Accounts of this differ hugely and there is hardly any reference to concrete cases. Therefore it seems that problems are dealt with on an incidental basis rather than in a structured way.

Officially the scheme is still owned by the government, but there is no longer a budget to run it. Only the water bailiff (WB), whose task it is to check the main canal for obstructions, keep it clear of rubbish, check the pipe outlets and open and close the outlet gate of the night storage dam, is still employed by the government.

Within the sub-canal units water is mostly shared informally. Although farmers often refer to an official schedule in which every plot receives water during one fixed day per week, exchanging shifts is common practice.

It was found that even though water shortages are often experienced, no reference was made to water theft within the sub-canal unit, while at the system level this was frequently mentioned as a problem. This perception is connected to a very clear tail-end problem at the system level: the sub-canals at the end of the main canal receive water less than six months per year, whereas the first sub-canals in the system have access to water from the main canal almost year-round. Although aggravated by unauthorized use, the causes of the problem are basically technical: too low an intake from the river, high leakages in the main canal and non-functioning pumps.

Key Issues for Revitalization

On the basis of my research I would conclude that in the Thabina irrigation scheme a reasonably successful rehabilitation was implemented. This, in turn, significantly increased water availability. Meanwhile, under the project, farmers received agricultural training, which many farmers mention as having significantly increased their yields. In addition, an MC consisting mainly of farmers has become the legal management body and is functional as such. Compared to rehabilitation processes of smallholder irrigation schemes that only focus on large infrastructure rehabilitation with a prescriptive, top–down implementation structure, the intervention at Thabina was clearly a step forward, although still largely focused on infrastructure repair in terms of budget allocation (some 90 per cent).[3]

However, the approach used in Thabina lacked a lot of the refinement that was developed in later phases of the revitalization exercise. Some of the issues

Table 5.2 *Responsibilities of three different organizational levels in Thabina: The sub-canal unit; the ward; and the scheme as a whole*

	Organization at the sub-canal unit	
Number of people	Up to 20	
Official organizational body	None	
Working area	The area served by one sub-canal	
Water management tasks	• Opening the piped inlets (connection to main level) • Scheduling irrigation within the sub-canal unit • Informal arrangements regarding irrigation shifts	
Maintenance tasks	• On individual basis cleaning a stretch of the main canal • Maintaining the sub-canal	
Organizational/ institutional tasks	None	
	Organization at the ward	
Number of people	Between 23 and 57	
Official organizational body	Ward meeting and ward committee	
Working area	A cluster of 3 to 5 sub-canals	
Water management tasks	None	
Maintenance tasks	None	
Organizational/ institutional tasks	• Information dissemination • Representation to the management committee • Enforcing rules and regulations	
	Organization at scheme level	
	Farmer management	Government management
Number of people	ca 160	2
Official organizational body	General meeting and management committee (MC)	Extension officer (EO) and water bailiff (WB)
Working area	The whole scheme area (200ha)	
Water management tasks	• Managing the pumps • Coordinating rotational water sharing among sub-canals in periods of little water	WB: Checking pipe outlets
Maintenance tasks	• Coordinating maintenance works • Maintaining major structures like the intake weir and pumps	WB: Keeping the canals free from obstructions

Table 5.2 *continued*

	Organization at scheme level	
Organizational/ institutional	• Rule making	EO: Support the MC
Tasks	• Marketing and strategic cropping advice • Information dissemination	

Source: chapter author

that have been addressed effectively in later approaches (Veldwisch and Denison, 2004), include the following:

- A lack of feeling of ownership over the project; farmers complain that incorrect priorities have been set, that money could have been spent better and that some tasks have never been properly finished.
- A lack of available credit or financial support.
- A big gap between the MC and the rest of the farmers.
- Problems centring around land tenure and the leasing of land; people not interested in farming holding on to their land.

There are, however, two issues in Thabina that have still not been brought up as specific points of attention in later phases: (a) the clear differences between the strategies and interests of commercial farmers and those of growers for home consumption and (b) the interaction between technical layout on the one hand and socio-political relations on the other.

These issues, despite their arising from an early case, are also highly relevant to the current revitalization approach; this section therefore describes the two issues in detail and gives recommendations at the end of each sub-section.

Commercial and Subsistence Farmers

For about 26 per cent of the PTO holders in Thabina, production for home consumption plays a significant role (Perret et al, 2003). These home-producers have smaller networks of contacts than the more commercial users of Thabina. Furthermore, their resources are limited in terms of money, educational background and other means. On the other hand, 32 per cent of the PTO holders are predominantly commercially oriented (Perret et al, 2003); they are characterized by an entrepreneurial attitude, using creative strategies to improve their socio-economic situation. In general, their positions are linked to a higher availability of, or better access to, resources such as knowledge, relations to supporting agents and directly available money and credits. Through this general advantageous position these people have a better chance of getting into the scheme's management, or at least close to it. Almost without exception the people within the MC are commercial farmers and to a lesser extent the same goes for the WCs. Parallel with this there is a huge difference in landholding sizes. People outside the MC have an average landholding of about 1.25ha whereas the average land holding of people in the MC is close to 3ha and that of its executive members is even close to 4ha.

The strong bias towards commercialization in the MC is to a great extent the result of its history in the revitalization programme. The objective of the management transfer was to make irrigation schemes financially self-sufficient. Management and maintenance expenses have to be carried by the users through the payment of user fees to a central management body. These user fees can only be paid if cropping within the scheme has enough economic benefits; this requires commercialization. Although it has never been a clearly expressed objective of the revitalization programme, in Thabina commercialization was clearly part of the design of the programme. It is therefore no surprise that commercial farmers currently dominate the MC.

On the other hand, it should be realized that many poor farmers engage in farming for the sake of food security rather than commercial gain. Time, energy and ability to partake in management tasks and extensive discussions, which are of no direct interest to survival, are often lacking. This does not mean that the large group of poor farmers has no opinion on the management of the scheme; sometimes they fiercely criticize the MC regarding its decisions and actions in water management. The most common issues of complaint are the lack of tractor services, unequal water sharing and non-functioning pumps.

Exemplary of the gap in information between the two groups is the understanding by many ordinary farmers that it is still the extension officer (EO)[4] who is responsible for managing the scheme, as it is the EO they go to when they have complaints about the sharing of water. On the other hand, people in or close to the MC hardly ever refer to the EO. Since the rehabilitation of the scheme and the installation of the MC there has arisen considerable vagueness about management responsibilities. The position of the EO is central to this vagueness. Officially his tasks have shifted so that the position is more of a support role for the MC.

Concerning these issues there are three recommendations for improvement. First, a clear choice has to be made about the objective of commercialization and the role of subsistence farming in smallholder irrigation. The implications of this choice have to be taken into consideration and dealt with in a participative formulation of the intervention plan. Second, after rehabilitation and the establishment of an MC, clarity needs to be established concerning the role of the EO, possibly by creating new concrete tasks and/or strategically replacing existing EOs with new ones. And third, more attention should be given to the foundation of MCs in order to root them in the entire community of users.

Socio-technical Interrelations

The physical layout of an irrigation scheme has consequences for both management possibilities and socio-political processes. In return, the social environment also shapes the scheme's physical layout. In the context of Thabina three examples are discussed in this sub-section: first, the tail-end problem is explained in relation to the layout of the off-take structures; second, it is explained why the ward is, from a hydraulic point of view, not a logical unit for organization; and third, a case is discussed in which individual power to a large extent determined the physical layout.

The tail-end problem

The tail-end problem in Thabina is such that the downstream area of the scheme only receives water during the rainy period, when it is least needed. The head-enders are by definition in a more powerful position, as the water first has to pass their area. The division structures at the off-takes to the sub-canals even add to the advantage of the head-enders. The discharge to sub-canals is independent of the discharge in the main canal, as it only depends on the head over the pipe, which is kept constant by the long crested weir (see Figure 5.2 above). Furthermore, the large number of off-takes connected to the main canal (18) makes it difficult for people in the tail-end to check whether all the upstream off-take pipes are closed. In this system water shortages are by default unequally spread – they solely hit the tail-enders. As an alternative the scheme could have been designed with a proportional division, which would not only proportionally divide the available water but also proportionally spread the impact of shortages throughout the scheme.

This tail-end–head-end contrast has affected the distribution of power within the scheme, even to such an extent that many people in the head-end have developed into relatively powerful commercial farmers, while in the tail-end people are mainly farming for subsistence. Furthermore, commercially-oriented newcomers often settled in the head-end area, while moves within the system have also taken place.

Wards as hydraulic units

In the analyses of the tail-end problem the sub-canal rather than the ward was taken as the unit of analysis, as these are the units separately connected to the main canal. The ward does not have a significant role in water division; it is more an organizational unit than a water management unit, as there is no common water entry point. A ward is a cluster of sub-canals, a cluster of separate water management units, therefore the ward is not seen as a water user. As an alternative each sub-canal could have been represented in the MC, but it would have been even better to physically make the wards separate water management units. The number of off-takes from the main canal could have been limited to four, one for each ward. The four secondary canals (or ward canals) would run parallel to the main canal, each supplying the three to five sub-canals of its ward (see Figure 5.5).

The single water entry point would make the ward a logical organizational and water management unit. It would make it easy for downstream wards to check the gate settings of the other wards. It would also be easy to put padlocks on these four structures and let them be managed by one or two elected farmers. At the same time farmers within a ward would become concerned with the water use of farmers in a neighbouring sub-canal. Currently, farmers in sub-canal 4 are not concerned with how much water is being used upstream, as usually there is

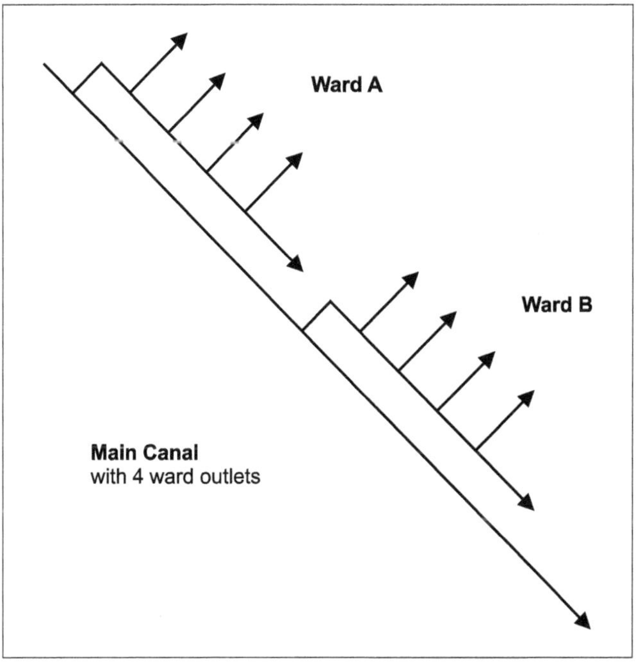

Source: Chapter author

Figure 5.5 *Ward canals as a technical solution for socio-organizational problems*

enough water in the main canal to reach off-take 4. In the proposed re-design, these canals would have to share a limited amount of water, which would ensure that people started to check up on each other.

An individual example

A good illustration of how one can read the socio-political situation of a scheme from its physical layout can be found in off-take 12 (see situation sketch in Figure 5.6). Where this sub-canal takes off there are two supply pipes that replenish the main canal with water pumped straight from the river with a maximum capacity of 380m^3/hr (about 100l/s). Currently it is the only functioning pump within the scheme. One of the pipes enters the main canal below the water surface so that the pipe is always filled with water, even when the pump is not running. Five hydrants (taps) have been constructed on pipe B so that at all times water can be taken, even when the pump is not functioning. It is the only place in the scheme with a pressurized off-take from the main canal and was constructed during the rehabilitation process.

One of the farmers who recently moved to sub-canal 12 is the vice-chairman of the WUA and crops 2ha, another is the chairman of the WUA with 5ha of land, the largest land holding within the scheme. Officially the MC decides on

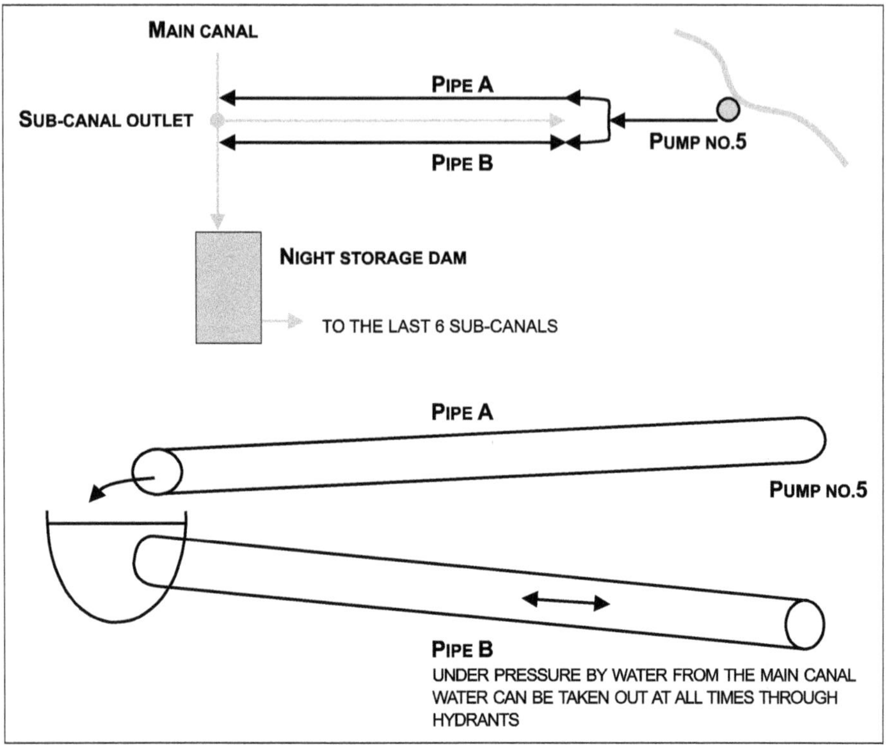

Source: Chapter author

Figure 5.6 *Sketch of the situation at sub-canal 12*

the operation of the pump, but in practice the chairman mainly controls it. He earned a lot of credit when he arranged for the pump to be repaired after it was vandalized; besides the people on sub-canal 12, people in the tail-end can also profit from the extra water. However, since excessive use of water at sub-canal 12 also represents a direct loss to the farmers downstream, they are critical of the pressurized pipe and the excessive use of water by the chairman. This criticism of the construction of the pressurized pipe is not openly expressed, as they know that criticizing the structure implies criticizing the people allied with the structure. It is the alliance of the chairman with this construction that gives it its stable character. Contesting the layout of the structure implies contesting the position of the chairman, who has a wide resource base in terms of both tractors and land, and both managerial and political influence. Most probably similar constructions at other places in the system would have been 'corrected' if similar circumstances did not prevail there.

The intricate interrelations between physical layout on the one hand and socio-political processes on the other have been insufficiently acknowledged in the design and rehabilitation of Thabina, and the same applies to many smallholder irrigation schemes around the world. Until this changes, irrigation reforms will continue to suffer from unintended side-effects, which are almost always obstacles to equitable development.

Conclusion

Right from the beginning, the revitalization approach was an innovative one and represented an improvement on mere infrastructure rehabilitation. In Thabina this approach has meant that the scheme now has a functioning MC and that infrastructure has improved on the basis of priorities set by water users. However, the scheme is far from problem-free: pumps have broken down, there are no clear rules and regulations, there is no functioning sanctioning system and the management body is only partly representative of the irrigator community and hardly accountable to them at all, to mention only a few of the difficulties.

The situation in which the intervention team started their work in Thabina was complex and in many respects a direct hindrance to establishing a productive scheme with collective management. First, agriculture was not the dominant source of income for many irrigators; second, the objectives and strategies of the users were very diverse. The fact that the scheme's land is spread over about ten different villages together with its problematic design makes it a complex system to manage, especially because water shortages are frequently experienced.[5] It is possible that if this situation had been seriously taken into account, Thabina would not have become a pilot project in the first place.

In later stages the revitalization approach has been improved. This has involved the shift of budget and effort from infrastructure-centred activities to soft development activities such as in-depth consultation, learnerships, institution building and agri-business development.

Although much has changed in the approach of the revitalization programme since the pilot project in Thabina, two aspects of the intervention that have had a strong negative effect on collective action were not resolved. These, described in the last section above, are:

1. An implicit objective of commercialization, which caused many subsistence farmers and non-farmers to stay within the system, putting a heavy burden on financial self-sufficiency. Assuming that a general tendency towards commercialization is necessary to justify the investment and operational costs of irrigation schemes, the challenge is to find ways to provide those people who are not willing or not able to commercialize a fair chance to continue farming in their preferred way (see also Veldwisch and Denison, 2004, p28).
2. The poor integration of socio-political and technical aspects of irrigation management. The approach towards improving smallholder irrigation systems has become more process-oriented and now also includes agricultural, empowerment and management training, yet a thorough understanding of the integrated nature of social, managerial and technical issues is arguably still lacking. The challenge here lies in building interdisciplinary teams; this goes a step further than a simply multi-disciplinary approach. Moreover, working in a process-oriented way, in which user experts have a substantial input,

overcomes part of the problem as experience shows that the interconnected understanding of an irrigation scheme usually comes naturally (see also Veldwisch and Denison, 2004, p26).

The early experiences in Thabina form a strong argument to make further adjustments to the rehabilitation and IMT processes that are currently starting in the 114 remaining smallholder irrigation schemes in Limpopo Province. It is important to look back at every stage and assess the effects of previous approaches. This makes it possible to learn from experience – to use solutions that worked for certain problems and to prevent making the same mistakes over and over again.

Acknowledgements

The author gratefully acknowledges the support of Dr Sylvain Perret (Cirad and University of Pretoria), who made this research possible by providing the basic funding. Additional funding came from the Water Research Commission through its Guidelines for Revitalization project, led by Jonathan Denison. Institutional support and office space were provided by the International Water Management Institute, South African office and representation.

But foremost I want to thank the people of Thabina, who welcomed me generously and offered me a glimpse behind the scenes of their irrigation scheme and their social life in its connection to irrigated agriculture. A special thanks to Mr Surprise Rasesemola for his important facilitation role during fieldwork.

Notes

1 Not in the formal South African policy sense in which a WUA involves a much larger area and in which the Thabina irrigation scheme collectively could be one member, but in the practical sense: all irrigating farmers are members of one organization that is primarily concerned with managing the irrigation system.
2 Currently all four ward leaders are men.
3 For a more elaborate analysis of the used approach see Veldwisch and Denison (2004, pp3–13).
4 Often he is even referred to as 'programme manager', as the function was called while the scheme was under government management.
5 Based on a list with enabling criteria drawn up by Agrawal (2001).

References

Agrawal, A. (2001) 'Common property institutions and sustainable governance of resources', *World Development*, vol 29, pp1649–1672

Callon, M. (1991) 'Techno-economic networks and irreversibility', in Law, J. (ed) *A Sociology of Monsters: Essays on Power, Technology and Domination,* Routledge, London, pp132–161

Chambers, R. (1997) *Whose Reality Counts? Putting the First Last*, ITDG Publishing, London

Hope, R. A. & Gowing, J. W. (forthcoming) *Does Water Allocation for Irrigation Improve Livelihoods? A Socio-Economic Evaluation of a Small-Scale Irrigation Scheme in Rural South Africa.* Centre for Land Use and Water Resources Research, University of Newcastle-upon-Tyne, UK

Latour, B. (1991) 'Technology is society made durable', in Law, J. (ed) *A Sociology of Monsters: Essays on Power, Technology and Domination,* Routledge, London, pp130–131

Long, N. (2001) *Development Sociology; Actor Perspectives.* Routledge, London and New York

LPDA (2002a) 'Water care programme (revitalization of smallholder irrigation schemes): Progress report', Limpopo Provincial Department of Agriculture, Polokwane, South Africa

LPDA (2002b) *Revitalisation of Smallholder Irrigation Schemes: A Business Plan for the Revitalisation of Smallholder Irrigation Schemes in the Limpopo Province,* Limpopo Provincial Department of Agriculture, Polokwane, South Africa

Mollinga, P. P. (2003) *On the Waterfront: Water Distribution, Technology and Agrarian Change in a South Indian Canal Irrigation System,* Wageningen University Water Resource Series 5, Orient Longman, Hyderabad, India

NPDALE (1998) 'Planning and implementation of irrigation schemes: Thabina Irrigation Scheme pilot project', report on the pre-development survey, Northern Province Department of Agriculture, Land and Environment, Polokwane, South Africa

NPDALE (2001a) *Request for Approval to Transfer Ownership and Management of Three Pilot Smallholder Irrigation Schemes From Government to the Beneficiaries Currently Utilising the Schemes,* NPDALE, Polokwane, South Africa

NPDALE (2001b) 'Water care programme: Progress report', Northern Province Department of Agriculture, Land and Environment, Polokwane, South Africa

Perret, S. (2002) 'Water policies and smallholding irrigation schemes in South Africa: A history and new Institutional challenges', *Water Policy,* vol 4, pp283–300

Perret, S., Lavigne, M., Stirer, N., Yokwe, S. and Dikgale, K. S. (2003) 'The Thabina irrigation scheme in a context of rehabilitation and management transfer: Prospective analysis and local empowerment', report no 46/03, Department of Water Affairs and Cirad-Tera, Montpellier, France

Radley, R., Crosby, C. T., Barnes, A., Fuls, J., Hattingh, C., Richters, E., and Pienaar, J. H. (1999) 'Improvement of water use efficiency on Thabina, Boschkloof and Morgan', ARC-IAE, Institute for Agricultural Engineering, Pretoria, South Africa

Shah, T., Van Koppen, B., Merrey, D., De Lange, M. and Samad, M. (2002) 'Institutional alternatives in African smallholder irrigation: Lessons from international experience with irrigation management transfer', IWMI research report no 60, Colombo, Sri Lanka

Veldwisch, G. J. A. (forthcoming) 'Running together or alone? Institutional analysis of water management at the Thabina Irrigation Scheme, South Africa', IWMI working paper, Colombo, Sri Lanka

Veldwisch, G. J. A. and Denison, J. (2004) 'From rehabilitation to revitalisation: The evolution of a small scale irrigation revitalisation approach in the Limpopo Province, South Africa', project no K/5/1463//4, report no 3, Water Research Commission

Vermillion, D. L. (1997) 'Impacts of irrigation management transfer: A review of evidence', research report no 11, IWMI, Colombo, Sri Lanka

6

Water Management on a Smallholder Canal Irrigation Scheme in South Africa

Simon S. Letsoalo and Wim Van Averbeke

Introduction

Management Transfer in South African Smallholder Irrigation

Before 1994 state organs were heavily involved in the establishment, operation, maintenance and management of smallholder irrigation schemes in South Africa (Van Averbeke et al, 1998, p5; Bembridge, 2000, p3; Merrey et al, 2002, p97). Since 1994 the government has implemented a policy of irrigation management transfer (IMT) to reassign these functions from the state or its agencies to farmers (Perret, 2002). IMT is a universal trend motivated by governments seeking to reduce recurrent expenditure on irrigation (Vermillion, 1997, p13; Shah et al, 2002, p64), and by the realization that irrigators must be the primary decision-makers in the management of scheme resources (Thompson, 1991, p15). Adoption of IMT policy in South Africa has been explicit in some parts, such as Limpopo Province (Limpopo Department of Agriculture, 2002, p7), and implicit in others, such as the Eastern Cape (Commission of Enquiry, 1995, piii). In Limpopo Province, IMT has been linked to the rehabilitation of smallholder irrigation schemes.

Social Capital and Institutions in Irrigation

It has long been realized that rehabilitation of irrigation schemes should proceed from a comprehensive understanding of both the physical apparatus and the associated social organization of existing irrigation activities (Bagadion and Korten, 1991, p123; Coward, 1991, p46). However, in the past the focus was primarily on the physical and technical aspects of these projects, while their social aspects were largely ignored (Bagadion and Korten, 1991, p123; Coward, 1991, p53; Ostrom, 1995, p125).

Social capital can be defined as the knowledge and skills individuals bring to the process of solving problems (Ostrom, 1995, p126). It is created when individuals spend time and energy working with others to find ways of achieving certain ends (Coleman, 1966, p98). To varying extents, participation in an irrigation scheme makes farmers interdependent, especially for water. As a result, to be successful, farmers have to work together.

Effective cooperation among farmers is dependent on the presence of functional institutions. The institution in this context is defined as a generic concept referring to a variety of rules that direct patterns of social behaviour (Bromley, 1982, p3; Swift and Hamilton, 2001, p85). In the management of water in canal irrigation, institutions stipulate how water is to be distributed among farmers and assign responsibilities for the care of that water (Ostrom, 1992, p46). Without enforceable rules that allocate both rights and duties, some individuals are tempted to take more than their share of water or to be opportunistic and free-ride on the efforts of others when it comes to care-taking duties (Ostrom, 2002, p464). Water management institutions may be established by farmers themselves, or in conjunction with others, such as government officials or irrigation agencies, and they may be formalized or remain informal (Coward, 1991, p46). The establishment and upholding of institutions are forms of creating social capital.

Institutions and Values

When deciding on the content of institutions people use their values and value systems as a frame of reference (Hubbard, 1997, p240). Values represent basic convictions that particular modes of conduct or end-states of existence are personally or socially preferable to others. Values influence attitudes and perceptions and are used to make judgements about social behaviour (Robbins, 2001, p62). Values have attributes of content and intensity. The content attribute of a value identifies a particular mode of conduct or end-state of existence as being important, while the intensity attribute specifies how important it is. Values ranked in terms of their intensity form a value system (Deacon and Firebaugh, 1981, p40; Robbins, 2001, p62).

Values may be absolute, normative or relative. Absolute values are firmly held regardless of surrounding factors or conditions. They are usually deeply rooted, resistant to change and often justified on spiritual or other fundamental grounds (Deacon and Firebaugh, 1981, p39). Normative values, like absolute values, tend to be prescriptive and binding. They are reinforced through people's own experiences or by the expectations of those around them. Often they have evolved from a consensus on how to deal with particular recurring situations (Deacon and Firebaugh, 1981, p39). Relative values are subject to interpretation and depend on an evaluation of circumstances (Ostrom, 2002, p462). The application of relative values is therefore subject to negotiation.

Institutions and Organizations

Linked to institutions are patterns of social interaction, which are referred to as social organizations (Hubbard, 1997, p240). Social organizations play a particularly important role in the way institutions are upheld. According to Coward (1985, p48) there is often inconsistency between what people accept as ideals of role expectations, which is the institutional dimension of a social system, and what actually occurs in terms of role performance, which represents the organizational dimension of the system. He points out that a lack of correspondence between role expectations and role performances is an important force for change in either the institutional or the organizational dimensions of a social system.

Conflict

Conflict is a general behavioural phenomenon of human existence and embraces a wide spectrum of social relations. Conflict is almost unavoidable in groups where members compete with each other for a scarce resource (Deutsh, 1973, p10). In many irrigation schemes water is a scarce resource, because demand for water exceeds supply (Chambers, 1988, pp158–163). Typically competition for water arises within a group of farmers served by a single supply furrow, or among groups of farmers on different irrigation blocks served by a single canal (Ostrom, 2002, pp466–467). A functional and robust social system is necessary to manage the interdependence among farmers so that conflicts may be avoided as much as possible and be resolved when they do occur (Bagadion and Korten, 1991, p74). Ineffective management of conflict usually has a negative effect on the overall performance of a scheme, and in severe cases may lead to the complete abandonment of irrigation activities, even when from a technical perspective the scheme is fully functional (Bembridge, 2000, p3).

Within a social system, such as a smallholder irrigation scheme, two types of interpersonal conflicts arise. The first consists of clashes between individuals over procedural or how-to-do-it problems. These form an 'affective' conflict. Such a conflict does not ordinarily stem from a disagreement on opinions or beliefs but from a struggle based on selfish or personal issues. Members of a group who refuse to follow the procedures of the group are called 'role deviants'; they are regarded as undesirable by others in the group, because the conflicts they create do not pursue group goals (Fischer and Ellis, 1990, p25). The second type of interpersonal conflict consists of intellectual opposition of group members to the content of ideas or issues pertinent within the group. This constitutes a 'substantive' conflict and involves 'opinion deviants'. Opinion deviants disagree with other group members about the content of ideas. Unlike role deviants they are tolerated and sometimes even admired by fellow members, because they are involved in a realistic conflict as a means to an end. The behaviour of opinion deviants is often intended to further the group's progress towards its goal (Fischer and Ellis, 1990, p26).

Objectives of the Study

IMT transfers power to farmers, including the power to develop or modify management systems. The current study sought to investigate what happened when a group of black South African smallholders acquired the power to transform or modify their water management system and to determine the probable reasons for any changes that were made. For this purpose the water management system in Dzindi, an irrigation project that has prevailed for five decades, was investigated. The objectives of the study were (a) to document the institutional and organizational dimensions of the system that have been used to manage water in Dzindi and any changes these dimensions were subjected to; and (b) to assess the functioning of the current water management system with special reference to its effectiveness under different circumstances. In the analysis of the water management system in Dzindi the institutional dimension was taken to refer to ideals of role expectations, including both normative (rules) and structural aspects, and the organizational dimension of the actual role performances, as proposed by Coward (1985, p49).

Methodology

Study Approach and Methods

The investigation used a case study approach and selected Dzindi as the case. Data were collected from different sources using a range of methods from January 2003 to October 2005. The first part of the study sought to document the institutional and organizational dimensions of the system used to manage water in Dzindi and to analyse its underlying values. Historical information on the water management rules and associated structural arrangements was obtained by constructing a time-line in which a collective of plot holders engaged in the participatory reconstruction of the history of the scheme with special attention to how irrigation water was managed. Additional detail on the history of the scheme was acquired by conducting focused interviews with individual participants, especially the five surviving plot holders who were present when Dzindi was established. The current situation was documented using written records held by the Scheme Management Committee (SMC) and through interviews with members of the SMC, the water bailiffs, the local extension officer and selected farmers in the different irrigation blocks.

During the subsequent phase of the study the practical application of the water management system was investigated using consecutive cycles of data collection, analysis and interpretation. Data collection methods involved participant observation of SMC meetings; field observations made during transect walks through the scheme, alone or in the presence of the bailiffs; and face-to-face interviews with a large number of farmers and other participants.

Study Site

To conduct the study, Dzindi (23°01'S; 30°26'E), a smallholder irrigation scheme in Limpopo Province (Figure 6.1), established in 1954, was selected because relative to other smallholder schemes in Limpopo Province and elsewhere in South Africa it has had limited state intervention and has not yet been revitalized.

Dzindi is a surface irrigation scheme covering an area of 136ha, which has been subdivided into 106 plots of 1.28ha (1.5 morgen) each. All plot holders in Dzindi farm single plots except for two who have double plots. Production in

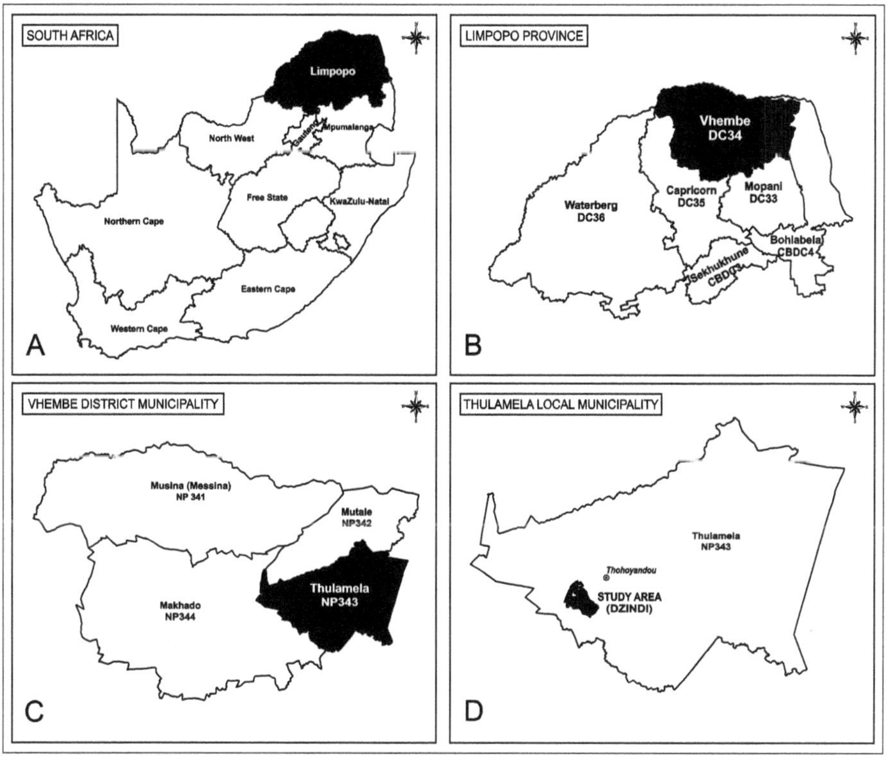

Source: Chapter authors

Figure 6.1 *Location of Dzindi*

Dzindi consists mainly of a rotation of maize in the summer and vegetables in the winter. Dzindi obtains its water from a weir on the Dzindi river. At the weir water enters the main concrete canal, which runs for about 14km and which conveys the water to the 4 irrigation blocks (Figure 6.2).

Farmers obtain water from secondary concrete canals, which lead the water from the main canal to the plots. Even when the flow in the river is adequate, the amount of water entering the scheme is sufficient only to allow farmers one irrigation per week. Low flow in the Dzindi river during winter and spring, deterioration of the conveyance system in the form of cracks in the secondary concrete canals and subsidence of parts of the main canal further limit the availability of irrigation water (Cadet et al, 2003).

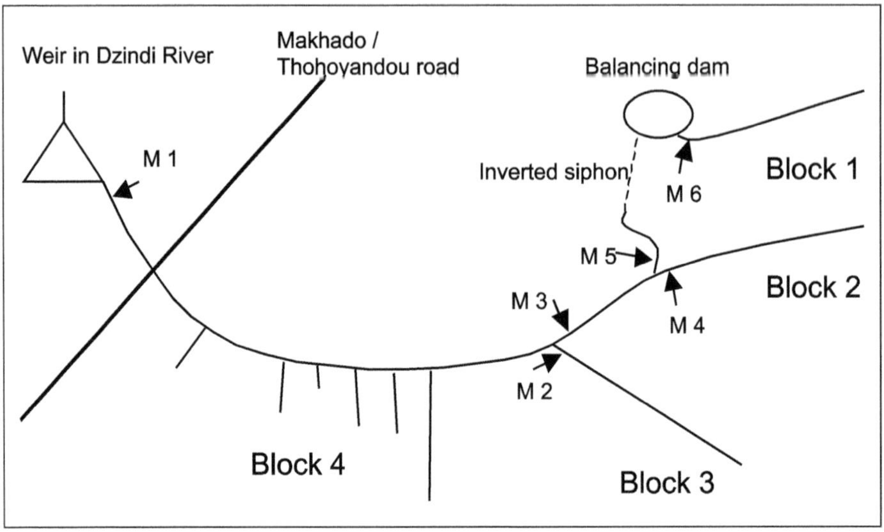

Source: Van der Stoep and Nthai (2005)

Figure 6.2 *Schematic layout of the water distribution network in Dzindi (not to scale)*

Results and Discussion

Introduction

Obtaining information about the formal aspects of the system that was used to manage water at Dzindi was relatively easy, because participants treated this information as public. For the same reason, recorded offences against the rules held by the SMC were discussed freely. However, the water management system also had an informal side, which was particularly relevant to the way plot holders interacted with each other in the sphere of water allocation. Collecting information on this informal, hidden side was difficult. Essentially participants were only prepared to discuss a particular element of the informal side when presented with evidence that the authors had already uncovered that element, and even then some elements were simply not talked about except in the most vague and general terms. Consequently, although considerable progress was made towards achieving the objectives of the study, the findings presented here must remain the result of work in progress.

The Formal Institutional Dimension of the Water Management System at Dzindi

When Dzindi started operating in 1954, the extension officer appointed to the scheme verbally introduced three rules to manage the use of irrigation water: (a) farmers were to irrigate in accordance with a timetable, which allocated irrigation time slots of half a day per week to each plot; (b) obstructing the flow in the main canal was not allowed; and (c) washing of one's body or laundry in the canal was forbidden. Two bailiffs in civil service were employed to police these rules. They reported offences to the SMC, which consisted entirely of elected plot holders, with the extension officer (EO) also in attendance. The SMC judged the cases brought before it and imposed monetary fines. Money collected from fines was banked in a scheme account and was used for the public good. When the SMC was unable to bring a case to its conclusion it referred the matter to the Chief, who, without exception, endorsed SMC recommendations. Involvement of the Chief resulted in the fine being doubled in order to acknowledge his contribution.

Cleanliness and equality in access to water were the values on which the water rules in Dzindi were based. Cleanliness is a normative value among the Venda people. In the traditional Venda belief system illness is caused by spiritual or personal agencies, such as ancestors and sorcerers (witchcraft), or by being in a state of pollution, which occurs at birth, after illness, after a crime and after a burial. Pollution is thought of as dirt, explaining why rituals of washing feature prominently in the purification process (Hammond-Tooke, 1993, pp178–179). Washing bodies or laundry directly in the canal causes body dirt to pollute the water. During their daily activities plot holders regularly come in contact with canal water, and they also use it to wash harvests of leafy vegetables. Consequently, they regard the idea of working with water that has been polluted with body dirt as repulsive.

Equality, on the other hand, is not a value local African society upholds. On the contrary, there is a general acceptance that individuals differ in terms of their position in society based on descent, age and gender (Hammond-Tooke, 1993, p99; Magubane, 2000, p179). Associated with position is the expectation of being shown respect, this including material expressions thereof. It follows that the value system that underpinned the water management rules in Dzindi was not entirely congruent with that held by the local society. Under such circumstances changes to the institutional or organizational dimensions of the water management system are likely to occur (Robbins, 2001, p67).

During its existence, Dzindi has lived through three major political eras. From establishment until about 1970 the scheme was under the control of the Department of Bantu Administration and Development of South Africa. This department pursued a vision of successful smallholder irrigation schemes with plot-holder households deriving full livelihoods from farming (Commission for the Socio-economic Development of the Bantu Areas within the Union of South Africa, 1955, p116). In Dzindi, the department introduced a range of support measures in pursuit of this vision, including a public land preparation service, but its stance towards farmers tended to be paternalistic and authoritarian. For example, the responsible Bantu Affairs Commissioner regularly visited Dzindi to verify that plot holders were sufficiently industrious on their plots, and according to surviving plot holders he did not hesitate to expel farmers who failed to display the necessary diligence. The commissioner derived his authority to remove plot holders from the scheme from the conditions of trust tenure and the permission to occupy (PTO) agreement with which farmers held plots in Dzindi. Effective utilization of the land at all times was one of the conditions farmers had to comply with in order to keep their plots.

From about 1970 until 1994 Dzindi experienced the homeland era, during which political authority was transferred from the Department of Bantu Administration and Development to the Government of Venda. In the case of Dzindi, the homeland administration, which was staffed by black people, was less authoritarian than the department. The regular assessment of farms was discontinued and with it the expulsions of plot holders for lack of industriousness. Direct operational support to farming, on the other hand, was maintained.

The democratic government that came to power in 1994 was of the opinion that smallholders needed to take control of their own resources and destinies and that direct operational support to their farming activities by the state was inappropriate (Department of Agriculture, 1995, p17). Implementation of this new vision soon followed; in Dzindi, for example, the public land preparation service was withdrawn in 1996.

Since inception, Dzindi has witnessed a gradual reduction in the influence of the state, especially from about 1989 onwards, when it became evident that democratic change in South Africa was inevitable. Declining state influence created room for the Dzindi community to modify its water management system. It also affected the roles of the civil servants appointed to the scheme. The two bailiffs became progressively less involved in policing duties. Their policing function had always exposed them to verbal abuse from the offenders they caught, and when neither the state nor the SMC insisted that they continue their watch over water use in the scheme, they gradually withdrew from patrolling the canals. Similarly, there was a steady reduction in the moral authority of the EO, a process accelerated by the adoption of the Participatory Extension Approach by the Limpopo Department of Agriculture.

With the exception of some minor modifications the Dzindi community did not tamper with the institutional dimension of the system that was used to manage water, even though they had the power to do so. The minor modifications it made to the rules consisted of granting plot holders permission to exchange irrigation time slots among each other, and authorizing selected farmers to cultivate parcels of land falling outside the scheduled irrigation area, as long as irrigating these parcels occurred during their allocated time slots. The absence of fundamental change to the formal institutional dimension of the system suggested that the water management rules and their underlying values, and the structure that was responsible for applying these rules, enjoyed widespread support within the group. However, a different picture emerged when the organizational dimension of the water management system was investigated.

The Organizational Dimension of the Water Management System in Dzindi

Keeping the canal water clean

Population growth accompanied by the settlement of large numbers of people on land surrounding the scheme has made it increasingly difficult for the Dzindi community to keep the water in the canal clean. Just below the weir, where the water emerges from the tunnel under the main road between Makhado and Thohoyandou, a carwash has been established. This enterprise extracts its water from the canal, which is not a problem, but the concrete surface on which the cars are washed drains into the canal, adding large amounts of soap to the water. Further down the canal, where it proceeds through residential areas, washing laundry in the canal, especially soiled baby nappies, has become common practice. These offences carry on unchecked because the Dzindi community lacks both the authority and the power to enforce the cleanliness rule on non-members of the scheme.

Distributing irrigation water among plot holders

The demand for irrigation water at Dzindi is a function of many variables, but climate is principal. On average, 80 per cent of the annual rainfall in the Dzindi area falls during the period from October to March (Weather Bureau, 1992). During the rainy season demand for irrigation water is often low, while supply is high, because of the strongly flowing river. During the dry period, which starts in April, demand for water tends to increase, reaching a maximum towards the end of the dry period in August and September. During the dry period flow in the river progressively declines, and once the flow rate drops below the threshold level of about $0.37m^3/s$, the volume of water entering the scheme is reduced (Van der Stoep and Nthai, 2005, p15). As a result, most offences against the water sharing rules occur during the dry winter and spring seasons.

In addition to seasonal variability in rainfall there are also substantial differences between years. On occasions the area is subjected to exceptionally high rainfall, such as in 2000 when farming activities in Dzindi had to be suspended for several months because of excess water. Dzindi also experiences periods of drought, such as in 1982–83 and again in 2004–05. Typically a drought develops when summer rains fail; during the subsequent winter and spring the flow in the Dzindi river drops below the threshold level much earlier than usual and continues to decline, causing acute water shortage.

In the ensuing text, the functioning of the organizational dimension of the water management system is described under two different sets of circumstances. The first set of circumstances, which occurred during the winter of 2004, more or less represented the norm at Dzindi in terms of water availability. Following good summer rains irrigation water in the scheme remained relatively abundant throughout the 2004 winter period. The second set of circumstances occurred during the winter of 2005 and represented a period of extreme water scarcity.

Circumstances of relative water abundance

In Dzindi, obstructing the main canal immediately below the distribution furrow turnout and irrigating during time slots allocated to other plot holders are the two main offences against the water sharing rules. Obstructing the main canal below the distribution turnout is done to increase the flow of water reaching one's plot. The higher the flow the less time it takes farmers to irrigate their fields.

When for some reason farmers are prevented from irrigating their plots during their allocated time slot, they seek to use a slot that belongs to someone else. The present system permits time slot exchanges among farmers subject to mutual consent, but some farmers act opportunistically and use the time slots of others without their consent. Another reason for stealing another plot holder's water is when crop water use is very high, and weekly irrigations are inadequate to fully meet crop demand.

Since the two bailiffs no longer patrol the canals, theft of water is usually discovered by plot holders. Usually, when plot holders become aware that the flow of water reaching their plot is less than expected, they go in search of the cause, which in nearly all cases is another plot holder breaking the rules. When an offence is discovered the aggrieved farmer confronts the offender. Usually, the offender immediately discontinues the offending action and pleads for forgiveness. If the apology is accepted, the conflict has been settled.

In cases where the aggrieved plot holder refuses to forgive the offender, the offender seeks assistance from a mediator to resolve the conflict. Old grudges between aggrieved and offending plot holders are typical circumstances for an offence not being settled amicably and immediately. People who are requested to mediate are usually elderly members of the scheme. Mediators are peacemakers; their role is to listen to the accounts of the affected parties without taking sides and to create the necessary conditions to restore the peace. Mediators help detect the underlying reason why the conflict was not settled immediately, and in this way steer the quarrelling parties towards reconciliation. Throughout the negotiations mediators avoid the use of harsh language. Instead they make use of reconciliatory idiomatic expressions and proverbs aimed at strengthening self-perception and rebuilding confidence of the people affected. In this way the mediator steers the offender towards a state of humbleness, remorse and willingness to cooperate and the aggrieved towards a preparedness to forgive.

When mediation fails to bring about a resolution of the conflict, the matter is reported to the block leader. Each of the four blocks has an elected leader; the position is an informal one of authority of indefinite duration, to be discontinued only for reasons of incompetence, illness or death. The main task of the block leader is exercising authority to help resolve conflicts over the sharing of irrigation water. The competence of the block leader is judged accordingly. When an aggrieved plot holder reports a conflict to the block leader, the block leader calls the offender to order, and urges for a resolution.

In some cases aggrieved plot holders by-pass the block leadership and report the problem directly to one of the bailiffs. As indicated earlier, the bailiffs no longer patrol the scheme, but they still become involved when a case of water theft is reported to them. Although the bailiffs are expected to report offences directly to the SMC, their first action is to enquire whether the parties involved have attempted to resolve the matter in amicable ways, using mediation or the intervention of the block leader. In instances where these avenues have not been explored, the bailiffs encourage the affected parties to do so. They only report cases to the SMC when all other means of achieving reconciliation have been exhausted. When an offence is brought before the SMC it enters the formal domain, as described earlier.

Table 6.1 *Relative importance of the different domains and levels at which conflicts between plot holders over irrigation water are resolved in Dzindi*

DOMAIN	LEVEL	PROPORTION OF CONFLICTS SETTLED (%)
Formal	Chief	1
	SMC	2
Informal	Bailiff	6
	Block leader	10
	Mediator	18
	Plot holder	63

Source: chapter authors

Structured interviews with a sample of ten randomly selected farmers using the list of Dzindi plot holders as the sampling frame provided an indication of the relative importance of the formal and informal domains in resolving conflicts among plot holders over water. Considering the sensitivity of the matter, the interviews avoided referral to conflicts in which respondents were personally involved. Instead, participants were asked to identify the different levels at which conflicts over irrigation water were resolved in their neighbourhood. The results of this assessment are presented in Table 6.1.

The data presented in Table 6.1 show that the vast majority (97 per cent) of conflicts between farmers over water in Dzindi are settled before they enter the formal domain. This demonstrates that without changing the formal rules and structure farmers did alter the water management system by modifying its organization. The key alteration was the decentralization of conflict resolution. Whereas the formal institutions assign responsibility solely to the SMC and the Chief for bringing conflict to an end, the organizational changes provided plot holders with the power to resolve disputes themselves.

Farmers perceived the organizational changes as desirable because they found the formal institutions and structure of the system too rigid. A multitude of valid reasons, such as family emergencies, other urgent livelihood activities and administrative obligations could cause them to miss their weekly time slot, and not watering a crop for two weeks usually spelt disaster. Similarly, most farmers had experienced circumstances that limited the time available for irrigation. Increasing the flow to their plot by depositing a large stone in the main canal just below the outlet was the obvious way to speed up the process. To create leeway for themselves farmers established a second organizational domain in which the formal water sharing rules were interpreted in relative terms. This domain was kept hidden from public scrutiny and was never openly discussed at plot holder meetings, even though all concerned knew about it. In this second domain a farmer who broke a rule when circumstances mitigated such action was in most cases forgiven by those directly affected, because they, in turn, expected to be forgiven when they needed to commit an offence. Roberts (1980, p187) refers to this social behaviour as the spirit and practice of mutual obligation.

The organizational changes that were made to the water distribution system indicate that the Dzindi community prefers to handle social relations and conflicts at an informal level. Under such circumstances there is no need for the active enforcement of the rules by a third party. This explains why the Dzindi community and its leadership did not insist that the bailiffs continue with their daily rounds. The adapted function of the bailiffs was to be on call for when the hidden domain was not capable of resolving a conflict situation and the formal domain needed to be invoked to bring about a resolution. Robbins (2001, p66) points out that locating conflict resolution at the informal, interpersonal level enhances uncertainty. The organizational changes the Dzindi community made to the water management system indicate that this group is not adverse to uncertainty.

The hidden organizational domain farmers introduced at Dzindi was intended to enhance flexibility, not to create a free-for-all. Guided by its own value system, it is largely self-regulating. Central to the effective functioning of this domain is the value of good social behaviour, which is based on respect, generosity, assistance to neighbours, kindness and forbearance as traditional social norms (Hammond-Tooke, 1993, p99). For a long time economic changes and exposure to other societies have influenced black South Africans, but traditional norms and values continue to play an important role in their worldview and patterns of social behaviour, especially in rural areas (McAllister, 2001, pp124–147; Ramose, 2002, p12). The current study showed this to also be the case in Dzindi.

Circumstances of extreme water scarcity

When the summer rains fail, the flow in the Dzindi river drops to very low levels during the subsequent winter and spring. This causes competition for water among farmers to rise. Such circumstances occurred during the winter and spring of 2005, following a very dry summer. The frequency with which the water distribution rules were broken increased rapidly, and a crisis developed. Farmers in the two tail-end blocks were particularly badly affected, because higher up farmers diverted most of the water into their turnouts by placing stones in the canal.

The water crisis accentuated the behavioural traits and personalities of different members of the group, and five broad behavioural categories could be identified. The first group, the 'bullies', openly broke the rules, thereby publicly dismissing the norms of good social behaviour. The second group, the 'opportunists', sneakily sought out opportunities to obtain water in excess of their legitimate share. By operating on the sly they tried to avoid social repercussions for their actions. Ostrom (2002, p464) refers to irrigators who actively pursue disproportionate advantages as 'rent seekers', and both the bullies and opportunists in Dzindi fitted this description. The third group, the 'righteous', abided by the rules, irrespective of the severity of the water crisis. They followed the irrigation timetable to the letter, and whenever they ran out of time to complete the irrigation of their crops during their allocated time slot, they resorted to night irrigation, which is permitted practice in Dzindi (Letsoalo and Van Averbeke, 2005). The fourth group, the 'evaders', fundamentally changed their farming activities in response to the crisis. They either reverted to dryland farming or they suspended farming completely, focusing on alternative livelihood activities instead. This behaviour enabled them to avoid being drawn into conflicts over water. The last group, the 'marginalized', was the group of plot holders that the bullies and opportunists took most advantage of. When the right to water of the marginalized was compromised, and they confronted the offenders, or complained to the bailiffs, the SMC, or the extension officer, their pleas were completely ignored. Typically, the marginalized were female plot holders who were widows or whose husbands were incapacitated by illness.

The differentiation in behaviour among plot holders was not restricted to the 2005 water crisis. Several role deviants displayed their particular behaviour also during times of relative water abundance. For example, for the past two years a farmer who purchased a pump has been extracting water directly from the main canal to irrigate a non-scheduled parcel of land he was permitted to cultivate. His pumping significantly reduces the stream of water that reaches lower-lying blocks, and plot holders farming in these blocks have repeatedly tried to convince him to stop pumping directly from the main canal. When their pleas were ignored, they complained to the SMC, which, in turn, pointed out to the farmer that his pumping from the main canal deprived tail-end farmers, but to no avail. The farmer argued that he only pumped water from the canal during his allocated time slot, and that this did not constitute a breach of the water management rules. Whereas his argument may be valid from a legal perspective, the practice of the farmer clearly violates the value of equality in access to water that underpins the management system in Dzindi; moreover it also blatantly disregards the norms and values that bolster the informal organizational domain.

Marginalization, particularly of female farmers who were *de jure* or *de facto* single, and whose plots were at the tail-end of a distribution furrow, also persisted during times of relative abundance of water. Typically plot holders higher up the furrow considered the time slots allocated to marginalized farmers as open time and used it for their own purposes whenever they felt like it, depriving the marginalized farmer of water. The reason for the marginalization of particular farmers, apparently with the consent of the group, has not yet been investigated. The observation that single female farmers were the prime target for marginalization may be linked to women occupying the bottom rung of the social ladder in traditional Venda society.

The social chaos experienced during the 2005 water crisis called for a ruling by the SMC to restore order, but the SMC did not intervene. Early in October, as the water crisis worsened, plot holders started to openly voice their opinions on how the problem needed to be addressed. One opinion was that the SMC needed to exercise its authority, and called for strict application of the rules and the systematic punishment of all offenders until such time that order was restored. The second was that the farmers in the block nearest to the weir needed to have their times reduced, because they were extracting a disproportionately large share of the water entering the canal. The third, which was ultimately sanctioned at a mass meeting, drew on experiences of the previous extreme drought in 1982–83. During that drought a bi-weekly irrigation timetable, which provided farmers with a full day of irrigation every second week instead of half a day every week, was introduced and adopted. Extending the duration of the time slot available to each farmer from half a day per week to a full day every fortnight is useful because low flow extends the time period required to complete the irrigation of a plot. Socially this particular solution was probably preferred because it did not seek to punish individuals, nor did it accentuate differences among groups of plot holders. Instead it focused on how the group as a whole could deal with the crisis in a harmonious way, until rain brought relief.

Conclusions

Summary of Findings

In Dzindi a water management system based on the values of cleanliness and equality of access was imposed by the state on establishment. Organizationally, the system relied on enforcement of the rules. Bailiffs in civil service employment patrolled the scheme and the elected scheme leadership (the SMC) judged and punished farmers that were caught breaking the rules. When the community acquired the power to modify the formal institutional dimension of the system, it refrained from doing so. Instead it elected to modify the system by making changes to the organizational dimension by injecting flexibility. This was achieved by creating a second organizational domain, which was kept concealed from public scrutiny. In this hidden domain distribution of water and conflicts arising from it were dealt with informally among farmers. This arrangement enhanced flexibility. Farmers could now break the water management rules when circumstances required without getting punished. The hidden domain was guided by its own norms and values, in which good social behaviour was central. The formal organizational domain characterized by centralization and rigid application of the rules was kept in place, to be invoked whenever a conflict could not be resolved in the informal domain. Consequently, the water management system in Dzindi was the product of a fusion of two value systems, one that was introduced to the community at the same time as irrigation, in which equality of access (to water) was a supreme and absolute value, and another, the traditional value system, in which the value of good social behaviour based on the norms of respect, generosity, assistance to neighbours, kindness and forbearance provided guidance.

Under conditions of relatively abundant availability of water the two-tiered organizational structure of the water management system functioned fairly effectively. The majority of offences against the rules were dealt with informally, limiting the time scheme management had to spend on maintaining order and resolving conflicts. However, two important weaknesses in the system were identified. The first related to the expectation of good social behaviour from all members of the group, which was shown not to be valid. Certain individuals blatantly disregarded the traditional value system and grabbed a much larger share of the water than they had a right to. Considerable evidence indicated that the system struggled to deal with such role deviants. The second weakness was that the system condoned the marginalization of certain members of the community, particularly single women.

When circumstances of extreme scarcity arose in Dzindi the weaknesses of the two-tiered water management system were accentuated. Farmers considered breaking the water management rules as justified because of the low availability of water, causing a rise in the frequency of offences. The hidden organizational domain was incapable of handling such circumstances, but when plot holders tried to invoke the formal organizational domain to restore a degree of order and equity, the scheme leadership in the form of the SMC did not assume its institutional role. Addressing the chaos required that the SMC take drastic action, including reinstating the patrolling mandate of the two bailiffs. However, fear of offending members of the community by making harsh decisions dominated the mood, causing the SMC to adopt a wait-and-see stance, which pleased no one but did not upset anyone either.

When elected or appointed to management positions in an organization, people are expected to make decisions in accordance with their mandate. In Dzindi, SMC members were generally reluctant to execute their management mandates. Instead they readily referred matters to general meetings for a decision in an attempt to achieve consensus. Delayed progress and inconclusiveness were key weaknesses of their management style. Both the community at large and the members of the SMC appeared to find it difficult to distinguish between the roles of SMC members as mandated managers and their roles as ordinary members of the group. The non-acceptance of role separation was a primary reason why the SMC tended to vacillate, preferring indecisiveness over the likelihood of slighting members of the group by ruling or deciding on controversial matters.

The water management system in Dzindi was also incapable of enforcing the water cleanliness rule. This was mainly due to the arrival of outsiders into the area, over which farmers and the SMC had little or no moral authority. In the absence of a broader institutional system and authority to enforce the cleanliness rule, the quality and safety of the canal water in Dzindi is bound to deteriorate further.

Lessons from the Dzindi Case

The Dzindi case shows that an indigenized community-based water management system is empowering for the group, because it enables different members to obtain experience in management. In Dzindi, where scheme management in the form of the SMC was elected regularly from among the plot holder community for a term of three years, new people regularly joined the management team. This included women, although in practice their portfolio was limited to that of secretary. Through their involvement in scheme management, plot holders in Dzindi built, sustained and reproduced a reservoir of empirical knowledge, which they tapped whenever the group needed to find solutions to challenges. The Dzindi study also indicates that indigenized water management systems can be maintained by participants themselves, at no direct cost to society at large.

The Dzindi case demonstrates that when a water management system is indigenized, social concerns can take preference, even when this brings about economic inefficiencies. At Dzindi the indigenized water management system pursued a sense of belonging among members of the group, which was achieved by encouraging generosity and compassion towards others, and forgiveness towards those that did wrong. When all members of the group subscribe to the values of such a system, both social and economic objectives can be realized effectively, but in Dzindi this was not the case. There were several role deviants who disregarded the traditional value system to persistently grab a disproportionately large share of the water. Their actions created a state of dissonance, and proponents of the system had to expend considerable amounts of time and effort to maintain concord, making the Dzindi management system expensive to maintain from an opportunity cost perspective. According to Ostrom (1995, p130) social systems need to be cheap to be sustainable, suggesting that the system in Dzindi is under strain.

By not explicitly adopting equality of access to water as an absolute value in the different dimensions of the water management system, scheme productivity in Dzindi was affected negatively. First, by allowing farmers to take non-scheduled land into cultivation, inequality in terms of plot size was created, resulting in less water being available per unit area. Second, by not dealing decisively with rent seekers using equality of access as the guiding principle, the management system caused other plot holders in the scheme to receive less water than was due to them, reducing the productivity of their farms.

In Dzindi the most obvious alternative to the indigenized water management system was the original system, which involved the absolute interpretation and strict enforcement of the rules to ensure that the underlying values of equality in access to water and cleanliness were upheld. However, evidence from Dzindi suggests that effective implementation of this system depends on the availability of an external agency to police and enforce the rules. The cost of such an agency would most probably have to be borne by society at large.

References

Bagadion, B. U. and Korten, F. F. (1991) 'Planning technical and social change in irrigated areas', in Cernea, M. M. (ed) *Putting People First: Sociological Variables in Rural Development,* 2nd edn, Oxford University Press, Ibadan, Nigeria, pp73–112

Bembridge, T. J. (2000) 'Guidelines for rehabilitation of small-scale farmer irrigation schemes in South Africa', Water Research Commission report no 891/1/00, Pretoria

Bromley, D. W. (1982) *Improving Irrigated Agriculture: Institutional Reform and the Small Farmer,* The International Bank for Reconstruction and Development and The World Bank, Washington, DC

Cadet, Y., Delcourt, K., Hoarau, L., Steinmetz, N., Ralivhesa, K. E., Letsoalo, S. S. and Van Averbeke, W. (2003) 'Losses in the distribution system of Dzindi irrigation scheme', poster paper presented at the Annual National Conference of the South African Association of Agricultural Extension, 19–22 May, Bela-Bela, South Africa

Chambers, R. (1988) *Managing Canal Irrigation: Practical Analysis from Asia,* Cambridge University Press, Cambridge, UK

Coleman, J. (1966) *Equality of Educational Opportunity,* Government Printing Office, Washington, DC

Commission for the Socio-economic Development of the Bantu Areas within the Union of South Africa (1955) *Summary Report,* Government Printers, Pretoria

Commission of Enquiry (1995) 'Ncora irrigation scheme', Agriculture and Rural Development Research Institute and the Faculty of Agriculture, University of Fort Hare, Alice, South Africa

Coward, E. W. Jr (1991) 'Planning technical and social change in irrigated areas', pp46–71 in Cernea, M. M. (ed) *Putting People First: Sociological Variables in Rural Development,* 2nd edn, Oxford University Press, Ibadan, Nigeria

Deacon, R. E. and Firebaugh, F. M. (1981) *Family Resource Management: Principles and Application,* Allyn and Bacon, Inc., Boston, MA

Department of Agriculture (1995) *White Paper on Agriculture,* Government Printers, Pretoria

Deutsh, M. (1973) *The Resolution of Conflict: Constructive and Destructive Processes,* Yale University Press, New Haven, CT

Fischer, B. A. and Ellis, D. G. (1990) *Small Group Decision-Making: Communication and the Group Process*, McGraw Hill Publishing Company, New York

Hammond-Tooke, D. (1993) *The Roots of Black South Africa*, Jonathan Ball Publishers, Johannesburg

Hubbard, M. (1997) 'The new institutional economics in agricultural development: Insights and challenges', *Journal of Agricultural Economics*, vol 4, no 2, pp239–249

Letsoalo, S. S. and Van Averbeke, W. (2005) 'Sharing the water: Institutional and organisational arrangements at Dzindi Irrigation Scheme in South Africa', *South African Journal of Agricultural Extension*, vol 34, no 1, pp34–43

Limpopo Department of Agriculture (2002) 'Revitalisation of smallholder irrigation schemes', Limpopo Department of Agriculture, Polokwane, South Africa

Magubane, B. M. (2000) *African Sociology – Towards a Critical Perspective*, Africa World Press Inc., Asmara, Eritrea

McAllister, P. (2001) *Building the Homestead: Agriculture, Labour and Beer in South Africa's Transkei*, African Studies Centre, Leiden, The Netherlands

Merrey, D. J., Shah, T., Van Koppen, B., De Lange, M. and Samad, M. (2002) 'Can Irrigation Management Transfer Revitalise African Agriculture? A review of African and International Experiences', pp95–104 in Sally, H. and Abernerthy, C. L. (eds) *Private Irrigation in Sub-Saharan Africa: Regional Seminar on Private Sector Participation and Irrigation Expansion in Sub-Saharan Africa*, International Water Management Institute, 22–26 October 2001, Accra, Ghana

Ostrom, E. (1992) *Crafting Institutions for Self Governing Irrigation Systems*, Institute for Contemporary Studies Press, San Francisco, CA

Ostrom, E. (1995) 'Constituting social capital and collective action', pp125–160 in Keohane, R. O. and Ostrom, E. (eds) *Local Commons and Global Interdependence*, Sage Publications Ltd, London

Ostrom, E. (2002) 'Crafting institutions', pp452–478 in Wolf, A. T. (ed) *Conflict Prevention and Resolution in Water Systems*, Edward Elgar Publishing, Northampton, MA

Perret, S. R. (2002) 'Water policies and smallholder irrigation schemes in South Africa: A history and new institutional challenges', *Water Policy*, vol 4, no 3, pp283–300

Ramose, M. B. (2002) *African Philosophy through Ubuntu*, Mond Books Publishers, Harare, Zimbabwe

Robbins, S. P. (2001) *Organizational Behaviour*, Prentice Hall, Upper Saddle River, NJ

Roberts, M. (1980) 'Traditional customs and irrigation development in Sri Lanka' pp186–202 in Coward, E. W. Jr (ed) *Irrigation and Agricultural Development in Asia: Perspectives from the Social Sciences*, Cornell University Press, Ithaca, NY

Shah, T., Van Koppen, B., Merrey, D., De Lange, M. and Samad, M. (2002) 'Institutional alternatives in African smallholder irrigation: Lessons from international experience with irrigation management transfer', research report 60, International Water Management Institute, Colombo, Sri Lanka

Swift, J. and Hamilton, K. (2001) 'Household food and livelihood security', pp30–56 in Devereux, S. and Maxwell, S. (eds) *Food Security in Sub-Saharan Africa*, ITDG Publishing, London

Thompson, J. (1991) *Combining Local Knowledge and Expert Assistance in Natural Resource Development in Kenya*, Acts Press, Nairobi, Kenya

Van Averbeke, W., M'Marete, C. K., Igodan, C. O. and Belete, A. (1998) 'An investigation into food plot production at irrigation schemes in central eastern Cape', WRC report 719/1/98, Pretoria

Van der Stoep, I. and Nthai, M. M. (2005) 'Evaluation of the water distribution system at Dzindi irrigation scheme', draft report on WRC project K8/585//4, WRC, Pretoria

Vermillion, D. L. (1997) 'Impacts of irrigation management transfer: A review of evidence', research report 11, International Irrigation Management Institute, Colombo, Sri Lanka

Weather Bureau (1992) 'Climate of South Africa: Climatic statistics up to 1990', report WB42, Department of Environmental Affairs, Pretoria

7

Emerging Rules after Irrigation Management Transfer to Farmers

Klaartje Vandersypen, K. Kaloga, Y. Coulibaly, A. C. T. Keïta, D. Raes and Jean-Yves Jamin

Introduction

In recent years, local users have been considered to be more effective managers of natural resources, especially considering the frequent failure of state-based resource management (Mosse, 1999). The current trend of transferring management responsibilities to users builds on this conviction. In particular, large-scale smallholder irrigation schemes have been subject to irrigation management transfer (IMT) in the past few decades. Farmers are now involved in the operation and maintenance of hydraulic infrastructure, financing and decision-making on the cropping pattern and the physical and institutional layout of the irrigation scheme (Groenfeldt and Svendsen, 2000). Most often, however, transfer of responsibilities has been imposed on them and it is not evident that farmers are able or willing to take on the complex management tasks (Le Gal et al, 2001; Meinzen-Dick et al, 2002). First, it requires a radical mentality shift from farmers who until then depended completely on the management for decision-making (Shah et al, 2002). Second, it demands a considerable understanding of the importance and functioning of the different management tasks and requires the availability of necessary information (De Nys, 2004). Third, the collective action problems inherent in such transfers also have to be surmounted through the drawing up of institutional arrangements. This is particularly so given the fact that individual farmers have an incentive to extract more water and invest less in maintenance than is optimal at the collective level (Tang, 1992).

The Office du Niger in Mali is an interesting case for evaluating some of these issues by considering the experiment of management transfer in collective irrigation schemes. The objective of this study is therefore to diagnose farmers' water management at the tertiary level, focusing on two principal activities of water management: water distribution and maintenance. The chapter first explains the setting and background of management transfer in the Office du Niger irrigation scheme and details the methods used in this study. Next, the rules in use for both activities are assessed, and their ability to resolve possible collective action problems is analysed. The relationship between internal and external group characteristics and the rules in use is also examined. Finally, the chapter looks at possible impediments to farmers' successful water management.

IMT: Setting and Background

The zone of the Office du Niger (14°18'N; 5°59'W) is characterized by a semi-arid climate. Yearly rainfall varies from 300mm to 600mm and is concentrated between July and September (Haefele et al, 2003). Reference evapotranspiration amounts to about 2500mm a year and exceeds rainfall in all months except August (Hendrickx et al, 1986). Soils are predominantly Fluvisols and Vertic Cambisols with a clayey texture (Haefele et al, 2003). The Office du Niger manages five administrative zones that together comprise 80,000ha, mainly destined for smallholder irrigation. The entire area is dominated by a dam on the Niger river from which irrigation water is conveyed through abandoned river channels to a hierarchic irrigation network composed of primary, secondary and tertiary canals. Field canals branching from the tertiary canals bring water to the rice basins and evacuate it to a drainage network (Figure 7.1). All irrigation and drainage canals are unlined. Each farmer owns several basins, constituting a plot with an average total surface of two hectares. Allocation of plots is such that field canals are often shared by two or three farmers. Farmers from the same secondary canal are accommodated in a nearby village. The area is mainly cultivated with flooded rice and yields of five tonnes per hectare are common (Kater, 2000). The main crop is grown during the rainy season. A second rice crop and vegetables are grown during the dry season on a limited area.

By the end of the 1970s, when its notoriously low agronomic and financial performance became unsustainable, the World Bank, together with other international donors, invested in the irrigation scheme while imposing reforms (Tall, 2002). The reform package included a comprehensive liberalization of production and marketing of crops and physical rehabilitation of the hydraulic infrastructure and institutional reforms, involving partial transfer of irrigation management

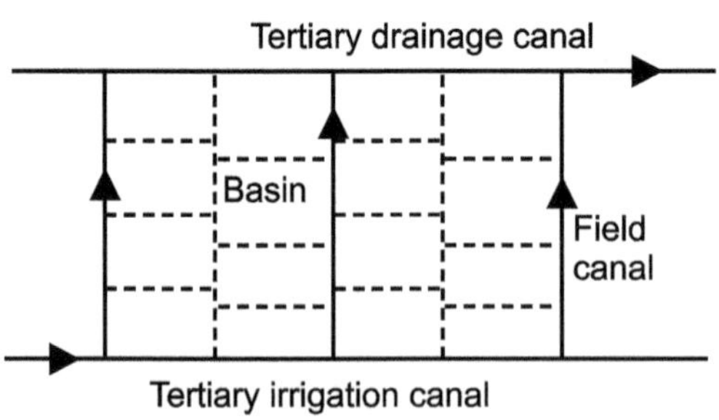

Source: chapter authors

Figure 7.1 *Physical design of a tertiary block*

responsibilities to farmers. This transfer has been more complete towards the lower levels of the canal system. Whereas the central management of the Office du Niger bears sole responsibility for water management at the primary canal level, farmers participate in decision-making at the secondary level through their representation. They are entirely responsible for water distribution and maintenance at the tertiary canal level (Touré et al, 1997).

While there is commitment to the policy of management transfer, the historical legacy and socio-economic and institutional setting of the Office du Niger irrigation scheme present several constraints for the practical translation of this policy. First, farmers lack management experience as nearly all decisions on the production and marketing of crops were taken at the central level, reducing farmers to mere labourers (van Beusekom, 2000). In view of this state of dependency, all knowledge and institutions for water management have to be created from scratch (Shah et al, 2002). Second, the irrigation scheme shows considerable heterogeneity in population in terms of socio-cultural background, endowments and interests. The area of the irrigation scheme, initially scarcely populated, was colonized through various waves of immigration from different regions of Mali and Burkina Faso (van Beusekom, 2000; Philipovich, 2001; Seebörger, 2003). Even though ethnic homogeneity was established in every village when settlers moved in, the high turnover of plot holding and reallocation of plots after rehabilitation mixed people with no common background in almost every village. It is often assumed that differences in socio-cultural background reduce the social capital of a group, which consists of trust, common rules, norms and sanctions, and feelings of connectedness, and hence reduce the scope for cooperation (Pretty and Ward, 2001). The economic growth triggered by the reforms went hand-in-hand with an economic differentiation between farmers. This differentiation has produced considerable variability in terms of plot size, agricultural equipment, access to credit and availability of family labour (Jamin, 1994). In addition, there is also a growing number of plot holders that are outsiders to the village and for whom rice cropping is only a secondary activity. In the rehabilitated parts of the irrigation scheme, these comprise some 10 to 40 per cent of the plot holders (Broekhuyse, 1987; Jamin and Doucet, 1994). Keita (2003) points out that in the Office du Niger, differences in endowments and interests cause farmers' priorities and capacities to diverge, causing constraints on cooperation. A third constraint on the practical translation of IMT policy is the absence of formal organizations for water management at the tertiary level, which adds an extra obstacle to cooperation (Ostrom, 1992; Meinzen-Dick et al, 2002). Farmers' organizations with competencies for water management are set up only at the village level, and these organizations are not effective in dealing with management issues at the tertiary canal level. Indeed, one village covers about 20 canals. Recent efforts to fill this void by organizing farmers for the maintenance of infrastructure at the tertiary level had only meagre results. Every canal has a chief (a position created as part of the reforms) who acts as an intermediary for passing on information between farmers and management agents and is supposed to organize and coordinate water management activities at the tertiary level. Lacking recognition by fellow farmers, however, the chief usually does not carry out any of his assigned duties.

The success of the reform package is beyond doubt. New settlers are attracted every year, the irrigated area is expanding quickly and yields have risen from little more than a tonne per hectare to five tonnes per hectare and more (Couture et al, 2002), resulting in a significant increase in farmers' income and economic activity in the region (Aw and Diemer, 2005). The success of farmers' water management instituted by the reforms, however, is less clear. Certain actions and decision-making on water distribution are effectively being taken over by farmers (Musch, 2001), but this is not necessarily coordinated at the level of the tertiary canal. On the other hand, maintenance tasks are generally considered to be neglected. Thus the ability of farmers to devise, monitor and enforce rules to resolve possible collective action problems in water management at the tertiary level remains in doubt. And inadequate water management might lead to unequal access to irrigation water, conflicts over water distribution and the deterioration of infrastructure.

Methods

Sample

This study is part of a broader research programme on strategies of water management at both the plot level and the level of the tertiary canal. The research took place in five villages selected in three administrative zones (Niono, N'Debougou and Molodo). In each village, about 20 farmers were chosen at random. For the purposes of this study, farmers farming alone on a tertiary canal were left out, leaving a final sample of 89 farmers from 59 tertiary canals. The villages are quite similar as regards total number of families, number of tertiary canals and average plot size. Key statistics about the villages are shown in Table 7.1.

Table 7.1 *Key statistics of the study villages and sample*

Particulars	*Name of village*				
	Coloni	*Foabougou*	*Banisiraela*	*Siengo*	*Siby*
Total number of secondary canals	1	2	1	1	3
Total number of tertiary canals	33	34	29	31	42
Total number of tertiary canals in the sample	12	15	11	12	9
Total number of farmers in the village	263	194	145	162	126
Total number of farmers in the sample	20	20	19	19	11
Average plot size in the village (ha)	2.1	3.1	3.4	2.4	2.9
Average number of farmers per tertiary canal in the sample	9	10	10	11	4
Average proportion of outsiders from the village per tertiary canal in the sample (%)	18	30	12	19	5

Source: survey data, land register of the Office du Niger irrigation scheme

The sample villages were chosen so that each represents a different method of rehabilitation or an area that has not yet been rehabilitated. The various donors who financed the rehabilitation have each followed their own philosophies with respect to the construction and operation of hydraulic infrastructure, resulting in different types of infrastructure with particular designs and dimensions of intakes, canals and basins. An overview of this is presented in Table 7.2. The main difference at the tertiary level is in the type of intake structure. In areas rehabilitated with Dutch cooperation, semi-modular sliding gates were used; these can be opened and closed by the farmers themselves. In other areas Neyrtec modules were used; these let through a water flow that is fairly constant with respect to fluctuations in the upstream water level. As a rule, the intake is secured with a lock and only water bailiffs employed by the Office du Niger can open or close the modules. In practice, though, the lock is often missing, or farmers have a copy of the key. Finally, one village has not yet been subject to rehabilitation and still has the original infrastructure with non-modular sliding gates at the intakes of all canals. Field canals were not provided in the original design, but farmers often created them themselves. Otherwise, plots were irrigated and drained from basin to basin.

Table 7.2 *Overview of differences in design and operation of the various types of infrastructure*

Donor and type of infrastructure	*Village*	*Properties*
French cooperation (Retail type)	Coloni	Sliding gates at the intake of secondary canals; Neyrtec modules at the intake of tertiary canals; intake at tertiary level secured; tertiary canals over dimensioned with respect to peak water requirements; one field canal per 2ha
Dutch cooperation (ARPON type)	Foabougou	Neyrtec modules at the intake of secondary canals; sliding gates at the intake of tertiary canals; intake at tertiary level not secured; one field canal per 3ha, additional field canals constructed by farmers
German cooperation (KfW type)	Banisiraela	Sliding gates at the intake of secondary canals; Neyrtec modules at the intake of tertiary canals; intake at tertiary level secured; one field canal per 2ha; a multiple of 7 field canals per tertiary canal in order to facilitate water distribution following a weekly rotation
World Bank (World Bank type)	Siengo	Neyrtec modules at the intake of secondary canals; Neyrtec modules at the intake of tertiary canals; intake at tertiary level secured; one field canal per 2ha
Not rehabilitated (not rehabilitated type)	Siby	Sliding gates at the intake of secondary canals; sliding gates at the intake of tertiary canals; intake at tertiary level not secured; no field canals, unless constructed by farmers

Source: chapter authors

Data Collection

The bulk of the data for this study were drawn from a questionnaire survey on the strategies of water management adopted at the tertiary level. The questionnaire contained a structured list of open questions in which farmers were asked to describe and evaluate the organization of water distribution and maintenance of the tertiary canal. In interviews there was further discussion of difficulties relating to water distribution experienced by other farmers of the tertiary canal, so that a diagnosis at the level of the tertiary canal can be made. When farmers answered that they experienced collective action problems with water distribution or maintenance, a follow-up question asked what they could do in order to alleviate those problems or prevent them from occurring. Another question, asked to all farmers, assessed which solutions they thought would work in order to improve their water management in general. Content analysis was used to analyse the questionnaire and categorize answers (Miles and Huberman, 1994). Additional information was gathered through field observations and informal discussions with farmers, water bailiffs and functionaries of the Office du Niger. Data on group characteristics at the level of the tertiary canals were obtained from the land register of the Office du Niger irrigation scheme and water bailiffs. The survey was conducted from June to October 2003. Descriptive statistics and correlation tests were generated using the SPSS computer package.

Results and Discussion

Water Distribution

Before management transfer, water distribution was supply-driven. Water was conveyed to the canals following an irrigation schedule established by the central management. Water bailiffs informed the farmers when their canal would be served and farmers were then required to irrigate their plots with the water available. However, since the management transfer, all irrigation canals are now continuously filled and water supply is on demand. Depending on whether the intake of the tertiary canal is locked, farmers can now either open the intake of the tertiary canal themselves, or communicate their water demand to a water bailiff who will open the intake of the tertiary canal accordingly. Thus, at every level of the canal system, the total demand of the downstream canals is satisfied. Taking turns in irrigation is only necessary between field canals. Indeed, the rehabilitated irrigation infrastructure is designed in such a way that not all field canals on a tertiary canal can be served simultaneously with an adequate flow rate. Irrigation manuals supplied with the rehabilitation prescribe that water distribution should follow a fixed rotation scheme between individual field canals on a weekly basis. It is, however, left to farmers to implement this rule in practice or to organize water distribution in some other way. Our first task, therefore, was to examine whether farmers established clearly defined rules to organize water distribution.

Results show that on only 30 per cent of the tertiary canals from the sample did clearly defined water distribution rules exist. A variety of rules were mentioned in the interviews, such as a rotation between upstream and downstream groups of field canals and preferential access to water for plots that are difficult to irrigate because of their topographic position. A fixed rotation between individual field canals on a weekly basis is rarely applied. However, the rotation schemes in use seldom follow fixed days and are frequently discontinued in periods of low water demand. On the other 70 per cent of the canals, no water distribution rules are recognized, so that anyone can irrigate at any moment, although informal consultation between farmers sometimes takes place in order to limit the number of farmers irrigating simultaneously or to prevent conflicts.

Subsequently, we investigated collective action problems of water distribution through the difficulties experienced by farmers due to fellow farmers' behaviour. Twenty per cent of the farmers interviewed reported having personally experienced such difficulties. These difficulties fall into three categories. The first category consists of fellow farmers impeding the water flow during irrigation by closing the gates of field canals or the tertiary canal, for example. Farmers would then be obliged to continue irrigation later on or to guard their plots continuously while irrigating in order to prevent such activity. The second category comprises the violation of rules by fellow farmers, causing one farmer's irrigation programme to be disturbed. A final category applies to farmers with plots that are difficult to irrigate and for which some coordination between farmers is necessary but impossible to achieve.

The research continued by examining whether there was a correlation between the application of clearly defined rules and the experience of difficulties. To this end, dichotomous variables were created for the existence of rules and the experience of problems and a Pierson chi-square test was carried out. The test shows that no such correlation exists (see Table 7.3). Clearly defined rules are thus not systematically established, even when there are problems to be solved, and furthermore establishing rules is no guarantee for solving problems.

A diagnosis at the level of the tertiary canal further clarifies these results. Clearly defined rules on water distribution are useful as they enhance the predictability of the water supply as regards quantity and timing (Ostrom et al, 1993). However, when water supply is plentiful and access is easy for all plots, the probability that irrigation activities of different farmers will hinder one another is small. Flooded rice has the additional advantage that water can be stored in the basins,

Table 7.3 *Association of the experience of problems with the application of clearly defined rules relating to water distribution*

Application of clearly defined rules	*Experience of problems*		*Total*
	Yes	*No*	
Yes	8 (26%)	23 (74%)	31 (100%)
No	9 (16%)	49 (84%)	58 (100%)
Total	17	72	89

Note: Pierson chi-square value = 1.384; df = 1; Significance = 0.239
Source: chapter authors

so that the timing of irrigation matters less. From the interviews, it appeared that on 58 per cent of tertiary canals in the sample, there is indeed no need to establish rules. Possible problems are easily resolved through informal consultation, so that difficulties are avoided. However, on 12 per cent of the canals establishing rules would have been useful at least for some farmers with plots that are difficult to irrigate, but this was not accomplished. On the other hand, the existence of clearly defined rules does not appear to guarantee avoiding difficulties. Only on 17 per cent of the canals have rules been devised that manage to avoid collective action problems. On a further 13 per cent, existing rules do not prevent such problems, most often due to violation by certain farmers. Indeed, formal monitoring and sanctioning mechanisms are completely lacking on all canals and there is no person or structure with the authority to establish them.

With no correlation between the experience of difficulties and the existence of rules relating to water distribution, other possible determining factors were investigated. It has been proposed that several physical and socio-economic factors might influence the organization of water management in irrigation (Aggarwal, 2000; Bardhan, 2000; Rinaudo, 2002; Meinzen-Dick et al, 2002); in order to test the relevance of these factors for distinguishing between canals on which water distribution rules are applied and those on which they are not, additional variables were created. The physical factors include group size, layout of irrigation infrastructure and level of water availability. The villages in this study are located near the head-end of the irrigation scheme, where water availability is ample. Due to design differences, however, the canal capacity in some villages is greater than in others, allowing for a larger proportion of field canals to take water at the same time with an adequate flow rate. This factor will therefore be analysed through the variable layout of the irrigation infrastructure. Among the socio-economic factors are social cohesion, dependency of group members on irrigation and group heterogeneity in terms of economic wealth and origin. Within the limits of this study, it was not possible to obtain detailed information about the farmers of the tertiary canals in the sample or their social environment. The socio-economic aspect of the research was therefore limited to the proportion of farmers from outside the village. This factor might influence the existence of water distribution rules in two ways: first, since they spend little time at their plots, and no time at all in the village, such farmers may impede the collective establishment of rules; second, with little time to devote to rice farming, such farmers may want to irrigate when visiting their plots without waiting for their turn. In other words they probably have no interest in rules limiting the number of days they are entitled to withdraw water. Pierson chi-square and Mann-Whitney tests were used to test for significance here.

Infrastructure was found to be the only variable significantly associated with the existence of clearly defined water distribution rules (Tables 7.4 and 7.5). On the canals of KfW type (Banisiraela village), rules are more often applied than elsewhere. Indeed, engineers wanted to facilitate the organization of a rotation schedule on a weekly basis by designing canals in such a way that each had a multiple of seven field canals. On canals of the Retail and not rehabilitated type, rules are seldom applied. As these have a large capacity compared with the others, there is often no need to establish rules. Another significant observation is that the existence of rules is average in both the canals of ARPON and World Bank

Table 7.4 *Application of clearly defined rules on water distribution by type of infrastructure*

Type of infrastructure	*Application of clearly defined rules*		*Total*
	Yes	*No*	
Retail	0	12 (100%)	12 (100%)
ARPON	6 (40%)	9 (60%)	15 (100%)
KfW	8 (73%)	3 (27%)	11 (100%)
World Bank	4 (33%)	8 (67%)	12 (100%)
Not rehabilitated	0	9 (100%)	9 (100%)
Total	18	41	59

Note: Pierson chi-square = 19.150; df = 4; Significance = 0.001
Source: chapter authors

Table 7.5 *Comparison of mean number of farmers on the canal and mean proportion of outsiders between canals on which there are clearly defined rules regarding water distribution and canals on which there are not*

Application of clearly defined rules	*Number of farmers* on the canal*			*Proportion of outsiders** to the village (%)*		
	Mean	*S.D.*	*n*	*Mean*	*S.D.*	*n*
Yes	9.83	6.22	18	12.13	16.16	15
No	8.71	6.59	41	22.03	29.96	37
Total	9.05	6.45	59	19.17	26.94	52

Note: * Mann-Whitney = 327.5; Significance = 0.492. ** Mann-Whitney = 249.0; Significance = 0.544
Source: chapter authors

types. Whereas the philosophy of ARPON was to work with cheap and locally constructed materials, the World Bank undertook a more costly rehabilitation using materials that are more expensive (this also applies to the KfW and Retail types). The price of rehabilitation per hectare therefore differed considerably, but this seems to have had no notable effect on farmers' water management.

To sum up, rules on water distribution have been established on less than one-third of the canals on average and are more common with certain types of infrastructure. However, problems occur as frequently on canals with rules as on those without. Indeed, given the lack of sanctioning mechanisms, rules are often not respected and are thus ineffective in solving collective action problems. On the other hand, the absence of rules does not necessarily lead to problems, as water supply is abundant.

Maintenance

The unlined irrigation canals of the Office du Niger are vulnerable to weed invasion and deterioration of the canal bed. Regular maintenance consists of removing aquatic plants from the canal and keeping the banks clear. This should be carried out once or twice per growing season. More thorough maintenance requires dredging of the canals and is necessary every two or three years. Neglect of these tasks eventually results in decreasing water flow and storage capacity. In addition, insufficient maintenance upstream also diminishes the water flow downstream, so maintenance should be a concern for all farmers. Apart from consequences for water flow, badly maintained canals also offer a habitat for several enemies of the rice crop, such as rats and granivorous birds. Before the restructuring of water management, farmers paid a yearly maintenance deposit for the tertiary infrastructure, which would be reimbursed when they carried out canal maintenance. Otherwise, the central management would use the money to pay a third party to do the work. In practice, however, farmers did not do the maintenance work themselves and the management did not employ others to do it, resulting in advanced degradation of the tertiary infrastructure. Upon restructuring, this system was abandoned; now farmers are simply required to carry out maintenance themselves. Organization and monitoring of this task are also left to farmers. The rules devised to organize maintenance of the tertiary irrigation canals are examined below. Only rules on regular maintenance are discussed.

Rules concern the timing of maintenance and the work required of each farmer; most frequently, individual farmers clean a section of the canal bordering their plot. The size of this section is then defined based either on equal effort on the part of all farmers or proportionately by plot size. On a few canals, maintenance of the whole irrigation canal is carried out collectively by all farmers. The timing of maintenance can be one particular day or a period of days defined according to the cropping season. It is important that maintenance is done in the same period on the whole canal, as parts of the canal not maintained are a bottleneck for the downstream parts and serve as a source for aquatic plants to reinvade the whole canal. Rules on timing and effort are thus complementary. On 24 per cent of the tertiary canals, rules exist on both aspects. On a further 61 per cent, rules have been devised only on the effort required of each farmer. Comparing the answers of farmers from the same tertiary canal given in the interviews, however, it appeared that there is often no consensus on the concrete arrangements prescribed by the rules. When no rules are applied, decisions on whether and when maintenance is carried out are left to individual farmers.

The research then examined collective action problems through farmers' appraisal of maintenance. Results show that 43 per cent of farmers are dissatisfied because maintenance of the various canal parts is not carried out in the same period, or because of shirking by fellow farmers, which means that they have to either carry out the share of maintenance of those farmers in addition to their own or suffer the consequences of insufficient maintenance. In the latter case, they moreover see the results of their own efforts spoiled by neighbouring parts of the canal not being maintained. Many farmers said they were discouraged by this and had therefore stopped maintaining their own section. Outsiders are especially accused of systematically shirking their maintenance duties and consequently discouraging others. It must be noted, however, that farmers have often a different perception of the importance of regular maintenance, especially since with the infrastructure being recently rehabilitated it has little impact on flow rates in the

Table 7.6 *Association of the experience of problems with the application of clearly defined maintenance rules*

Application of clearly defined rules	*Experience of problems*		*Total*
	Yes	*No*	
Rules on both timing of maintenance and effort required of each farmer	4 (16%)	21 (84%)	25 (100%)
Rules on timing of maintenance only	28 (53%)	25 (47%)	53 (100%)
No rules	6 (55%)	5 (46%)	11 (100%)
Total	38	51	89

Note: Pierson chi-square = 10.138; df = 2; Significance = 0.006
Source: chapter authors

short term. For that reason, some farmers are dissatisfied, even though the maintenance level is sufficient in the view of others, while others may be satisfied with what may be deemed insufficient by some.

The relationship between the type of maintenance rules applied and the problems encountered by farmers has also been examined using the Pierson chi-square test. To this end, a categorical variable was created for rules applied (categories are 'rules on both timing of maintenance and effort required of each farmer', 'rules on timing of maintenance only' and 'no rules'), with a dichotomous variable for dissatisfaction with maintenance ('yes' if problems are experienced, 'no' otherwise). Results show that a correlation indeed exists (Table 7.6). More specifically, when rules on both timing of maintenance and effort required by each farmer are applied, significantly fewer problems are cited. On the other hand, problems are about as frequent on canals with only rules on effort required as on canals with no rules at all.

Again, further analysis of the interviews sheds light on these results. As with water distribution, virtually no formal monitoring and sanctioning mechanisms are in place to enforce the rules in use. Instead, rules on the timing of maintenance have been simply abandoned when they were broken too often. This means that the rules remain in place only in cases where maintenance is carried out by all farmers and everyone is satisfied. On the other hand, most farmers think everyone should contribute fairly to maintenance, so rules on the effort required of each farmer more often remain in place, even though many shirk their duties. As stated above, this does not automatically lead to dissatisfaction since not all farmers attach the same importance to regular maintenance. In a limited amount of cases, farmers are dissatisfied about maintenance without being able to establish rules.

Following the analysis of the association between the application of rules on maintenance and the occurrence of problems, we took a closer look at the physical and socio-economic factors that might be involved. Relevance was tested for the same variables put forward in the analysis of water distribution rules (number of farmers on the canal, type of infrastructure and proportion of outsiders). The significance of correlations with the application of clearly defined rules was again tested using the Pierson chi-square and Kruskal-Wallis tests. Regarding maintenance, it was proposed that the proportion of outsiders might have an influence since, in addition to their absence impeding the collective negotiation of rules, they most often do not rely on rice farming as they have other sources of income. Their motivation for maintenance and thus for rules on maintenance is therefore less. Furthermore, it appeared to be the case that outsiders frequently broke the rules and that rules on maintenance were often abandoned when not respected.

Results indicate that the proportion of outsiders indeed negatively influences the application of rules, whereas the influence of the number of farmers on the canal is only marginally significant (Table 7.7). The type of infrastructure plays no significant role in this context (Table 7.8). These results contrast with those

Table 7.7 *Comparison of mean number of farmers on the canal and mean proportion of outsiders between canals on which there are clearly defined rules regarding maintenance and canals on which there are not*

Application of clearly defined rules	*Number of farmers* on the canal*			*Proportion of outsiders** to the village (%)*		
	Mean	*S.D.*	*n*	*Mean*	*S.D.*	*n*
Rules on both timing of maintenance and effort required of each farmer	10.64	7.08	14	9.92	17.71	13
Rules on timing of maintenance only	7.28	4.01	36	15.27	24.98	30
No rules	13.67	10.34	9	45.56	30.25	9
Total	9.05	6.45	59	19.17	26.94	52

Note: * Chi-square = 5.87; df = 2; Significance = 0.053. ** Chi-square = 10.54; df = 2; Significance = 0.005
Source: chapter authors

Table 7.8 *Application of clearly defined rules on maintenance by type of infrastructure*

Type of infrastructure	*Application of clearly defined rules*			*Total*
	Rules on both timing of maintenance and effort required of each farmer	*Rules on timing of maintenance only*	*No rules*	
Retail	0	10 (83%)	2 (17%)	12 (100%)
ARPON	2 (13%)	10 (67%)	3 (20%)	15 (100%)
KfW	3 (27%)	7 (64%)	1 (9%)	11 (100%)
World Bank	4 (33%)	6 (50%)	2 (17%)	12 (100%)
Not rehabilitated	5 (56%)	3 (33%)	1 (11%)	9 (100%)
Total	14	36	9	59

Note: Pierson chi-square = 10.97; df = 8; Significance = 0.204
Source: chapter authors

obtained on water distribution rules, where rules were established only when necessary. In principle, maintenance is always necessary, so rules are more common and the social capital of the group of farmers starts to play a role.

To summarize, maintenance rules on effort required of each farmer exist on 61 per cent of canals, while a complete set of rules including timing and effort required is found on only 24 per cent. The absence of a complete set of rules is made more likely by the presence of outsiders and gives rise to complaints about insufficient maintenance by fellow farmers. Once again, the absence of sanctioning mechanisms makes it hard to impose rules, which therefore are frequently abandoned.

Impediments to Successful Organization of Water Management

From the results, it appears that many groups of farmers have been unable to devise, monitor and enforce rules to resolve collective action problems. Possible impediments to successful organization of water management at the tertiary level are thus investigated in this section by looking at the strategies adopted by farmers in order to alleviate collective action problems or prevent them from occurring. Farmers' ideas to improve water management in general are also discussed.

With respect to water distribution, some farmers have adopted individualistic strategies to reduce difficulties. These include advancing or postponing sowing and planting dates with respect to their neighbours in order to spread irrigation requirements, deepening the rice basins and reshaping the irrigation infrastructure so as to divert more water to their plots. These individualistic strategies do not only come at a personal cost but also inflict additional problems on fellow farmers. Furthermore, they do not solve all the problems; indeed with respect to maintenance no individualistic solution exists. Very few farmers (less than 8 per cent of those citing problems) said they confronted fellow farmers about the problems they created, and even those admitted they did not always succeed in convincing them to change their behaviour. From such confrontations, moreover, conflicts often arise. Most farmers say they can do nothing to solve their problems, since they have no influence over their colleagues and want to avoid conflicts. This leads us to a first possible impediment to successful organization of water management in the area: peer pressure, which can be a very powerful enforcement instrument (Aggarwal, 2000), is often ineffective; outsiders are especially immune to it as they share no other activities with the local farmers of the tertiary canal.

As to the general improvement of water management, in only 14 per cent of cases was more coordination of individual decisions and actions considered desirable. Another solution often proposed is that individual responsibilities are defined and enforced not by the farmers themselves, but by central management. This is a remarkable result, since one of the main goals of the management reforms in the Office du Niger irrigation scheme was to transfer responsibilities towards farmer groups. A second impediment might thus be the incomplete mentality shift towards assuming collective responsibility.

Conclusions and Perspectives

Twenty years after the first reforms, farmers' water management is still immature. Clearly defined rules on water distribution and maintenance at the tertiary level are difficult to establish when necessary, and, once established, difficult to enforce. The historical and socio-economic conditions of the irrigation scheme do not favour collective action. Indeed, two major impediments seem to be the ineffectiveness of peer pressure among farmers and their incomplete mentality shift towards assuming collective responsibility. In order to cope with problems ensuing from the lack of coordination, farmers rather adopt individualistic strategies that come at a personal cost and possibly exacerbate fellow farmers' problems. However, since water supply is abundant and the infrastructure recently rehabilitated in a large part of the irrigation scheme, the absence of rules does not necessarily lead to problems, meaning organization of water management at the tertiary level is not always required.

The analysis of the Office du Niger irrigation scheme illustrates some of the flaws of IMT. First, it has been too abrupt. Given the irrigation schemes' history of farmer dependency, time is needed for farmers to fully adopt their new responsibilities and grow in their new roles. Second, it was assumed that farmers' organization of water management would arise spontaneously, thereby overlooking social impediments to collective action. In many smallholder irrigation schemes, management transfer was accompanied by imposing institutional and organizational arrangements on farmers. These cannot guarantee that collective action will emerge, as shown by Veldwisch (2004), but, when adapted to local realities, they might greatly facilitate collective action for water management (Letsoalo and Van Haverbeke, 2004).

The current state of affairs in the Office du Niger irrigation scheme is not deemed sustainable in the long term. One of the main political goals for the Office du Niger is to further expand the irrigated area without a substantial increase in total water consumption. Water will thus become scarcer and an equitable distribution will become indispensable. Furthermore, as the irrigation infrastructure ages, maintenance will also gain importance. At that point, effective organization of water management at the tertiary level will become crucial. Measures of sensitization and group empowerment accompanying the process of management transfer will therefore be desirable. To further promote collective action, it might also be necessary to provide farmers with institutional arrangements which are adapted to the complex social context.

References

Aggarwal, R. (2000) 'Possibilities and limitations to cooperation in small groups: The case of group-owned wells in Southern India', *World Development*, vol 28, pp1481–1497

Aw, D. and Diemer, G. (2005) *Making a Large Irrigation Scheme Work: A Case Study from Mali*, World Bank, Washington, DC

Bardhan, P. (2000) 'Irrigation and cooperation: An empirical analysis of 48 irrigation communities in South India', *Economic Development and Cultural Change*, vol 48, pp847–866

Broekhuyse, J. Th. (1987) *De Verbetering van de Rijstverbouw en van de Samenwerking van de Rijstverbouwers van het Office du Niger*, Koninklijk Instituut voor de Tropen, Amsterdam

Couture, J-L., Delville, P. L. and Spinat, J-B. (2002) 'Institutional innovations and water management in Office du Niger (1910–1999): The long failure and new success of a big irrigation scheme', Groupe de Recherche et d'Echanges Technologiques working paper 29, GRET, Paris

De Nys, E. (2004) 'Interaction between water supply and demand in two collective irrigation schemes in North-East Brazil: From analysis of management processes to modelling and decision support', PhD dissertation, FLTBW, Katholieke Universiteit, Leuven, Belgium

Groenfeldt, D. and Svendsen, M. (2000) *Case Studies in Participatory Irrigation Management*, World Bank, Washington, DC

Haefele, S. M., Woporeis, M. C. S., Ndiaye, M. K. and Kropff, M. J. (2003) 'A framework to improve fertilizer recommendations for irrigated rice in West Africa', *Agricultural Systems*, vol 76, pp313–335

Hendrickx, J. M. H., Vink, N. H. and Fayinke, T. (1986) 'Water requirement for irrigated rice in a semi-arid region in West Africa', *Agricultural Water Management*, vol 11, pp75–90

Jamin, J-Y. (1994) 'De la norme à la diversité: l'intensification rizicole face à la diversité paysanne dans les périmètres irrigués de l'Office du Niger', PhD dissertation, Institut National Agronomique Paris-Grignon, Paris

Jamin, J-Y. and Doucet, M. J. (1994) 'La question foncière dans les périmètres irrigués de l'Office du Niger (Mali)', *Les Cahiers de la Recherche Développement*, vol 38, pp65–82

Kater, L., Dembélé, I. and Dicko, I. (2000) 'The dynamics of irrigated rice farming in Mali', *Managing Africa's Soils 12*, Russell Press, Nottingham

Keita, A. (2003) 'Gestion sociale de l'eau au niveau tertiaire en zone Office du Niger: Cas du kala inférieur', Institut Polytechnique Rural, Katibougou, Mali

Le Gal, P-Y., De Nys, E., Passouant, M., Raes, D. and Rieu, T. (2001) 'Intervention et décision collective dans les périmètres irrigués', in Malézieux E., Trébuil G. and Jaeger M. (eds) *Modélisation des Agroécosystèmes et Aide à la Décision*, Cirad/INRA, Montpellier, France, pp351–369

Letsoalo, S. S. and Van Averbeke, W. (2004) 'When water is not enough: Institutions, organisations, and conflicts surrounding the sharing of irrigation water at a smallholder scheme in South Africa', in *Proceedings of the Workshop on Water Management for Local Development*, WRM 2004, Pretoria, pp362–373

Meinzen-Dick, R., Raju, K. V. and Gulati, A. (2002) 'What affects organisation and collective action for managing resources? Evidence from canal irrigation systems in India', *World Development*, vol 30, pp649–666

Miles, M. B. and Huberman, A. M. (1994) *Qualitative Data Analysis: an Expanded Sourcebook* (2nd ed), SAGE Publications, Thousand Oaks, CA

Mosse, D. (1999) 'Colonial and contemporary ideologies of "community management": The case of tank irrigation development in South India', *Modern Asian Studies*, vol 33, pp303–338

Musch, A. (2001) 'The small gods of participation', PhD dissertation, Universiteit Twente, Enschede, The Netherlands

Ostrom, E. (1992) *Crafting Institutions for Self-Governing Irrigation Systems*, Institute for Contemporary Studies, San Francisco, CA

Ostrom, E., Schroeder, L. and Wynne, S. (1993) *Institutional Incentives and Sustainable Development: Infrastructure Policies in Perspective*, Westview Press, Boulder, CO

Philipovich, J. (2001) 'Destined to fail: Forced settlement at the Office du Niger, 1926–45', *Journal of African History*, vol 42, pp239–260

Pretty, J. and Ward, H. (2001) 'Social capital and the environment', *World Development*, vol 29, pp209–227

Rinaudo, J-D. (2002) 'Corruption and allocation of water: The case of public irrigation in Pakistan', *Water Policy*, vol 4, pp405–422

Seebörger, K-U. (2003) 'Recent changes in migration patterns, agricultural development and strategies to reduce poverty in the area of the Office du Niger, Republic of Mali', International Workshop on Migration and Poverty in West Africa, Sussex Centre for Migration Research, Falmer, UK

Shah, T., van Koppen, B., Merrey, D., de Lange, M. and Samad, M. (2002) 'Institutional alternatives in African smallholder irrigation: Lessons from international experience with irrigation management transfer', research report 60, IWMI, Colombo

Tall, E. H. O. (2002) 'La restructuration de l'Office du Niger. L'adoption d'un cadre institutionnel propice au développement', in Kuper M. and Tonneau J-P. (eds), *L'Office du Niger, Grenier à Riz du Mali, Succès Economiques, Transitions Culturelles et Politiques de Développement*, Cirad, Montpellier/Karthala, Paris

Tang, S. Y. (1992) *Institutions and Collective Action: Self-Governance in Irrigation*, Institute for Contemporary Studies Press, San Francisco, CA

Touré, A., Zanen, S. and Koné, N. (1997) *La Restructuration de l'Office du Niger: Contribution de ARPON III Coopération Néerlandaise*, Office du Niger, Ségou, Mali

Van Beusekom, M. M. (2000) 'Disjunctures in theory and practice: Making sense of change in agricultural development at the Office du Niger, 1920–60', *Journal of African History*, vol 41, pp79–99

Veldwisch, G. J. (2004) 'Local governance of Thabina irrigation scheme: Three years after rehabilitations and irrigation management transfer (Limpopo Province, South Africa)', in *Proceedings of the Workshop on Water Management for Local Development*, WRM 2004, Pretoria, pp335–352

8

Crafting Water Institutions for People and Their Businesses: Exploring the Possibilities in Limpopo

Felicity Chancellor

Introduction

Pockets of indigenous, and often long established, irrigation all over Africa testify to smallholders' past successes. Irrigation raises agricultural productivity and rural income. However, smallholder irrigation as we think of it now, with designated plots within a system that distributes water under imposed rules and conditions, developed largely in the early 20th century. Administrations usually built the systems, subsidized the running costs and managed the water and farming activity at a detailed level. In Southern Africa particularly, where political objectives dwarfed economic considerations, irrigation costs were typically high relative to returns from irrigated farming.

Small-scale irrigation was adopted as a development strategy. Governments, donors and NGOs invested in ambitious schemes justified by optimistic feasibility studies. Investment was undertaken with a minimum of local participation and often fitted poorly into the local economy; even projects that did not fail were characterized by lack of enthusiasm and low economic impact. Management was 'top–down', aimed at moulding farmer behaviour to fit the irrigation system, an approach that came under increasing criticism and led later projects to incorporate some degree of participation. However, lack of skilled participatory techniques, rushed implementation and poor design combined to give poor results. Even 'participatory projects' failed to meet the needs of users and ticked-over lamely achieving disappointing returns and often existing on subsidy. By the start of the 21st century, large numbers of smallholder farmers were struggling with inadequate irrigation infrastructure, flawed management regimes and poor services; government departments and NGOs struggled to find effective ways out of the impasse and improve irrigation performance; while global market liberalization consigned producers to unfavourably competitive markets (IFAD, 2001).

The development climate has changed, however: programmes and projects now adopt people-centred approaches and give careful consideration to economic viability and entry to global markets. Better-informed and more focused participation is offered, featuring empowerment and promoting equity and entrepreneurship. The change is significant: governments are embarking on programmes of Irrigation Management Transfer (IMT), demolishing the old systems, informal networks and practices that grew up in the past (Hedden-Dunkhorst et al, 2002; Gapko et al, 2001; Shah et al, 2002). Creating and adopting new mind-sets, new practices and robust, transparent processes is a new challenge, and one that must be met to enable smallholder-irrigators to survive the new economic climate and achieve local food-security and development (Hodgson, 2003; ICID, 2004).

Most governments are committed to empowering irrigators to run their own businesses and irrigation systems, and to follow their own development paths. They are also generally committed to equitable laws, protecting human rights and eradicating poverty and must seek processes to achieve their multiple objectives. Subsidies must go; therefore government must improve understanding of the constraints to profitability and sustainability, to achieve balance and provide acceptable compromises as the urgency to withdraw financial and managerial support gathers momentum.

This chapter considers the key area of the nature of institutions appropriate for owning and managing smallholder irrigation schemes following IMT. It explores how institutional changes can achieve sustainable smallholder irrigation development, foster the achievement of irrigators' ambitions, promote equitable distribution of benefits and improve the situation of the poor.

The reality that good institutions alone will not guarantee success is recognized, but it is contended that the part they play in sustaining entrepreneurship and the wider support system, which will be required by irrigators for many years, is key to the long-term success of farmer-managed smallholder irrigation.

The Study

This chapter is based on the findings of a Department for International Development (DFID), UK research study carried out between 2000 and 2003 on creating sustainable smallholder businesses in Southern Africa (Chancellor et al, 2003). The study was carried out in two phases. The first was a broad investigation covering 15 smallholder irrigation schemes in Zimbabwe, South Africa and Swaziland, which represented a range of situations from small subsistence food plots to larger smallholder schemes potentially suited to entering commercial markets. The study used interviews with samples of farmers, including many different types of household to reflect the needs of the poor and not so poor, young and old, and men and women. This survey was supplemented by open-ended key-informant interviews and a series of participatory focus groups based on an initial analysis of the interview data.

The main issue that emerged was the need for a strong commercial focus. Discussions considered the important issues that might change existing expectations and behaviour among farmers to foster the development of a business mindset. Participation and marketing consistently came to the fore. Lack of clear objectives to provide an organizational framework for effective collective decision-making was a significant issue. Schemes where farmers participate regularly exhibited more involvement in management, greater understanding of the reasons for decisions and more support for group decisions resulting in better performance. In addition it was noted that a reliable system to communicate decisions from one level to the next is almost as important as the decisions themselves (Chancellor et al, 2003, Annex 1).

Following on from these organizational and marketing concerns, the second phase focused on smallholder schemes in Limpopo Province in South Africa, in particular the New Forest and Dingleydale scheme, which is part of the pilot programme for the revitalization of smallholder irrigation. The research focused on the role of marketing in releasing money to finance and sustain infrastructure in the IMT process and grappled with the simultaneous processes of establishing water user associations (WUAs) responding to the market situation while managing water and accessing services, inputs, credit and labour.

Findings from this investigation identified significant concerns about participation, marketing and institutions. Specifically, it is argued that the transfer process has not yet established an appropriate institutional framework to define the basis for ownership and management by members. There is therefore a poor basis on which to seek and develop commitment from the farmers and other stakeholders. Lack of commitment inhibits the establishment of appropriate institutional arrangements and undermines the legitimacy of the new institution and the authority of its leaders and officers. The interactive process to establish the management committees (MCs) and other structures currently used on the schemes highlighted the need for the department of agriculture and support agencies to take a more proactive role in clarifying the institutional basis of the transferred schemes. In addition, new management must address planning ahead, tapping commercial and provincial expertise to provide a range of solutions relating to self-financing at the scheme level. Developing finances and on-going support mechanisms is also crucial to reducing the risk of failure for farmer managed schemes (Chancellor et al, 2003, Annex 2). The new water act and the consequent emergence of the institution of the WUA have increased the urgency for this clarification.

The Context

Recent theories and processes of development have focused on the individual and local or communal levels, and on the place and role of the state (Ashley and Maxwell, 2001). Improvements in understanding about how adults learn, and lessons from failed technology-based modernization, have led to recognition that change must be driven from the perspective of stakeholders. Development is no longer identified and legitimized by an external agency and adoption of foreign technologies, but as support and facilitation of local internal capacity. This focus on social development leads to a widespread demand for participatory approaches, shared analysis and planning between intervention agencies and client groups. Recognition of the legitimacy of different, and multiple, stakeholder views and needs has broadened and enriched understanding of the meanings and nature of community and how communities and groups interact with the state and market, particularly in the maintenance and management of common pool resources. Empowerment of the disadvantaged to articulate their views and needs to influence policy formulation is seen as the very stuff of development. However, agencies of the state still tend to implement government policies through authoritative and interventionist approaches. These and other concerns have led to considerable changes in perceptions about the role and place of the state in general, and as an agency for change in particular.

Development is now recognized as a process of structural change. All states and societies across the world, developed and developing, are grappling with this change without clear blueprints to draw on to formulate new relationships between state, market and civil society. Many persuasive arguments are put forward for transferring some, or all, public services and publicly owned resources to the private sector and the discipline of the market place. The Republic of South Africa (RSA) is implementing enormous changes in the role of the state, market and civil society (van Zyl et al, 2001). Not least are the changes in institutional frameworks, organization for the planning, management and distribution of water through the National Water Act (NWA), and the transfer of state-run and owned smallholder irrigation schemes to farmer management (Perret, 2002). These changes are pursued through processes of participation and are intended to create structures to promote greater participation of citizens in service provision and wider processes of development decision-making.

The NWA breaks new ground in its attempt to involve the population in water management decisions through the establishment of WUAs. It demands a great deal from both the administration and the population in making decisions and taking responsibility for development of South Africa's water resource, which is arguably the scarcest of all resources in the country, and one with very high opportunity costs. If the act is to work as intended, it is crucial that participation is espoused by all the stakeholders; since water has a unique role in rural development, both people and government are concerned that it is well managed. Those who already use water for production, often through investment in infrastructure, are anxious to participate to protect their positions – water-users who are short of water, for example, are keen to participate to ensure a better supply – but many individuals and groups have little idea of how they might use water productively, or little confidence in their own ability to participate, and thus their participation is in doubt (Jaspers, 2001; Mackay, Rogers and Roux, 2003; Nemarundwe and Kozanayi, 2003; Svendson, 2002).

Two separate ministries are currently concerned with smallholder irrigation. The provincial Department of Agriculture (DOA) has a responsibility for members as new or emerging scheme 'owners' and managers, and as agricultural producers, while the Department of Water Affairs and Forestry (DWAF) has a responsibility for members as water managers/users and members of WUAs. The agricultural and economic imperative is to create farmer-managed schemes that operate to source services and inputs, and improve productivity and market ability in order to become self sustaining. The idea is for the farmer-managed scheme to manage water within the overall scheme, operate and maintain the system and develop the irrigation infrastructure in the future. Farmers must establish businesses with sufficient potential for profit to sustain the schemes. This is the only way that the farmers can continue to receive a supply of water. The DOA obeys an economic imperative: efficient use of resources entails payment of the costs associated with each resource. And the cost of water can no longer be ignored: infrastructure, operation and maintenance all cost. Whether the full costs can be met by smallholders is not certain at this stage but it is clear that they must make a significant contribution. They must then address fundamental issues about the nature of property ownership, economies of scale and market access, in addition to improved production practices. The DWAF obeys a political imperative: efficient use of water and equitable development based on water allocation necessitating change in the structure of water distribution. The department aims to make these changes through catchment management agencies (CMAs) and WUAs.

The ownership and agricultural economic imperatives on the schemes and the water distribution structural and political imperatives for the WUAs both require participation as the cornerstone of the changes required (Merle et al, 2000). However, the nature and processes of these engagements vary in complexity and requirement. The schemes must have identity and a framework for ownership and management, but the terms of ownership and the rules and conditions for collective decision-making are not yet clear. A WUA will manage water in its area, but the rules and conditions of participation in a WUA are onerous, demanding voluntary participation far in excess of worker availability in most smallholder schemes.

Institutions

The justifications for reducing the role of the state in water management typically include arguments associated with reducing costs, increasing efficiency and improving the quality and choice of services on the one hand, and arguments associated with improvements in equity, local control and ownership and stakeholder empowerment on the other. The overall objective is for communities to become the drivers of development. In the RSA under the NWA, the nation is engaged in pursuing these objectives, at the national level, through changes in the nature of public ownership and management of the water system as a whole through CMAs, which are responsible for the country's 19 catchment areas and for allocating water to WUAs.

Each WUA is responsible for representing and engaging with all the water users of the area (the members) on the management and distribution of water. At each level (member to WUA and WUA to CMA) payment for water use will have to be made and WUAs will be able to market water supplies between them. A WUA provides all citizens in its area with access to water for whatever purpose; it is a public system in the sense that it is controlled by the public membership (subject to approval by the CMA) but is also commercial in the sense that it markets water. In terms of institutional frameworks and linkages between state enabling systems and delivery and user systems (Hobley and Shields, 2000), the institutional architecture may be expressed as in Figure 8.1 below:

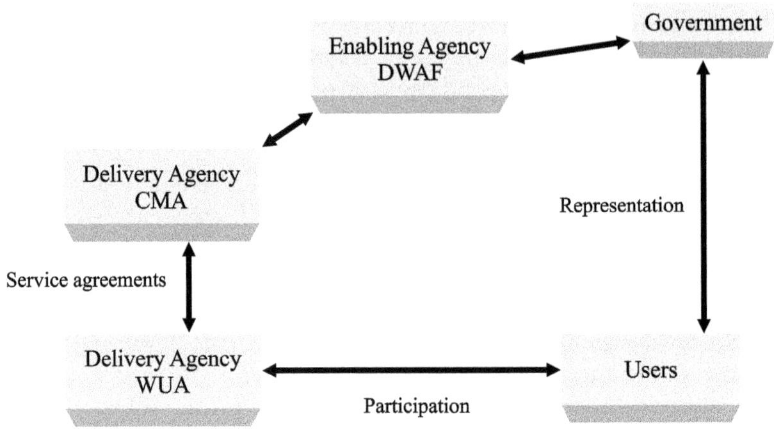

Source: After Hobley and Shields (2000)

Figure 8.1 *Water institutional architecture*

Both CMAs and WUAs are delivery agencies. As a state agency the CMA has a direct influence on what a WUA can do, while users influence the planning and nature of services provided by a WUA as participating members. Users are not simply receivers of water, as is the case with traditional command-based public service providers, nor are they in a market relationship as with a private delivery agency. Under the terms of the NWA the relationship between the delivery agency (the WUA) and the users is intended to be interactive, based on membership voice through active participation in planning, distribution and use. Poverty alleviation and empowerment are close to the goal of efficient water management among government priorities for a WUA.

Elsewhere in the world, in the Philippines, Tunisia, Morocco, Turkey, Bangladesh and Nepal, for example, a WUA has a more restricted role and is normally associated only with managing transferred irrigation schemes (Angood et al, 2003; ICID, 2004). These examples suggest that resource and management transfer to an appropriate WUA is a positive thing not only in terms of sustainable local water management but also in a general development sense. Therefore a WUA is in the private sector in the sense that it is for production purposes and belongs to its users/members. In contrast, a WUA in South Africa is a public institution in that it has statutory and legal responsibilities for the provision of water to all citizens, albeit being membership-based. Essentially, then, it has different purposes from those associated with WUAs elsewhere.

The question then is how these multiple objectives can be achieved within the WUA framework established in the NWA. To answer this, we need to understand what a transferred smallholder irrigation scheme will be like. Smallholder irrigation schemes in the former homelands were essentially state systems, with land holdings allocated and held through traditional rights-to-occupy/use systems, imposed in that members were there through force of circumstances. For now it must be assumed that the rights to occupy and use the land will continue to reflect the traditional mechanisms of the past. The transferred 'common' resource is then the irrigation system and infrastructure. However, IWMI Research Report 60 (2002) argues that IMT alone will not be enough to achieve viable sustainable smallholder irrigation schemes. What then would an appropriate institutional framework for collectively owning and operating it look like?

Smallholder farmers can be as efficient as larger farmers, but, if production is to provide a reasonable return, then economies of scale through collective association are crucial. IMT experts generally agree that only smallholder schemes that implement similar forms of integration as the ones of large commercial farms are successful. This can be achieved by developing organizational and management systems to handle input, credit and market constraints as well as the coordination and delivery of irrigation water. Such a view is fully endorsed by the present DFID research.

There are then three interlinked elements to be considered in seeking to develop appropriate organizational and institutional frameworks for smallholder schemes:

1 The identity and status of the scheme system and infrastructure as a common pool resource.
2 The need and options for formal and informal production, purchasing and marketing collaborative associations.
3 Organizational and institutional links to associate with a WUA to meet both their goals and those of the WUA, as defined in the NWA.

Identity and Status of the Schemes

Generally, voluntary schemes created by farmers for themselves are successful, which strongly influenced the adoption of IMT policy (Vermillion and Sagardoy, 1999). However, the fact that farmers/occupiers become owners does not provide the same basis, or motivation, for creating a relationship between members as when schemes are created by voluntary self-interested collective action. In the case of transferred ownership, collective action is an imperative not a matter of choice. The farmers are there, and unless they chose and are able to leave, which is uncommon, they are obliged to become joined together in some entity to collectively own and manage the irrigation scheme system and infrastructure.

The transfer of a scheme as a 'commons' resource involves rights and responsibilities. Following Agarwal and Gibson's (2001) discussion of common property, the notion of a scheme as a 'commons' means that it is an institution that establishes rights of access and use and some degree of cooperation or communal constraint on individual behaviour. It is not clear what the nature and extent of the rights and responsibilities are, or how the commons will be owned. The state, as previous owner, must clarify this situation and the irrigators who will take over this commons can learn what it is they will own, and how it will work. This understanding is key to identification of the right kind of support for management committees and appropriate participatory processes to empower scheme members and enhance their decision-making capacity.

The options for a scheme as a legal entity range from a legally constituted community-based organization to a registered shareholder company or cooperative. In the former, ownership is expressed as rights of access and use; in the latter, ownership is expressed by holding shares. In either case ownership is restricted to those who already have rights to occupy and use the land. How individual rights or shares can be transferred to others must be clarified in the legal conditions of the entity, but, given how the land is presently held, scheme membership rights or shares could only be held in conjunction with land rights, and as such would have to be given up if the land rights were given up.

To maintain a system and develop it as agreed by its members/shareholders, the framework adopted must provide for the scheme to trade, seek and acquire resources, such as credit, as a single legal entity. It would thus be then essentially 'private' systems that served the needs and aspirations of the members/shareholders. The scheme would have to operate as a sound business to cover costs and develop and maintain the system. The constitution of the scheme would clarify the obligations and responsibilities of the members or owners, mechanisms for participation and services to be provided. A clear understanding of the basic terms of ownership and control, and associated responsibilities is essential to participatory interaction and decision-making. If this is not the case, we would argue that there is a poor basis on which to build a sustainable or effective institution.

Schemes can only function if they have means to constrain individual behaviour that might otherwise affect their viability. The most obvious threat must come from those who do not pay their water dues (however these are finally determined) and/or those who do not make appropriate use of their land. Non-payment raises the costs for all other irrigators and, if sufficient numbers default, may lead to the collapse of the scheme, or at least a particular part of it. It may be argued that appropriate land use is a matter for the individual, as long as the water fees are paid (possibly by selling on the water), but this would defeat the development objectives of the investment, i.e. increased production and productivity through irrigation.

In either case, the scheme must possess the legal authority to ensure the degree of cooperation required by, if necessary, removing rights of access and use. Appropriate procedures in the case of non-payment for irrigation water may have to be determined with the WUA, but in the case of land, under the current dispensation this will have to be determined with the tribal and/or municipal authorities. It is important, therefore, that all relevant institutions are made fully aware of the institutional arrangements to be achieved in the transfer process. They must be fully committed to supporting this process – particularly with respect to establishing the legitimacy and authority of the managing committees and the officers of the institution.

Agricultural and Marketing Associations

The economic imperatives of irrigated production suggest that farmers on smallholder schemes must find ways to collaborate in production and marketing. These collaborations can be of many different kinds. Small schemes may develop fully integrated production and marketing systems, but on larger schemes these functions may remain separate. A point to be emphasized here is the need to recognize the distinction between requirements for identity of the scheme as a commons and formations for group production and marketing. The irrigation infrastructure is a common resource and scheme communities have to get together so that scheme control can be transferred to them. This transfer imperative requires an agreed understanding of the organizational and institutional framework, while the form that production or marketing associations take is entirely a matter of choice for the individual irrigator. Departments of agriculture may assist farmers to appreciate the benefits that collaboration or collective action might bring, but the extent and direction of any association should reflect the opportunities and choices available.

A close relationship between schemes, or scheme-based businesses, and the local WUA is a key condition of success. Not only do businesses rely on adequate and timely water, but a WUA relies on the profitability of schemes, individuals, groups, co-ops, companies and trusts for secure, sustainable finances. User-fees will finance water management by operating and maintaining infrastructure and development for equitable access where required. Funds will also be needed to ensure that the poor can exercise their water rights, which may initially involve training and capacity building for which additional support will be available to the WUA from central government.

Organizational and Institutional Arrangements

We argue therefore that it is inappropriate for a smallholder scheme to be a WUA because a WUA in South Africa is constituted to manage local water infrastructure and to ensure fair and reliable water supply to all its members, which is incompatible with the institutional arrangements necessary to establish a collective but essentially private, independent and business-based irrigation system. Nor is it compatible with the collective bodies formed for agricultural production and marketing purposes, or underwriting collateral arrangements for purchasing and marketing contracts that do not apply to all irrigators. A WUA sits best where users share a water resource. This can only be determined on a case basis, according to unique local features; and yet it is usually beyond scheme boundaries and requires involvement of adjoining schemes, informal or non-agricultural users. For example, Thabina already has a WUA; it is relatively small (234ha), with some 150 members. As depicted in Figure 8.2, the primary system comprises a single main canal from an upstream dam, a storage dam and secondary canals draining back to a river. This means that adjacent users below the dam, but upstream of the scheme, are outside the WUA-managed area. These domestic users and informal irrigators, however, depend on the same water source as scheme members and have a legitimate right to be members of a WUA, but not necessarily to participate in the scheme. In theory, they could form a separate WUA but the result would be several WUAs sharing the resource. Conflict resolution or cooperative action would then become the responsibility of the higher management level, the CMA, and such an arrangement is not what was envisaged. Decision-making at the CMA level would erode the empowerment and local decision-making expectations associated with WUA establishment.

The Thabina Management Committee is burdened by the WUA requirements to report, develop a business plan, monitor water and address issues of inclusion, equity and poverty alleviation in the wider community. These tasks add to the committee's already considerable effort in improving the performance of farmers through better input deals and better services. Not only does the scheme have limited access

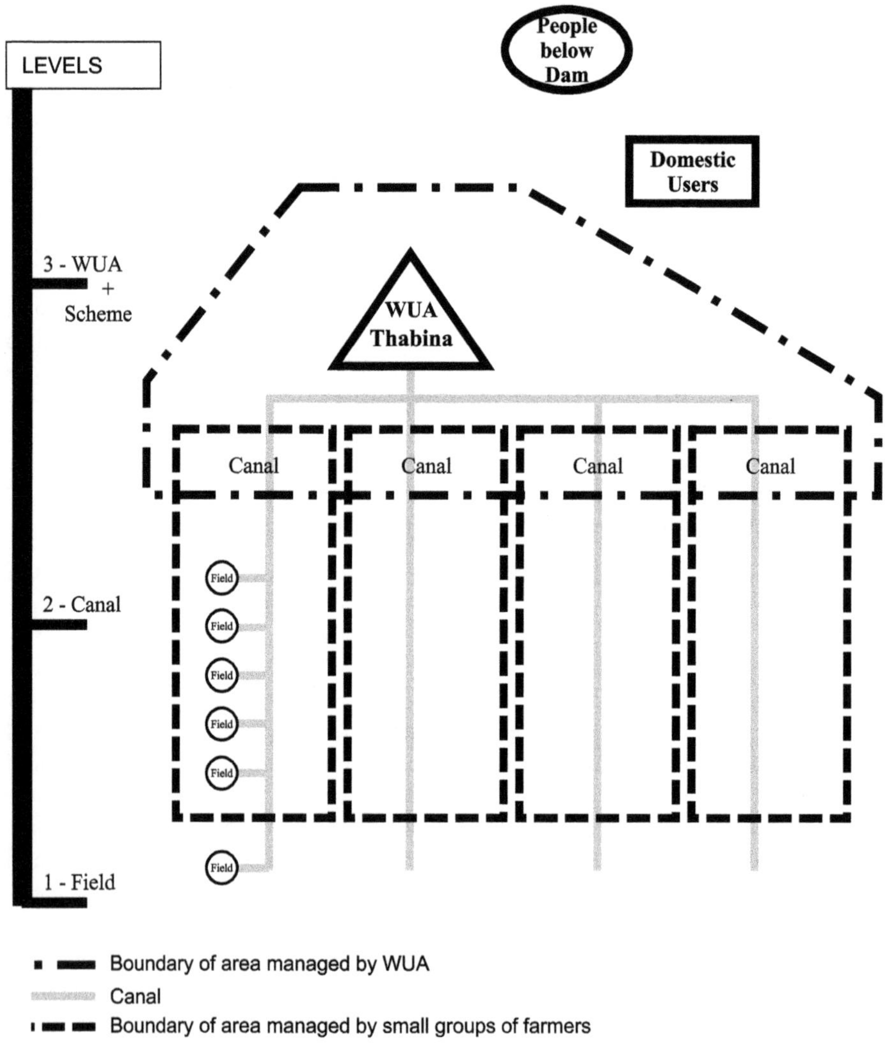

Source: chapter author

Figure 8.2 *Thabina: Existing WUA*

to resources, it also has a weak commercial basis with only a small proportion of the farmers regularly returning a comfortable profit. There is a long-term problem within the scheme relating to equity between these and poorer farmers.

We argue the need to redefine the boundary of the WUA, as illustrated in Figure 8.3. The WUA-managed area is now 'elevated' to outside the scheme with rep-

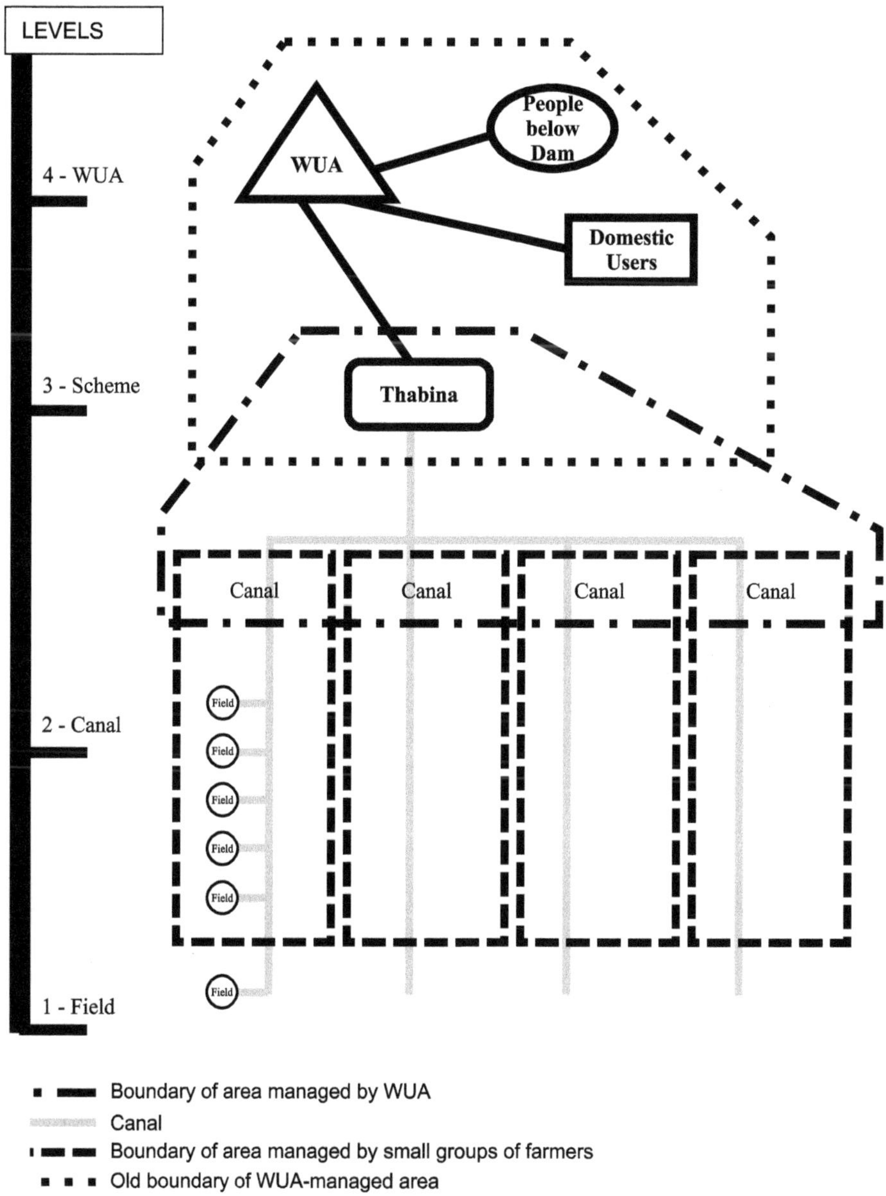

Source: chapter author

Figure 8.3 *Alternative structure*

resentation from the scheme, the irrigators below the dam and the domestic users. The scheme may choose to continue with the joint roles of system management and agricultural production and marketing, given its small size. Alternatively, farmers might commit to the scheme but prefer other groupings to follow market and production strategies. Either way, the question remains of how a scheme and separate production and/or marketing groups should be represented on a WUA, other than as individuals in their capacity as members of a WUA.

Dingleydale/New Forest

Organizational and institutional issues are more complicated on larger schemes comprising members from different communities, other water-using businesses within the scheme, and complex water supply, such as at Dingleydale and New Forest (DD/NF). Figure 8.4 shows the physical layout illustrating the interconnected water supply through the Orinoco Dam to separate parts of the scheme. The commercial ex-ARDA (Agricultural and Rural Development Association) farm Champagne irrigates from the same source, which also supplies domestic water to surrounding villages.

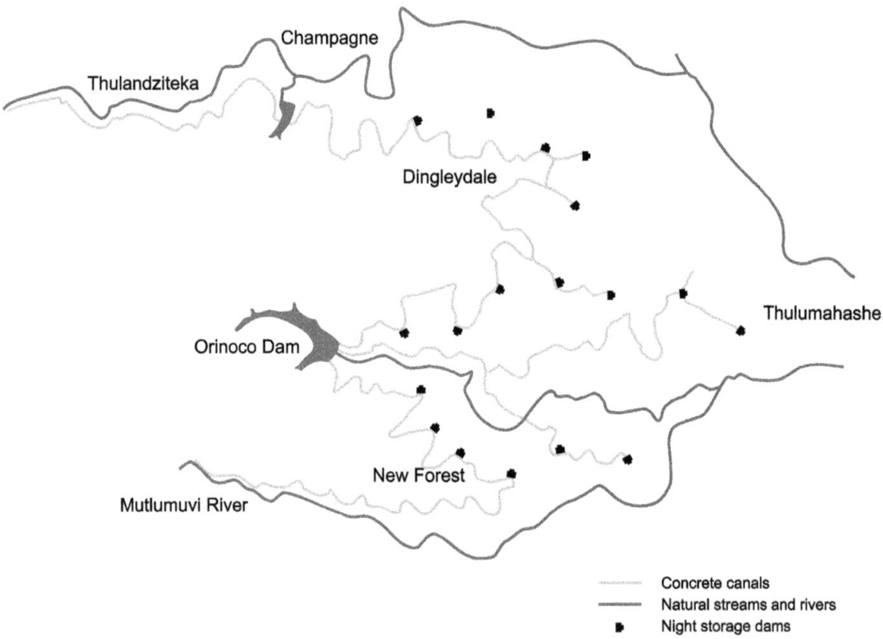

Source: chapter author

Figure 8.4 *Map of Dingleydale/New Forest*

DD/NF totals some 1600ha with some 5000–8000 people dependent on it for all or part of their livelihoods. Given that each scheme is large, and shares water resources with a large commercial enterprise as well as with domestic users, there appears to be an overwhelming case to separate the schemes from the WUA.

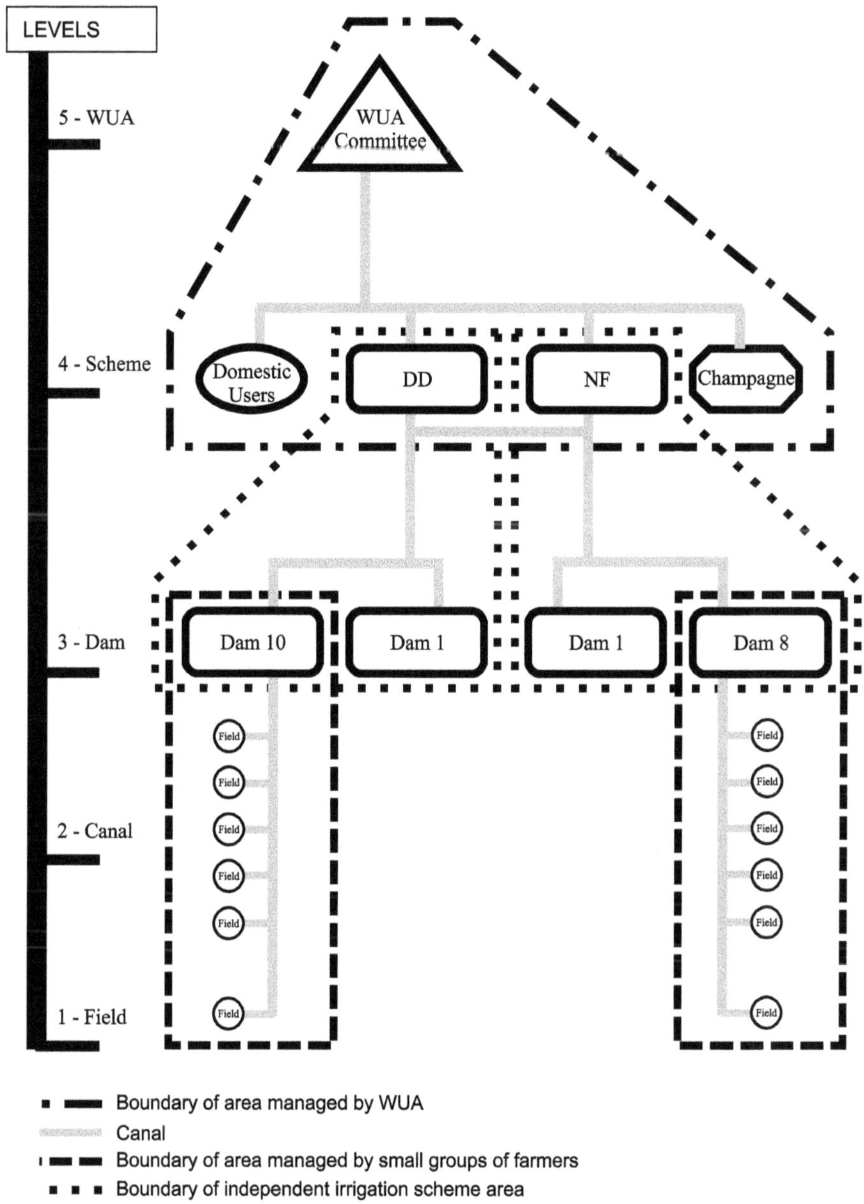

Source: chapter author

Figure 8.5 *Diagrammatic scheme levels, Dingleydale/New Forest*

A WUA could then be established as the appropriate body to manage the resource for all users. Figure 8.5 shows Dingleydale and New Forest as independent irrigation schemes which would, with Champagne and the domestic user's representatives, be members of the WUA. As discussed above, this assumes that group representatives could sit as WUA members.

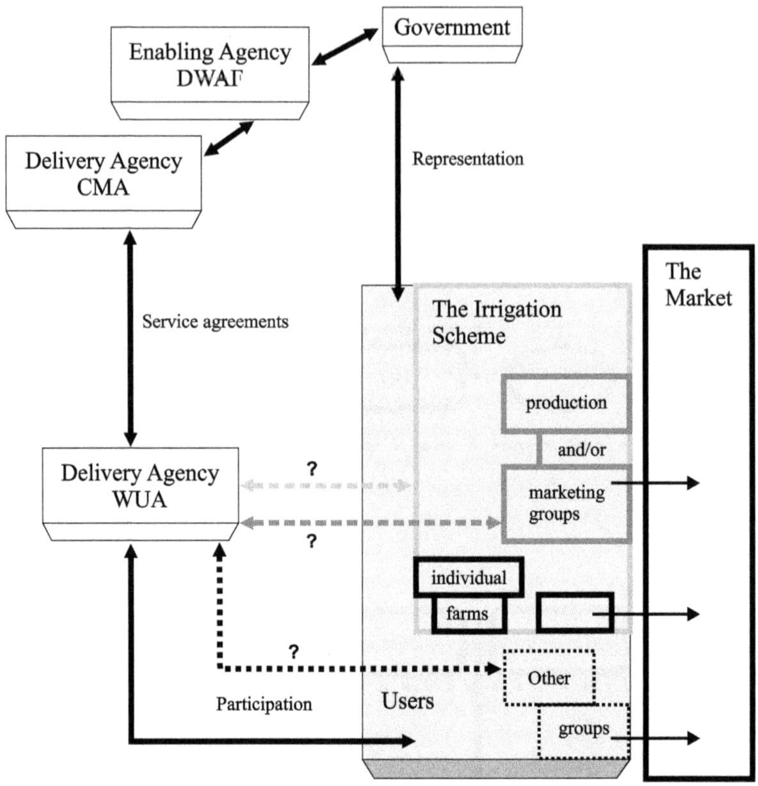

Note: ? indicate issues about group and legal entity representation on a WUA.
Source: chapter author

Figure 8.6 *Water institutional architecture, with user associations (in this case one irrigation scheme, and other groups inside or outside the scheme)*

The size of DD/NF also suggests a need for business groupings below the level of the two sub-schemes. For instance, production and/or marketing business associations might develop (some already exist) at balancing dam level (see Figure 8.4).The scheme level (level 4) management committees would be responsible for distribution between balancing dams and any other functions required and agreed by the second level dam groups or associations to be undertaken at scheme level.

The place and boundaries of a WUA in relation to schemes and farmers' production/marketing groups can be summarized by extending the institutional diagram presented in Figure 8.1 to the form illustrated in Figure 8.6.

In this illustration, there is only one scheme but there could be two or more. The WUA members are those in the shaded area. The scheme is a separate legal entity belonging to scheme members. The scheme members are water users and as such are individual members of the WUA, as are water using members of other groups, whether these are within the scheme or outside the scheme. Question marks indicate issues about group and legal entity representation on a WUA.

These, either formally or informally, are de facto likely to be represented as entities through their members, but presently membership has only been defined formally as that of individuals in the WUA area.

Discussion

The action of the DOA in the existing and future revitalization programme will culminate in handing over the schemes to farmer management – IMT. Irrigators must then operate in a framework relating to the irrigation scheme in which they farm and be concerned with the survival of the infrastructure and the level of water delivery service needed to sustain production. This framework does not need to be the same as production and marketing groups, but it must be supported and its rules must be kept if water is to continue to be available. The scheme in this sense must have authority in the farmers' eyes and they must commit to it.

The small scale of individual irrigators' farming enterprises makes it very difficult for them to compete in the market without some element of collective or cooperative activity unless individuals are able to add value or create highly differentiated products. Options for collective action are strongly influenced by the institutional framework in which irrigators operate. The commercial context requires them to create flexible, or even temporary, production or market related associations to react quickly and appropriately to changing market conditions. Individuals may want to operate alone, to operate together with nearby irrigators or to join others producing the same crop. They may find it advantageous to belong to more than one group. They need to be free to join and leave on agreed conditions, probably on an annual or seasonal basis.

The DWAF is largely concerned with efficient management of water but the institutional framework that it promotes has a wider remit. A WUA is a statutory body. It will manage water and collect fees just as will be done on irrigation schemes but its additional functions serve to save and reallocate water to ameliorate the present imbalance in water use. In the long run, it is expected to promote more efficient water use, protect the environment and prevent waste. All citizens will have a voice in this body to ensure water allocation and management that is appropriate to the physical conditions and demographic character of the area. In a WUA participants will come from a wide range of backgrounds and carry with them the interests of groups undertaking very different activities that compete for water. A wide range of power will also characterize the members and shape relationships between them.

There is clearly commonality between the activities and objectives of a farmer-managed smallholder scheme and a WUA, and for a time it seemed that one farmer committee might fulfil both roles. However, this represents a daunting change from the pre-1994 situation. The practicality for smallholders of dealing with such a heavy load of participation is limiting (Cleaver, 2000 and 2001). It is a myth that labour and energy abound in rural areas. For most smallholder irrigators, labour and management skills are scarce because irrigation is labour-intensive and management training has not been available. Time spent in voluntary participation has high opportunity costs, yet participation is essentially voluntary. (Lane, 1995)

In addition, there is an inherent contradiction in objectives: whereas an irrigation scheme needs to follow profit motives, a WUA has a statutory obligation to pursue equality and poverty eradication, which suggests that a dual purpose committee would not work.

Modern participation is bounded by frameworks that purport to encourage, focus and accelerate the spontaneous people-driven nature of old-style community activity. While participation is not a new concept invented by aid agencies and governments but a traditional coping strategy (Curtis, 1995), the term has been widely used and has become genuinely difficult to define. Chambers (1995) suggested three uses of the word. First, he identified a cosmetic label to make proposals look good; second, a co-opting process to mobilize labour and reduce costs; and third, an empowering process. There is difficulty, however, in achieving truly empowering participation. Many factors determine the potential for empowerment, such as the personal history of stakeholders, human capital and other resources of both stakeholders and development agencies, and the social capital of the community.

One legacy of the old South Africa is significant lack of experience in handling participation, both among the majority and among people working in development. People are rooted in the top–down approach, often justified by the perception that it allows immediate action and that details of participation and majority decision-making can be sorted out later, 'once a meal is on the table'. There are circumstances in which such views seem reasonable, thus the mantra is often accepted, even by stakeholders for whom empowerment is the main objective.

The relative power of stakeholders varies enormously. This is particularly the case in relation to water, as the contrast between the large commercial farmer and the smallholders exemplifies. In addition, activities that seem transformative in terms of local voice and decision-making may, on closer inspection, be highly inequitable for women and poor and disabled people. The quality and nature of participation and whether it is seen as a 'means' to enhance a project or increase efficiency, or 'an end in itself' are issues still under discussion in development circles (Lane, 1995; Kabeer, 1994; Meinzen-Dick, 2000; Nicol and Mtisi, 2003). Interpretation of the NWA in this respect reveals elements of both. On the one hand, a WUA will become the means of managing water, reducing the financial and development responsibility of government; on the other, communities will be empowered, by the activities and decision-making in managing water, to direct development in their locality.

In the past, participation was relatively superficial in smallholder irrigation schemes. The role of a farmer committee was not decision-making, but liaison, conveying management decisions and receiving farmers' views. It was very different from the participation of large-scale farmers in the former irrigation boards, where farmers made decisions, investments and controlled development. IMT has led to the formation of new committees that will undertake these roles. However, although the process initially, albeit unwittingly, concentrated activities in the hands of the better-off and more confident farmers, bypassing important subgroups, it was soon recognized that all irrigators had to be represented and that committees should be broad-based and democratic. This is essential to ensure the commitment of the vast majority of members.

Another significant factor affecting participation is the attitudes different people bring to it. Some carry conviction as to how a specific task should be approached and carried out; others come with deep uncertainty and a predetermined tendency to accept the suggestions of others. Skilled facilitation is therefore needed to prevent the powerful from swamping the less powerful. Within smallholder irrigation schemes, emerging farmers are powerful people, demonstrably more so than subsistence plot holders. Yet, all have significantly less power than any other competitors for water resources.

The DFID research study explored the characteristics of participation in one of the revitalized schemes, DD/NF (which might be regarded as two separate schemes but for the complex interdependency of water supply). Formerly belonging to different homelands, much of the present networking in the DD/NF scheme reflects its duality, and revitalization has progressed differently in the two sub-regions, reflecting not only the differences between the two committees but also between the two communities.

The participation level achieved on the scheme appeared relatively good and indicated that the revitalization process was to some extent successful. However, poor communications – not only bad roads and lack of phones, but also poor literacy, unreliable verbal communication, social and ethnic barriers and failure to understand existing networks and how people receive and filter information – resulted in as many as 60 per cent of people in one part of the scheme not understanding what was going on in relation to IMT despite the high level of activity all around them. At best 30 claimed they were unable to grasp the proposed changes, and many people did not even know of the existence of the new management committee. This implies there is a serious communication problem in a scheme expected soon to register as a WUA, when all will be contributing members.

This situation can hardly be explained only as a farmers' problem. The difficulty the development side has in identifying a clear purpose and plan of action also seems to be a major impediment to effective participation, which suggests that the two branches of government must give more attention to presenting a united and coherent plan to drive the participation process upon which both the farmers and future water users depend. Presently farmer participation is deflected by the perception that the proposed institutions will not serve their purposes. Meetings that fail to reach a consensus, insufficient response to the concerns voiced by farmers and participation fatigue, exacerbated in South Africa by the disadvantages of the past, fuel farmers' doubts as to how they will sustain participation in the future. There are uncertainties over whether members' payments will really fund the activity needed, how payment levels will be set and payments enforced, and whether, indeed, the majority of farmers will be able to pay.

Conclusion

In many of South Africa's new WUAs the viability of smallholder irrigation schemes will be central to sustainability and will be the basis on which the WUA will grow to address the wider issues of water management and productive water use for all local people. It is therefore important that the institutional arrangement supports irrigators over the wide spread of their activities. Not all irrigators will be able to embrace commercial farming and those who cannot must be enabled to profit at some level in order to contribute to management costs. Specialist market strategies will be required, and this chapter identifies three important imperatives.

Economic Imperative

Attention must be given to marketing and market support for irrigated produce in order to ensure that newly-established smallholder farmer-managed irrigation can achieve economic survival.

This should include institutionalizing services for the following:

- Provision of market information to publicize location of markets, seasonal patterns of demand and continuing trends, competition from other suppliers and costs and benefits of processing, packaging, storage and transport, supplemented by advice on accountancy business management, institutional frameworks for group activity and contracting.
- Provision of extension advice and training courses on marketing through a government-assisted separate and specifically trained marketing advisory service.
- Brokering supply agreements (eg via a marketing agency) to assist farmers in negotiating contracts with potential commercial buyers and ensuring compliance with the terms of agreements.

Institutional Imperative

Institutional arrangements in relation to smallholder irrigators must achieve clarity on the separation of institutions, defining the scope and purpose of each in a way that people can understand and therefore know their role in relation to it.

There need to be separate institutions to manage irrigation systems to which the smallholders as individuals must commit; institutions to assist in production and marketing, to which commitment is a matter of agricultural choice and management; and institutions to oversee equity, efficiency and empowerment in relation to future water management in the wider environment, to which all citizens will belong.

These separations broadly relate to irrigation scheme management, production and marketing groups and the activities of a WUA. There is a need for close liaison and for flexibility. Individuals will be members of some or all of the institutions.

Participation

Participation is the way in which farmers and other rural people will engage with the water-related development. Participation involves understanding the rights and responsibilities of engaging in people-centred development and is worthy of significant investment. The impact of past disadvantages must be understood specifically in relation to water and development. Different sorts of engagement will be appropriate for different groups. However, it is important to achieve clarity on the objective of participation in the different institutional settings needed for successful smallholder irrigation and other water uses. Objectives and expected outcomes must be clearly communicated to a much greater proportion of people than is now the case, both inside and outside of existing irrigation schemes, so that water users can be confident that they understand what is going on.

References

Agarwal, A. and Gibson, C. C. (eds) (2001) *Communities and the Environment: Ethnicity, Gender and the State in Community-Based Conservation*, Rutgers University Press, Vale, NJ

Angood, C., Chancellor, F., Morrison, J., and Smith L, (2003) *Contribution of Irrigation to Sustaining Rural Livelihoods: Bangladesh Case Study OD/TN 114.* HR Wallingford Ltd, Wallingford, UK

Ashley, C. and Maxwell, S. (2001) 'Rethinking rural development', *Development Policy Review*, vol 19, no 4, pp395–425

Chambers, R. (1995) 'Paradigm shifts and the practice of participatory research and development' in N. Nelson and S. Wright (eds) *Power and Participatory Development: Theory and Practice*, IT Publications, London

Chancellor, F., Shepherd, D. D. and Upton, M. (2003) 'Towards sustainable smallholder irrigation businesses', DFID Project OD 149, HR Wallingford, Wallingford, UK

Cleaver, F. (2000) 'Moral ecological rationality, institutions and the management of common property resources', *Development and Change*, vol 31, no 2, pp361–383

Cleaver, F. (2001) 'Institutions, agency and the limitations of participatory approaches to development', in Bill Cooke and Uma Kothari (eds) *Participation: The New Tyranny?*, ZED Books, London, pp36–55

Curtis, D (1995) 'Owning without owners, managing with few managers: Lessons from Third World irrigators' in N. Nelson and S. Wright (eds) *Power and Participatory Development: Theory and Practice*, IT Publications, London

Gakpo, E. F. Y., du Plessis, L. A. and Viljoen, M. F. (2001) 'Towards institutional arrangements to ensure optimal allocation and security of South Africa's water resources', *Agrekon*, vol 40, no 1, pp87–103

Hedden-Dunkhorst, B., Machethe, C. L. and Mollel, N. M. (2002) 'Smallholder water management and land use in transition: A case study from South Africa' *Quarterly Journal of International Agriculture*, vol 41, nos 1–2 (Special Issue: Agricultural Water Management and Land Use in Relation to Future Water Supply), DLG-verlag, Frankfurt, pp59–76

Hobley, M. and Shields, D. (2000) *The Reality of Trying to Transform Structures and Processes: Forestry in Rural Livelihoods.* Overseas Development Institute, London

Hodgson, S. (2003) 'Legislation on water users' organizations: A comparative analysis', FAO Legislative Study, FAO, Rome

ICID (International Commission for Irrigation and Drainage)(2004) 'Socio-economic sustainability of services provided by irrigation, drainage and flood control schemes in the water resources sector', position paper, ICID, New Delhi, India

IFAD (International Fund for Agricultural Development) (2001) *Rural Poverty Report 2001: The Challenge of Ending Rural Poverty.* Oxford University Press, Oxford

Jaspers, F. G. W. (2001) 'The new water legislation of Zimbabwe and South Africa: comparison of legal and institutional reform' in *International Environmental Agreements: Politics, Law and Economics*, vol 3, no 5, Kluwer Academic Publishers, Dordrecht, The Netherlands, pp305–325

Kabeer, N. (1994) 'Empowerment from Below' in N. Kabeer (ed) *Reversed Realities: Gender Hierarchies in Development Thought*, Verso, London

Lane, J. (1995) 'Non-governmental organizations and participatory development: The concept in theory versus the concept in practice' in Nelson, N. and Wright, S. (eds) *Power and Participatory Development: Theory and Practice*, IT Publications, London

MacKay, H. M., Rogers, K. H. and Roux, D. J. (2003) 'Implementing the South African water policy: Holding the vision while exploring an uncharted mountain', *Water SA*, Water Research Commission, Gezina, South Africa, vol 29, no 4, pp353 358

Meinzen-Dick, R. (2000) 'The need for irrigators' organisations', *Zeitschrift fur Bewasserungswirtschaft*, DLG Verlag, Frankfurt, Germany, vol 35, no 2, pp129–146

Merle, S., Oudot, S. & Perret, S. (2000) *Technical and Socio-economic Circumstances of Family Farming Systems in a Small-scale Irrigation Scheme of South Africa (Northern Province)*, PCSI report no 79/00, Cirad-Tera, Montpellier, France

Nemarundwe, N. and Kozanayi, W. (2003) 'Institutional Arrangements for Water Resource Use: A Case Study from Southern Zimbabwe', *Journal of Southern African Studies*, vol 29, no 1, pp193–206

Nicol, A. and Mtisi, S. (2003) 'Politics and water policy: A southern Africa example', *IDS Bulletin*, Institute of Development Studies, University of Sussex, Brighton, UK, vol 34, no 3 (Special Issue: Livelihoods in Crisis? New Perspectives on Governance and Rural Development in Southern Africa), pp41–53

Perret, S. R. (2002) 'Water policies and smallholding irrigation schemes in South Africa: A history and new institutional challenges', *Water Policy*, vol 4, no 3, pp283–300

Shah, T., van Koppen, B., Merry, D., de Lange, M. and Samad, D. (2002) 'African smallholder irrigation: Lessons from international experience with irrigation management transfer', IWMI Research Report 60, Colombo, Sri Lanka

Svendsen, M. (2002) 'The role of irrigation in reducing poverty and shrinking income disparities', *Integration and Management of Irrigation, Drainage and Flood Control* (Volume 1B), 18th International Congress on Irrigation and Drainage, Montreal, Canada, International Commission on Irrigation and Drainage (ICID), New Delhi, India, pp1–8

Van Zyl, J., Vink, N., Kirsten, J. and Poonyth, D. (2001) 'South African agriculture in transition: The 1990s', *Journal of International Development*, vol 13, no 6, pp725–740

Vermillion, D. L. and Sagardoy, J. A. (1999) 'Transfer of irrigation management services: Guidelines', FAO Irrigation and Drainage Paper no 58, FAO, Rome

Part III

9

Conflict Analysis and Value-focused Thinking to Aid Resolution of Water Conflicts in the Mkoji Sub-catchment, Tanzania

Leon M. Hermans, Reuben M. J. Kadigi, Henry F. Mahoo and Gerardo E. van Halsema

Introduction

International policy documents on water resources management often stress the crucial importance of water to sustain life and development (eg SIWI, 2000; WWAP, 2003). The true meaning of these words is reflected in the harsh realities faced by water users in drought-prone regions throughout the world. In regions that suffer from occasional or chronic water scarcity, the distribution and use of scarce water resources requires hard choices and, as a result, conflicts between water users easily erupt.

Within today's paradigm of integrated water resources management (IWRM), the management of water resources is seen as a process that aims at coordinated development and management of water and other natural resources (GWP, 2000). In closed basins, in which there are no utilizable outflows in the dry season (Molden, 1997), conflicts over water are likely to be part of this process. The sustainable implementation of IWRM principles is therefore only possible if the involved stakeholders find a way to resolve their conflicts; sustainable water resources management requires stakeholders to jointly manage their water resources, rather than fight over them.

Unfortunately, resolving conflicts in closed or closing basins is not an easy task. All utilizable water is committed to present uses and making water available for additional uses requires a transfer of water or an increase in productivity (Molden et al, 2001). Furthermore, at the local level the constraints posed by higher level institutions and external events further restrict the room for negotiated agreements.

The pervasiveness of conflicts over water in closing basins and the difficulties in resolving them make a proper understanding of conflicts a prerequisite to the formulation of an effective IWRM strategy. Understanding the specifics of a conflict, including its dynamics and its evolution, provides an essential basis for subsequent efforts towards conflict resolution. Since conflicts are driven by the interests and values of stakeholders, insight into the values underlying conflicts may further help the understanding of conflicts and the identification of promising ways out. And such identification is of crucial importance, as while jointly constructed win–win solutions are often heralded as ideal in conflict resolution, their identification still remains more of an art than a craft (GWP, 2001).

In this chapter the use of two analytical frameworks – conflict analysis (Fraser and Hipel, 1984) and value-focused thinking (Keeney, 1992) – that support a better understanding of local conflicts over water is discussed. The use of these frameworks is illustrated in a case of a closed water basin in the United Republic of Tanzania to show how they support a better understanding of local conflicts. Furthermore, the case illustrates how the frameworks aid the identification of promising solutions, beyond the traditional solutions in water management that either imply 'transferring water' or 'increasing water productivity' (Molden et al, 2001). Although the discussed analytical frameworks have been used for policy analysis and decision analysis for quite some time, they have so far received little attention in the IWRM community. This chapter aims to illustrate that these frameworks deserve a wider use in the field of water resources management.

Analytical Frameworks to Aid Understanding and Resolution of Conflicts

Understanding Conflict ...

Before addressing a water conflict, a proper understanding of both the conflict and its history is considered essential. Generally, frameworks for conflict analysis are built around a core that addresses the key issues over which the conflict arises, the parties involved in the conflict, the parties' main interests or objectives and the resources or means available to them to control issues and to influence the course and outcomes of the conflict (Fraser and Hipel, 1984; Howard, 1989; Fang et al, 1993; Timmermans and Beroggi, 2000; Obeidi et al, 2002).

Depending on the specific interests of the analysts, other aspects might also be covered, such as the perceptions of the parties in the conflict (Bennett et al, 1989), the arguments used in conflicts (Horita, 2000; Hermans, 2003), the coalition building (Kilgour et al, 1996) or the resolution strategies employed by the parties involved (Castro and Nielsen, 2003).

... and Moving Towards Conflict Resolution

The conflict analysis framework helps to build understanding of a conflict, which is the first step towards conflict resolution. However, in many cases, it is not easy to identify a way to end the conflict. In these cases, a good understanding of the interests and values of parties in the conflict is likely to help the identification of creative ways out of the conflict. Although values are incorporated in conflict analysis frameworks through the interests and preferences of the involved stakeholders, the focus in conflict analysis is much more on options than on underlying values. However, a key to effective negotiation is 'creating value' (Sebenius, 1992), which suggests that the underlying values of stakeholders merit a more detailed analysis.

Focusing on stakeholders' values can help in supporting conflict resolution by suggesting directions for new alternatives (Keeney, 1992; Gregory and Keeney, 1994). Therefore, a value-focused thinking approach, as elaborated by Keeney (1992), is used to derive the stakeholders' perspectives on the underlying values of water. This not only offers a further understanding of the conflict but, more importantly, also stimulates the identification of possible directions out of it (McDaniels and Trousdale, 1999; Gregory et al, 2001). This framework puts stakeholders and their fundamental objectives and values at the centre and analyses them before addressing the more specific measures and alternatives available to stakeholders. Using this approach for water related conflicts implies a broad interpretation of the values of water, which may include, but are not confined to, an economic valuation or a financial cost–benefit analysis of predefined alternatives.

Introduction to the Case Study: Conflicts Over Water in the Mkoji Sub-catchment, Tanzania

The uses of the above-mentioned analytical frameworks for conflict analysis and value-focused thinking are illustrated in a case of local water conflicts in the Mkoji sub-catchment (MSC) in southwest Tanzania. This case study is based on material collected as part of a larger project for which several activities were undertaken, including an extensive household survey, a comprehensive analysis of available monitoring data and past research findings, focus group discussion and a participatory planning workshop (FAO, 2005).

The MSC is a rural area with a relatively low population density of about 146,000 people on an area of about 3400km^2. It drains into the Great Ruaha river, which in turn is part of the larger Rufiji river basin (see Figure 9.1). Water scarcity in the area has resulted in shrinkage of the Usangu wetlands, drying up of the Great Ruaha river in the Ruaha National Park and problems with power generation at the two main hydroelectric facilities, Mtera and Kidatu. Downstream water users are furthermore experiencing drinking water shortages during the dry season, deficits of water and pasture for livestock, reductions in areas suitable for fish breeding, and less land suitable for wildlife, while tourism in the Ruaha National Park also is affected by the drying up of the Great Ruaha river. As a result, there is increasing pressure on the communities in the upper sub-catchments of the Rufiji river basin to release more water to meet downstream water needs.

The MSC is one of the upper sub-catchments from which more water could be released to benefit downstream users in the Rufiji river basin. However, the MSC is also a closing sub-catchment and suffers increasing water shortages. Dur-

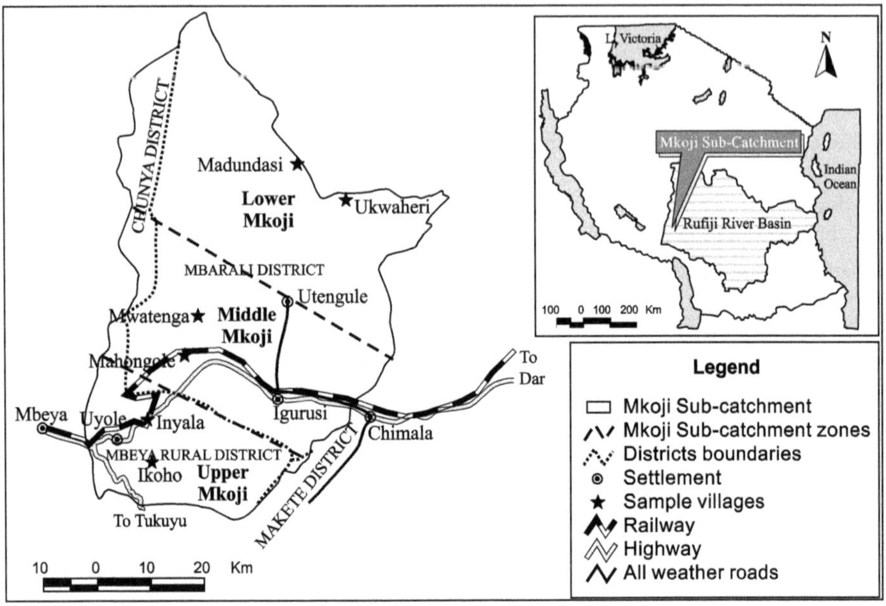

Source: FAO (2005)

Figure 9.1 *Location map of the Mkoji sub-catchment in Tanzania*

ing the dry season, all the rivers dry up a few kilometres downstream of the highway that runs through the sub-catchment, which leaves the lower part without water. In fact, the communities in the lower parts of the MSC are experiencing similar problems to those further downstream: shortages in domestic water supply, lack of water and pasture for livestock, reduced breeding grounds for fish and shrinking wetlands.

Within the MSC, three different agro-ecological zones can be distinguished; these have different climatic conditions and also differ in terms of the availability of land and water resources:

1. The Upper Zone, which mainly falls within the Mbeya Rural District and which is characterized by a mountainous landscape and a semi-humid to humid climate that allows year round cultivation of crops.
2. The Middle Zone, which has a less favourable climate, but where the land and water resources enable rice cultivation and irrigation in parts of the zone.
3. The Lower Zone, which covers the plains that extend into the downstream sub-catchments and which has a semi-arid climate and a low population density. This zone is inhabited by pastoralists who raise their cattle in the plains of the Lower Mkoji and neighbouring areas.

Competition for water concentrates in the Middle and Lower Zones, as these are the areas where water courses run dry during the dry season. Here, competition for water is fierce and easily escalates into serious conflicts. Two types of conflicts that frequently occurr in this area are:

- upstream/downstream conflicts within a water using community: within the MSC there are important conflicts within the irrigation schemes that are used for rice cultivation in the Middle Zone; and
- conflicts between water-using sectors: another important conflict in the MSC is that between livestock keepers and agricultural irrigators. This is a multiple use conflict with an upstream/downstream dimension within the sub-catchment, as livestock keepers are usually located in the Lower Zone while irrigation occurs in the Upper and Middle Zones.

In addition to these conflicts, other conflicts also occur in the MSC. However, to allow for a good illustration of the two analytical frameworks, this chapter is limited to the above two types of conflicts, which can also be found in many other closing catchment areas.

Local Conflicts Over Water in the Mkoji Sub-catchment

Introduction to the Conflict Analysis Framework

As stated above, various methods for conflict analysis have been developed to help gain insight into the development and possible course of conflicts, and the options different actors have to influence these. These methods are generally based on game theory or related approaches and have also proven their use in the field of water resources management, as illustrated by applications for water-related conflicts in Canada, the US and the Philippines (Obeidi et al, 2002; Hermans, 2003; Hermans et al, 2003).

Within a conflict analysis framework, the interactions between actors are modelled as strategic games. Actors are assumed to be more or less rational agents whose behaviour is guided by a combination of their objectives and the actions under their control, i.e. options. The interactions of actors revolve around certain issues and each actor has different options to influence these issues. Often, actors will have several options, which can be combined into strategies, based on different combinations of options. Each actor will choose his own strategy, and together the individual strategies can be combined to describe the possible outcomes of the conflict (Howard, 1989). Modelling the actors, issues, options and objectives thus enables an insight into the range of possible outcomes of conflict.

The analysis of options structure developed by Howard (1971, 1989) provides the conceptual basis for the majority of the more elaborated methods for conflict analysis. The basic steps of such an analysis of options are to:

1 Review the issues to be decided, ie over which conflict arises;
2 Review who controls the issues, either directly or indirectly; and
3 Ask how actors control these issues, resulting in an inventory of options.

A logical next step in conflict analysis then is to review the possible outcomes by systematically identifying feasible combinations of options that actors could implement. As part of this, the dependencies between options also have to be formulated, for example, 'option X can only be implemented if option Y is also implemented', or 'options Y and Z are mutually exclusive'. In theory, in a game with N stakeholders $s_1, \ldots, s_N$ who have O_i options ($i = 1, \ldots, N$), there are $O_1 \times \ldots \times O_N$ possible outcomes. As the number of stakeholders and the number of options available to them increase, the number of possible outcomes will increase sharply due to the amount of possible combinations. Conversely, the dependencies between options will reduce the number of possible outcomes, because they rule out those containing logically or physically impossible combinations of options.

Once options and possible outcomes are identified, preferences of different actors for each of the possible outcomes can be ranked, using ordinal preferences. The result is a fully quantified model of the conflict, which can be mathematically analysed for equilibria. The latter refers to a stable outcome, ie one from which none of the parties in the conflict can move to a more preferred outcome on his own by changing his selection of individual options without expecting a sanction from one of the other stakeholders.

These last steps are mentioned for a better understanding of the full theoretical framework underlying conflict analysis, but they are not pursued here. However, even without the last calculation steps, framing a conflict in terms of the framework's basic concepts can help increase understanding of water conflicts at the local level, even when information is less detailed or complete (Howard, 1989).

Understanding the Conflict Over Irrigation Water for Rice

Background, evolution and dynamics

Paddy cultivation is the main source of income for households in the Middle Zone, as can be seen from Figure 9.2. This figure provides an overview of the main sources of income in each of the three agro-ecological zones of the MSC. It is based on the results of a household survey carried out in 2003 covering 246 households in 6 villages (2 villages in each of the 3 zones) throughout the sub-catchment (for details see FAO, 2005).

An important water requirement for paddy cultivation occurs early in the growing season, when the establishment of nursery fields and the subsequent transplanting of paddy takes place. Most farmers attempt to transplant their paddy early in the season, because this will enable them to take advantage of seasonal dynamics in market prices, whereby rice that is marketed early in the season fetches a higher price.

Following the dynamics of the rice markets, the competition for water among paddy farmers reaches its annual peak at the onset of the wet season. Every year, during the start of the wet season, there is a fierce struggle among farmers to obtain water for early transplanting, even more so in years when the onset of the wet season is delayed. This may result in conflicts, whereby competing farmers

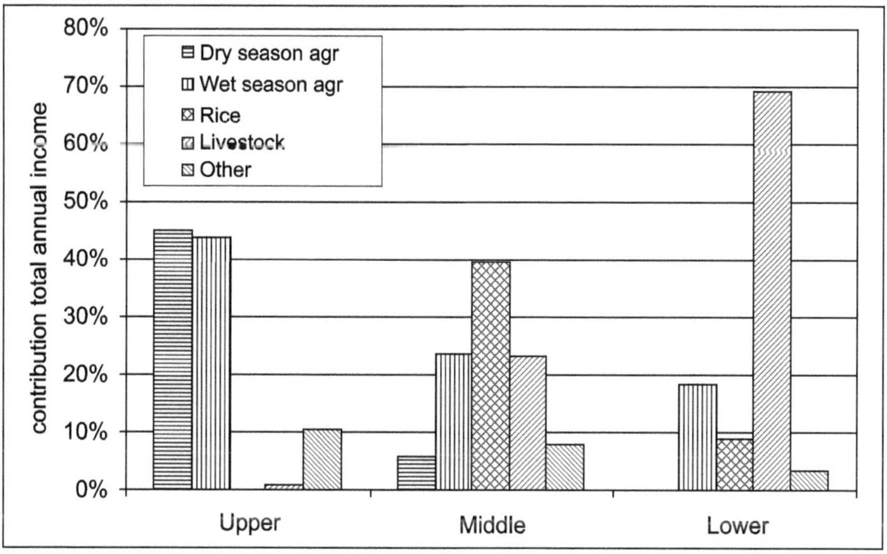

Source: FAO, 2005

Figure 9.2 *Sources of income for households in the three zones of the MSC*

destroy water canals and intakes to allow water to flow to their own fields. These conflicts may even erupt in violent fights and sometimes result in court cases, as shown by examples before the Igurusi Primary Court (Maganga et al, 2003).

Furthermore, conflicts seem to have worsened over the past years, probably due to the increase in irrigation activities in the Upper and Middle Zones. In both zones, there has been a trend to modernize irrigation infrastructures and to increase the capacity to abstract stream flows for irrigation. This has resulted in the construction of modern (concrete) intake and diversion structures and realignment of main and secondary canals, together with the establishment of operation and maintenance committees and new guidelines for water scheduling between farmers. Irrigation improvement projects have benefited some of the villages in the Middle Zone, but generally the modernization activities appear to be more widespread in the Upper Zone villages. As a consequence, upstream abstraction capacities have increased, leaving less water available for irrigators in the Middle Zone. Whereas the irrigation schemes in the Upper Zone have water flowing in almost all intakes, rotation in the Middle Zone involves inter-intake allocation.

Framing the situation with conflict analysis concepts

Using the conflict analysis framework, conflict can be summarized in terms of issues, parties and parties' options and preferences. The issue can be identified as access to irrigation water within the Middle Zone irrigation schemes. The irrigating paddy farmers are the main parties involved, and they all prefer to have sufficient irrigation water for their paddy, and as early in the season as possible. One can further distinguish top-end and tail-end irrigators within the schemes, as well as farmers that have joined the irrigation associations and those that have not.

One of the obvious options that are available to the irrigators is to jointly establish and enforce the water rotation schedules. However, while this would help to resolve conflicts *within* the irrigation schemes, it would not make much difference to the general trend of decreasing water availability in the Middle Zone. Therefore, it is likely to result in winners and losers, dividing slices of an ever decreasing cake and leaving some farmers with less water than in the current situation. Furthermore, enforcement is already problematic, as illustrated by the fact that many of the control gates within the improved irrigation systems have been removed or damaged (Lankford, 2004).

Another option would be to improve water use efficiency, by using irrigation improvement programmes, improved crop varieties or other water saving measures. However, although there is ample room to increase water use efficiency, this has proven problematic in the past. Irrigation improvement programmes are generally expensive and often require external funding sources. The irrigation improvement projects that have been undertaken in the past in several villages within the MSC have not always had very positive results. They mainly benefited the upstream communities at the first improved intake point along the river, at the expense of downstream communities (Lankford, 2004). Moreover, improvement projects often reduced the need for joint maintenance activities of the irrigation schemes by farmers, in this way contributing to an even less cooperative climate (Lankford, 2004). As for the other options: seeds for improved crop varieties are expensive and involve recurring expenses for seed purchases, while improved on-field water management practices often require training, initial investments and a transition period. This quickly puts these options out of reach for the average farmer.

In the end, the easiest option for the top-end irrigators is just to take the water they need, leaving downstream irrigators deprived of water. There is not much downstream irrigators can do about this, unless the upstream irrigators choose to cooperate, which means that this is a more or less stable outcome. Unsurprisingly, this is exactly the outcome that is currently occurring during the onset of the wet season in the MSC.

Understanding the Conflict Over Water for Livestock, Irrigation and Wildlife

Background and evolution of the conflict

Whereas rice is the main source of income for the farmers in the Middle Zone, livestock is the main source of income in the Lower Zone (see Figure 9.2 above). During the wet season livestock keepers graze their herds in the Lower Zone of the Mkoji, while in the dry season, when the Lower Zone does not provide enough pasture, they move their herds to other seasonal grazing lands. Normally, about 75 per cent of the cattle in the MSC are herded outside the MSC during the dry season (FAO, 2005).

Traditionally, livestock keepers grazed their herds on the pastures around the Ihefu perennial swamp in the neighbouring Usangu Plains, until the government of Tanzania decided to declare a considerable portion of this area the Usangu Game Reserve, where livestock keepers would no longer be allowed to take their animals. Although this decision was announced back in 1998, the restricted access to the grazing grounds around the Ihefu perennial swamp was only recently enforced by the district government officials.

Of course, this loss of access to grazing lands forced livestock keepers to find other suitable lands for their livestock during the dry season, which resulted in a move towards the irrigated areas in the Middle Zone. However, in these areas the livestock keepers easily come into conflict with the irrigating farmers, for instance when their cattle damage irrigation intakes and canals or graze irrigated fields.

Framing the situation with conflict analysis concepts

The issue in this conflict is access to land and water during the dry season, competing over these resources for livestock, irrigation and wildlife. The main parties are the livestock keepers, the irrigators and the Game Reserve officials, who all prefer to use as much water as they can access from the scarce dry season water resources.

The livestock keepers basically have two main options: they drive their cattle either illegally to the Usangu Game Reserve or to other nearby grazing lands, including those in the Middle Zone. The former option drives livestock keepers into conflict with the Game Reserve officials, which means that they risk fines as well as being chased away; the second drives the livestock keepers into conflict with the Middle Zone farmers, who do not have many options other than to protect their property by persuasion or by force. The result is often escalation of the conflict or, in some cases, formal settlement of conflicts through village leaders and courts. It is easy to understand that this situation satisfies no-one, as imposing fines and court settlements does not represent a very structural solution, especially not for the livestock keepers.

A more drastic option for livestock keepers would be to reduce their livestock keeping activities and develop other livelihood activities. To some extent this option is also being implemented. The loss of natural grazing grounds pushes the livestock keepers towards intensification and expansion of their wet season cropping activities, while diminishing floods allows for the reclamation of flood plains into agricultural land. The use of draught animal power enables the farmers in the Lower Zone to cultivate larger areas on heavy clay soils and as a result a shift towards rainfed agriculture can be observed. However, this shift is only taking place to a limited degree. Most livestock keepers prefer to hold on to their livestock herds as their traditional source of income and, furthermore, rainfed agriculture is limited to those places where conditions allow for it. The potential of rainfed agriculture to provide a sustainable livelihood for all Lower Zone households remains unknown. This means that under the current conditions and preferences, the conflicts over water in the Middle Zone between livestock holders and irrigators are likely to continue.

Insight Gained from Conflict Analysis: An Impasse in Local Conflicts

The descriptions of both the conflicts above paint a picture of conflicts between different parties that are to an important extent driven by external forces, such as increased irrigation activities in the Upper Zone and the closing off of grazing lands in the Usangu plains. So far the local stakeholders have mainly simply responded to these external forces and they have done so with rather limited scope, looking for options in their immediate environment.

Framing both conflicts using the concepts of conflict analysis shows that in both cases, the conflicts have reached an impasse. Paddy farmers argue among each other for the irrigation water that reaches their irrigation schemes, while livestock keepers look for the nearest place to graze their livestock, entering into conflict with irrigators and government officials. In conflict analysis terms, these situations are more or less stable outcomes, even though this leaves all the involved parties dissatisfied.

The conflict analysis clarifies the fact that within the current set of options available to the local stakeholders, none of them can improve their position without the help of others. This means that, in order to get out of the impasse, they can either hope for a positive external intervention from national or international organizations, which may have new and more effective options at their disposal – although the past does not hold too much promise here – or they can try to cooperate, finding local solutions by starting up a local dialogue.

The process of a local dialogue seems the more promising way to go, with or without external support and facilitation, but even this does not offer an obvious way out. A dialogue needs perspective; without this the MSC would not be the first place where dialogue ends in deadlock.

Values Involved in the Water Conflicts

Introduction to Value-focused Thinking

When using conflict analysis to frame the conflicts in terms of access to water, both conflicts seem stuck. And offering a perspective to resolve these conflicts is not easily done using the conflict analysis framework. The framework is useful to understand the conflict and the severity of the current impasse, but it does not offer an easy way out. For possible solutions it is necessary to take a closer look at the values that underlie these conflicts. In looking at values, identifying *why* water is valued so highly by the local stakeholders, there might be more possibilities to identify ways out of the conflicts and to provide a perspective for local dialogues.

Analysing the values that drive stakeholders in conflict situations can be done using value-focused thinking, as developed by Keeney (1992). Conflict analysis focuses on actors and their options and as such it represents 'alternative-focused thinking'. However, alternatives are only relevant as a means to achieve values and therefore values, not alternatives, should be the primary focus of analysis, as Keeney argues. Although it is useful to combine and implement iterations between both approaches of articulating values and creating alternatives, values should come first, resulting in 'value-focused thinking' (Keeney, 1994).

Articulating and structuring objectives is central to value-focused thinking. Each objective is a statement of something that one wants to strive towards and thus an insight into objectives helps to clarify values. Fundamental objectives concern the ends that stakeholders value in a specific decision context; means objectives are methods to achieve ends (Keeney, 1994). From the perspective of an individual water user, fundamental objectives may be related to economic efficiency, to maximize income from water-related production activities, or to social and sustainability issues such as meeting basic food requirements and ensuring a better future for their children.

Only once objectives have been articulated and structured do alternatives enter the analysis. Often, alternatives are already identified early on in a process, as with alternatives that have been used in the past for similar situations or alternatives that are readily available (Keeney, 1994). However, these obvious alternatives are not necessarily the most promising ones. Focusing on objectives helps to uncover additional alternatives that may offer new and more promising ways to realize fundamental objectives.

Value-focused thinking fits well in the tradition of analytical frameworks that have been in use in management science and operations research, but its innovation is in its focus on values rather than alternatives. Furthermore, value-focused thinking offers an operational framework that can be used as a basis to identify and structure objectives, measure the achievement of objectives, quantify objectives with a value model or utility function, guide creativity in identifying alternatives, and execute a formal evaluation of the identified promising alternatives (Keeney, 1992).

In the next sections, the use of the framework for moving towards conflict resolution is illustrated for the conflicts over water for rice and water for livestock, irrigation and wildlife. A complete and detailed description of the application of value-focused thinking would be beyond the scope and limitations of this chapter and therefore the discussion here is limited to the mere way of thinking, illustrating how even employing the general principle of focusing on values can help to identify new ways out of a conflict.

The Value of Water for Rice

In the description of the yearly returning conflicts over water for rice it has already been mentioned that the fluctuations in market prices are an important driving force in the conflicts. Therefore, the underlying values and the associated fundamental objective are the same for all parties in the conflict. The conflict is not over water per se, but over securing a good income: it is not over obtaining cubic meters of water but over obtaining Tanzanian Shillings.

One can easily see the reason to fight over water from looking at rice prices during the season: market prices for rice vary considerably, with prices that are higher early and late in the season. As already mentioned, early rice fetches an especially good price at the market. When marketing rice at the regular time, when the bulk of the crop reaches the market, a household with average rice production can earn an income of US$107 for its harvest, whereas it can earn as much as US$309 if the harvest can be marketed early. Although the actual difference in income will be a bit smaller due to reduced yields for early rice, the difference will still be considerable, taking into account that average household incomes in the Middle Zone are around US$300 per year (Kadigi et al, 2003; FAO, 2005).

Realizing the importance of the price dynamics in the local markets also means realizing that these price dynamics are related to scarcity, as basic market economics teaches us that scarcity of goods drives up prices in the market. And most probably the scarcity of rice on the local markets is due to the scarcity of water to produce this rice. The result is a cycle in which the scarcity of water drives up the price of rice, which in turn reinforces the demand for water as more farmers want to take advantage of high market prices, which then in turn aggravates the conflicts over water.

So far, the quest for solutions to the conflicts of the paddy farmers has focused on water: improving water use efficiency or changing de facto water allocations by enforcing rotations. Even though there is ample room for water savings and re-allocation of this saved water, however, past experiences show that this is difficult and also likely to increase tensions because water savings may only benefit upstream farmers, while re-allocations will result in winners and losers.

Focusing on underlying values opens the way for other alternatives to enter the debate, focusing not on water but on income generation. In this context a promising way out of this vicious cycle of conflict would seem to be the establishment of a joint management and marketing system to address the marketing of rice. So far, farmers have not been able to mobilize themselves for collective or coordinated marketing of their produce (FAO, 2005), but doing so would enable them to set up a fairer mechanism to share the benefits of early rice. Furthermore, it would allow them to construct shared storage facilities that would increase their control over the timing of marketing and could increase their bargaining position in negotiations with other market players. In this way, it could *create* value by increasing the shared income from rice marketing for the Middle Zone farmers.

The Value of Water for Livestock, Irrigated Agriculture and Wildlife

In the conflict between livestock keepers, irrigators and game reserve officials, again securing income is an important underlying objective, but here another factor is the use of water to secure basic livelihoods and jobs, regardless of whether the produce is marketed or not. For those farmers that can access irrigation water during the dry season, this water provides them with the ability to raise cash crops in a period when there are no other local livelihood options available to them. For livestock keepers, access to dry season water and grazing grounds for their livestock is crucial simply to sustain their agro-pastoral livelihoods.

Figure 9.3 illustrates the economic values involved in this conflict using the average economic water productivities of different water using sectors.

The economic water productivities in Figure 9.3 were estimated using different methods for different uses, although uniformity has been strived for in the application of methods. Crop water productivity was based on reported farm gate

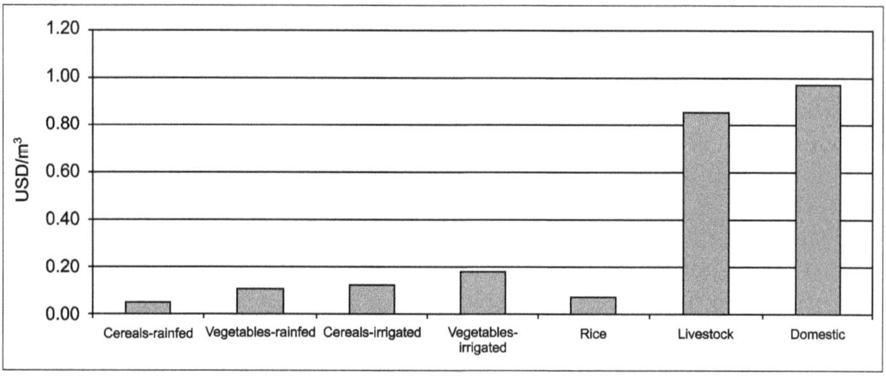

Source: FAO (2005)

Figure 9.3 *Economic water productivity of different sectors in the MSC*

prices, actual evapotranspiration (ETa) and calculations using the FAO's CROPWAT model. Livestock water productivity was estimated using reported household incomes from livestock combined with literature values for livestock water consumption. In both cases, production costs were not taken into account, as significant production inputs (eg labour) could not be valued in monetary terms; thus, reported values are upper bounds. Domestic productivity was estimated using contingency valuation and observed water prices for commercially vended water (see FAO, 2005 for further details).

Figure 9.3 suggests a high value of livestock water productivity. This is explained by the high market value of cattle and the fact that the estimates are based only on water withdrawal for *direct* consumption, excluding the water needed to produce the food for the cattle. The exact economic water productivity for livestock in agro-pastoral farming systems is difficult to estimate due to the relations and overlaps between different water-using activities. Livestock consumes a considerable amount of water through the water contained in its fodder, but simply accounting for all this water as livestock water consumption misses the point that livestock may graze on crop residues that would otherwise be lost. Or livestock may graze grasslands, feeding on crops that would otherwise be unfit for human consumption, thus perhaps competing with environmental water uses, but not with other human related uses. Complications here are not confined to the MSC, as cattle herds are taken outside the area during the dry season to graze in wildlife parks, where they compete for water with the wildlife. This also means that the cattle import a considerable amount of 'virtual' water from outside the sub-catchment.

Despite the limitations in establishing these economic water productivity estimates, the considerable differences in Figure 9.3 between values for livestock and for crop production indicate that livestock keeping is quite profitable in comparison with the cultivation of crops. This opens up some room to exchange 'water for income', i.e. a local taxation system of some sort where livestock keepers gain access to grazing land in the upper parts of the MSC in return for some financial compensation or investments that can be used to start up other activities for Middle Zone households. Such a system may enable some of the Lower Zone households to continue livestock keeping, supplementing the transfer to rainfed agriculture that is already ongoing.

Initially this solution may not receive a warm welcome from the side of the livestock keepers, as it means they will have to pay for their water and pasture. However, the livestock keepers that are illegally driving their herds to the Usangu Game Reserve are already paying for access to dry season land and water, although they are paying fines rather than taxes. In the long run the establishment of a local taxation system or payment scheme would seem to be worthwhile for the livestock keepers, taking into account that such a system may be one of the few options to cope with the serious threats to their very livelihood base. There is a lot at stake for them.

Just as in the paddy irrigation conflict, here too this value-focused option may open the way to creating value. It offers Lower Zone livestock keepers a way to sustain access to dry season grazing lands and at the same time generates a source of income for the Middle Zone irrigators.

Benefits and Limitations of Value-focused Thinking

The framework for value-focused thinking has been used here to illustrate how the approach can bring new alternatives to help resolve conflicts. In fact it has been employed to a very limited extent, focusing only on underlying *economic* values. For a full application of value-focused thinking a more detailed analysis would be needed, identifying, structuring and assessing the full range of values involved, beyond only economic values, and using this to identify additional alternatives and to evaluate them for their expected impacts on the range of objectives. Nevertheless, even this limited application of value-focused thinking helps to illustrate how value-focused thinking can help to move towards conflict resolution.

The conflict analysis indicated an impasse in both local water conflicts discussed here, but focusing on the underlying values pointed to some new solutions, in addition to the options identified under conflict analysis. The solutions offered through value-focused thinking are a local taxation system for the conflicts over grazing land for cattle and the establishment of farmers' cooperatives for the conflicts over water for rice. Although these solutions are neither of groundbreaking originality nor the definitive answer to all the problems, it can be argued that they offer a useful addition to the set of alternatives that are generally considered in working towards conflict resolution.

Value-focused thinking did not lead to the identification of revolutionary innovations and in this sense the outcome may be disappointing to some. Nevertheless, one should bear in mind that, until now, the solutions identified through value-focused thinking have remained outside the scope of the irrigation and water specialists that have been looking at the water problems in the closing Ruaha river basin. So far, farmers' cooperatives and livestock taxation systems have not been part of the mainstream water debate, which rather focused on irrigation modernization, *water* user associations, *water* rights and *water* user fees.

It will be clear that the solutions offered through value-focused thinking are not the complete answer, that they are not easy to implement and that they do not render useless the other options to improve water use efficiency and water allocation. Nevertheless, the value-focused thinking solutions are promising in the sense that they enable the parties in the conflicts to *create* value, while staying within reach of what can reasonably be expected from the local stakeholders.

In situations like that in the MSC, straightforward solutions or 'silver bullets' are unlikely to be found. The conditions in these areas, characterized by poor populations, marginal livelihoods and limited natural resources, usually only leave room for piecemeal or incremental improvements. Anything else is likely to be out of reach for local stakeholders, and even when introduced by external parties – for instance through national interventions or through international development assistance – it is still likely to exceed the absorption capacity of local communities, making local ownership and long-term sustainability doubtful. Therefore solutions that enable incremental steps forward and that are within reach of the local stakeholders offer a more promising perspective for lasting conflict resolution.

Conclusions

This chapter has illustrated the use of two analytical frameworks, conflict analysis and value-focused thinking, to understand local conflicts over water and to identify ways for their resolution. This was done using two local conflicts in the MSC that are exemplary for water conflicts in various other river basins.

The conflict analysis enabled a description of the evolution and present state of the conflicts in the MSC. It indicated both the seriousness of these conflicts and the inadequacy of the options that local stakeholders currently have to resolve them. In terms of conflict analysis, the current situation is stable and therefore conflicts are likely to continue unless new options are identified or introduced by national or international stakeholders.

For this reason the identification of additional solutions was crucial, and to this end value-focused thinking was used. Value-focused thinking offered another perspective on water conflicts, opening space for new solutions to enter the debate. This was not a broadening of *scale*, scaling up water problems to the regional, national or international levels that are beyond the control of local stakeholders, it was much more a broadening of *scope*, including other areas that may offer room for new trade-offs and even the creation of value, while still remaining within reach for local stakeholders.

Through the application of these analytical frameworks, a better understanding of the seriousness of the current situation has been gained and two promising solutions were identified that might help in moving towards conflict resolution. These solutions are not the complete answers to these conflicts, but they illustrate that there are options outside the traditional water engineering solutions which are within the control of local stakeholders and which can help them to create value in closing river basins. This approach may not lead to groundbreaking solutions, but neither can its contribution be dismissed.

Acknowledgements

This chapter is based on work funded through the FAO – Netherlands Partnership Programme. We would like to acknowledge the contribution of all those involved in this project. Thanks are especially due to the members of the Soil Water Management Research Group in Igurusi and Morogoro and the local stakeholders in the MSC who participated in focus group discussion and stakeholder meetings.

References

Bennett, P., Cropper, S. and Huxham, C. (1989) 'Modelling interactive decisions: The hypergame focus', in Rosenhead, J. (ed) *Rational Analysis for a Problematic World: Problem Structuring Methods for Complexity, Uncertainty and Conflict*, J. Wiley & Sons, Chichester, UK, pp283–314

Castro, A. P. and Nielsen, E. (2003) 'Overview', in Castro and Nielsen (eds) *Natural Resource Conflict Management Case Studies: An Analysis of Power, Participation and Protected Areas*, FAO, Rome, pp1–17

FAO (2005) *Water Productivity and Vulnerable Groups in the Mkoji Sub-Catchment. A Local Case Study in Integrated Water Resources Management in the United Republic of Tanzania.* Project report, prepared by FAO and Sokoine University of Agriculture, Rome

Fang, L., Hipel, K. W. and Kilgour, D. M. (1993) *Interactive Decision Making: The Graph Model for Conflict Resolution.* John Wiley & Sons, New York

Fraser, N. M. and Hipel, K. W. (1984) *Conflict Analysis: Models and Resolutions.* North-Holland, New York

Gregory, R. and Keeny, R. L. (1994) 'Creating Policy Alternatives Using Stakeholder Values', *Management Science*, vol 40, no 8, pp1035–1048

Gregory, R., McDaniels, T. and Fields, D. (2001) 'Decision aiding not dispute resolution: Creating insights through structured environmental decisions', *Journal of Policy Analysis and Management*, vol 20, no 3, pp415–432

GWP (Global Water Partnership) (2000) 'Integrated water resources management', TAC Background Papers No.4, GWP Technical Advisory Committee, Stockholm, Sweden

GWP (2001) *IWRM ToolBox Version 1 – Policy Guidance, Operational Tools, Case Studies*, GWP, Stockholm, Sweden

Hermans, L. M. (2003) 'Agenda setting in policy analysis: Exploring conflict for a case of water resources management in the Philippines', *Proceedings of the IEEE International Conference on Systems, Man and Cybernetics*, Washington, pp3314–3321

Hermans, L. M., Beroggi, G. E. G. and Loucks, D. P. (2003) 'Managing water quality in a New York City watershed', *Journal of Hydroinformatics*, vol 5, no 3, pp155–168

Horita, M. (2000) 'Folding arguments: A method for representing conflicting views of a conflict', *Group Decision and Negotiation*, vol 9, no 1, pp63–83

Howard, N. (1971) *Paradoxes of Rationality: Theory of Metagames and Political Behavior*, MIT Press, Cambridge, Massachusetts

Howard, N. (1989) 'The manager as politician and general: The metagame approach to analysing cooperation and conflict', in Rosenhead, J. (ed) *Rational Analysis for a Problematic World*, J. Wiley & Sons, Chichester, England, pp239–261

Kadigi, R. M. J., Kashaigili, J. J. and Mdoe, N. S. (2003) 'The economics of irrigated paddy in Usangu Basin in Tanzania: Water utilization, productivity, income and livelihood implications', *Proceedings of 4th WaterNet/Warfsa Symposium*, Gaborone, Botswana, pp79–84

Keeney, R. L. (1992) *Value-Focused Thinking: A Path to Creative Decision Making*, Harvard University Press, Cambridge, Massachusetts

Keeney, R. L. (1994) 'Creativity in decision making with value-focused thinking', *MIT Sloan Management Review*, vol 35, no 4, pp33–41

Kilgour, M. D., Hipel, K. W., Fang, L. and Peng, X. (1996) 'A new perspective on coalition analysis', *Proceedings of IEEE Conf. on Systems, Man and Cybernetics* (vol 3), Beijing, pp 2017–2022

Lankford, B. (2004) 'Irrigation improvement projects in Tanzania: Scale impacts and policy implications', *Water Policy*, vol 6, no 2, pp99–102

Maganga, F. P., Kiwasila, H. L., Juma, I. H. and Butterworth, J. A. (2003) 'Implications of customary norms and laws for implementing IWRM: findings from Pangani and Rufiji Basins, Tanzania', *Physics and Chemistry of the Earth*, vol 29, no 15–18, pp1335–1342

McDaniels, T. and Trousdale, W. (1999) 'Value focused thinking in a difficult context: Planning tourism for Guimaras, Philippines', *Interfaces*, vol 29, no 4, pp58–70

Molden, D. (1997) 'Accounting for water use and productivity', SWIM Paper 1, IWMI, Colombo, Sri Lanka

Molden, D., Sakthivadivel, R. and Habib, Z. (2001) 'Basin-level use and productivity of water: Examples from South Asia', Research Report 49, IWMI, Colombo, Sri Lanka

Obeidi, A., Hipel, K. W. and Kilgour D. M. (2002) 'Canadian bulk water exports: Analysing the sun belt conflict using the graph model for conflict resolution', *Knowledge, Technology, and Policy*, vol 14, no 4, pp145–163

Sebenius, J. K. (1992) 'Negotiation analysis: A characterization and review', *Management Science*, vol 38, no 1, pp18–38

SIWI (2000) 'Water and development in the developing countries – A study commissioned by the European Parliament', SIWI Report 10, Stockholm Timmermans, J. S. and Beroggi, G. E. G. (2000) 'Conflict Resolution in Sustainable Infrastructure Management', *Safety Science*, vol 35 no 1–3, pp175–192

WWAP (World Water Assessment Programme) (2003) *Water for People Water for Life. The United Nations World Water Development Report*. UNESCO Publishing and Berghahn Books, New York

10

Determinants of Quality and Quantity Values of Water for Domestic Uses in the Steelpoort Sub-basin: A Contingent Valuation Approach

Benjamin M. Banda, Stefano Farolfi and Rashid Hassan

Introduction

This chapter estimates the value that domestic users in the Steelpoort sub-basin (SPSB) attribute to improved quantity and reliability of water supply and to improved quality of water. Apart from estimating the willingness to pay for improved quantity and quality of water, the chapter describes the pattern of water use and its distribution in rural and urban parts of the SPSB.

Water resource allocation is an important issue in present day South Africa. The policy framework for water allocation and management includes the Water Services Act (WSA, 1997) and the National Water Act (NWA, 1998), both of which place on government the responsibility for, and authority over, water resource management, including its efficient and equitable allocation and distribution.

The NWA provides the framework for a decentralized water resource management based on 19 Water Management Areas (WMAs), where the emerging catchment management agencies (CMAs) will be in charge of local policies for water allocation to competing users.

The SPSB, situated in the Olifants WMA, is a good case study for the analysis of water demand management and allocation at catchment level. The sub-basin includes water users ranging from domestic households to industries, mines and commercial farms. It is expected that these users place different values on quantity and quality of water depending on their socio-economic characteristics and, for the economic sectors, their production functions.

There are important gains from water resource valuation for policy-makers. First, placing values on water from various sources including surface water allows policy-makers to decide whether costs can be recovered from investing in water resource development, including quality control. Second, the values placed on water can be used as input into a cost–benefit analysis to identify institutional and infrastructural changes that may improve water resource management. Third, failure to recognize the economic value of water has led to wasteful and environmentally damaging uses of the resource. Managing water as an economic good is an important way of achieving efficient and equitable use and of encouraging conservation and protection of water resources (Dublin Statement, 1992). Last, many studies that attempt to value water recommend that charge and subsidy levels should be supported by empirical analysis (Halpern et al, 1999).

A contingent valuation survey was carried out in the SPSB to collect factual information necessary for conducting the intended analysis. Consistent with the new policies on water allocation, the study identifies Steelpoort as one of the strategic catchment areas already under stress from the increasing demand for water by different users. While the Department of Water Affairs and Forestry (DWAF) has a goal of using water pricing, limited-term allocations and other administrative mechanisms to bring supply and demand into balance in a manner that is beneficial and in the public interest (Thompson et al, 2001), the study models willingness to pay (WTP) for water not as a measure of price but of values attached to the resource by users.

Goldblatt (1998) used WTP to assess the effective demand for improved water supplies in informal settlements in Johannesburg. The justification for estimating demand functions in most studies is that policy-makers would like to have some efficiency criteria for investing in the public good, especially in the water sector, which is judged by most as the least financially autonomous sector.

The South African experience is particularly remarkable in that there is a huge demand for new infrastructure given the inequality inherited from the Apartheid era. There is much public pressure on government to redress the inequality from the national statistics as of 1994 of about 99.9 per cent of white South Africans having access to formal water services as opposed to 43.3 per cent of black South Africans. On the other hand, the public investment required to achieve equity in water access has no corresponding pricing mechanism to ensure efficiency since pricing water at the margin (equating prices to the marginal cost of water provision) would be regressive. In addition, South Africa is recovering from a 'political game of passive resistance' that took the form of non-payment of rents and service charges to government during the Apartheid era (Goldblatt, 1998).

The WTP method can estimate the benefit that people attach to the provision of water. The estimation of WTP is based on the Hicksian compensated demand function that, for all practical purposes, assumes utility maximization and the related axioms about preferences of consumers are satisfied. This is justified since water is provided at a fee in the study area, although some sources of water are available at no fee.

The chapter is organized in five sections. Following this introduction section 2 sets the theoretical basis and specifications of the empirical model; section 3 describes the data used to calibrate the empirical model; section 4 presents and discusses the empirical results; and the conclusion and implications of the study are drawn in section 5.

The Contingent Valuation Framework

Contingent valuation (CV) is an important survey-based procedure for eliciting the economic value of the quality and availability of non-market commodities (Niklitschek and Leon, 1996). The method is particularly attractive because of its simplicity and flexibility and is commonly applied to cost–benefit analyses and environmental impact assessments.

The CV framework is based on maximizing utility from the consumption of market and non-market goods such as environmental quality, Q.[1] Q is therefore used as an argument in the individual's utility function, $U(X, Q)$, where X is a vector of market goods. The individual's problem is to maximize $U(.)$ subject to the budget constraint $PX = M$, where P is a vector of market prices, and M is income. When Q is given, the solution of the utility maximization problem is a set of Marshallian Demand functions, $X_i(P, M, Q)$.

The Contingent Valuation Methodology (CVM) builds on the above framework adopting indirect utility functions, $V_i(P, M, Q); \forall\, i$, where $i = 1, \ldots, N$ denote individuals.

Since a market does not exist for the environmental good Q, its value is inferred from survey data reporting households' willingness to pay (WTP), or willingness to accept compensation (WTA) for a change in its quantity or quality (Kuriyama, 1998).

When a policy change is implemented so that quantity or quality of Q improves, i.e. from Q_0 to Q_1, the CVM survey measures the compensating surplus an individual is willing to pay to enjoy the improvement; that is, remain at the same (compensated) utility level.

$$V_i\left(P,Q_0,M\right)=V_i\left(P,Q_1,M-WTP\right) \tag{1}$$

Where WTP is individual *i*'s stated willingness to pay, P, M and Q as defined above.

The individual's willingness to pay for the change could formally be represented by the change in expenditure resulting from a change in Q while holding the utility level constant at U_0.

$$WTP_i = m\left(P,Q_1,U_0\right)-m\left(P,Q_0,U_0\right)= P\left\{H\left(P,Q_1,U_0\right)-H\left(P,Q_0,U_0\right)\right\} \tag{2}$$

Where $m\ (.)$ is the expenditure function, and $H\ (.)$ is the Hicksian (compensated) demand function.[2]

Until recently, stated preference data analysis focused mainly on WTP measures and their implication for environmental decision-making (Ryan and Wordsworth, 2000). The major advantage of the WTP approach is that anything of value to people can be translated into utilities in a framework that operates in financial terms (Anand, 2000). As illustrated above, underlying the WTP method is the assumption of utility maximization and related axioms about preferences of consumers.

Questions have been raised in the literature on reliability and appropriateness of WTP measures for resolving social choice problems. These include the reliability and validity of the survey instrument, the rationality of responses, and the sensitivity of the results to sequence, context and delivery of the questionnaire (Ryan and Wordsworth, 2000). WTP values are subjectively held to be valid to fill gaps caused by non-existent markets in situations in which we can only rely on responses to hypothetical questions. Assigning hypothetical or actual WTP to environmental options requires that the completeness axiom holds, yet the hypothetical nature of the CVM-WTP survey implies that preferences do not exist until the time an individual is invited to make a choice.

Anand (2000) further argues that WTP should not be used as a criterion for social choice because incompleteness of preferences is fundamental to any social choice problem; in addition, a normative analysis of social choice problems is logically not constrained by the same axioms of completeness and transitivity of preferences as assumed by the WTP approach. Individual preferences will in general be relevant even though decisions made on behalf of the citizenry are not the same as those made by the market, since among other things the relations between beneficiaries, providers and other stakeholders, including the state, are quite different in the two cases. Equating social and individual choice either specifies the decision problem wrongly or uses wrong preferences to solve it.

Consumer theory gives a positive relationship between WTP and household income (Deffar, 1998). The household's socio-economic characteristics, such as household size, education and municipal location, and the resource characteristics, such as the source of water and the distance to the source, are then assumed to account for variations in WTP responses (Whittington et al, 1990; Hokby and Soderqvist, 2001).

The 2-Step Model

In implementing the CVM survey, a respondent is presented with questions on whether or not he or she is willing to pay either for improved availability or quality of the commodity in question. In the first step the researcher is interested in the probability of an outcome of a dichotomous choice problem (yes/no) measured by the latent response variable (*wi*), ie:

$$\left.\begin{aligned} P(wi=1) &= F(X,\beta) \\ P(wi=0) &= 1-F(X,\beta) \end{aligned}\right\} \tag{3}$$

Where X is a vector of explanatory variables explaining the individual's choice of whether or not to pay and β is the set of parameters. The marginal effects on $P(.)$ are derived from the cumulative density function. Assuming a logistic distribution, the standard logistic model is expressed as the odds, i.e.:

$$\frac{F(X,\beta)}{1-F(X,\beta)} = e^{Z_i} \tag{4}$$

The log of odds is linear in Z_i, for all individuals in the sample, where $Z_i = \beta'X$.

The probability that an individual would be willing to pay is given by:

$$P(wi = 1) = \frac{e^{Z_i}}{1 + e^{Z_i}} \tag{5}$$

For simplicity it is usually assumed that the same variables that influence the value of the stated willingness to pay determine whether or not an individual would be willing or not willing to pay in the first place.

The regression model for equation (5) is found by taking the mathematical expectation of the latent variable, given the observed component X:

$$E(w \mid X) = 0[1 - F(\beta'X)] + 1[F(\beta'X)] = F(\beta'X) \tag{6}$$

The first order conditions are given by:

$$\frac{\partial E(w \mid X)}{\partial X} = \left(\frac{dF(\beta'X)}{d(\beta'X)} \right) \beta = f(\beta'X)\beta \tag{7}$$

Where $f(.)$ is the probability density function corresponding to the cumulative density function (cdf), $F(.)$.

The second step consists of using the estimated models to calculate the expected mean WTP. The general formulation of the empirical Tobit model is given as:

$$WTP_i^* = \beta' X_i + \varepsilon_i \tag{8a}$$

$$WTP_i = \begin{cases} 0 \text{ if } WTP_i^* \leq 0 \\ WTP_i^* \text{ if } WTP_i^* > 0 \end{cases} \tag{8b}$$

Where X_i' is, for individual i, a vector of explanatory factors in the regression; β is a vector of coefficients; and WTP_i^* is the stated willingness to pay for individual i.

The vector X_i' referring to WTP for improved availability of water was assumed to include the following variables: frequency of availability of water, per capita water consumption, household monthly income, age of the household head, source of water and the maximized WTP probability for improved quantity of water from step 1. To estimate the WTP for improved quality of water, the vector X_i' was assumed to include the maximized WTP probability for improved quality from step 1, the household monthly income, the amount of water consumed, the water user's appreciation of quality and the source of water.

The Tobit specification for estimating WTP for both improved quality and quantity was preferred over the ordinary least squares (OLS) (Tables 9.8 and 9.9).

The OLS assumes that those who are not willing to pay for water in the logistic regression would not participate in the market for improved availability of water, even if more water was to be provided to them. The Tobit regression assumes that the OLS is a misspecification, since an a priori filter for those with $WTP = 0$ amounts to sample selection bias, and the least squares estimators are as a result biased and inconsistent.[3]

The Tobit model was estimated using maximum likelihood to generate estimators that are consistent and asymptotically normal, although these properties are highly sensitive to specification errors (Amemiya, 1973; Judge et al, 1987). The Tobit model is less useful for predicting fitted WTP. The standard Tobit may give predictions that fall outside the domain of permissible values. Inclusion of observations for which $WTP = 0$ causes the function to swivel towards negative values, although the WTP function has a lower limit of zero. Values falling below the lower limit would not make any economic sense, although the underlying model would remain statistically valid.

The complication with predicted values from a Tobit also arises because there are three possible regressions depending on the nature of the study (Greene, 2003). The first regression is similar to the OLS when the dependent variable is largely uncensored:

$$E\left(WTP_i^*\right) = X_i'\beta \tag{9}$$

The second and third, and their corresponding marginal effects are given below:

$$\left.\begin{aligned} &E\left(WTP_i \mid WTP > 0\right) = X_i'\beta + \sigma\frac{f_i}{F_i} \\ &\frac{\partial E\left(WTP_i \mid WTP > 0\right)}{\partial x_j} = F_i\beta_j \end{aligned}\right\} \tag{10}$$

$$\left.\begin{aligned} &E\left[WTP_i\right] = F_i E\left[WTP_i \mid WTP > 0\right] \\ &\frac{\partial E\left[WTP_i \mid WTP > 0\right]}{\partial x_j} = \beta_j\left[1 - (f_i X_i'\beta)/(\sigma F_i) - (f_i / F_i)^2\right] \end{aligned}\right\} \tag{11}$$

Where f_i and F_i are respectively the probability distribution function and cumulative distribution function of a standard normal distribution evaluated at $Z_i = \dfrac{X_i'\beta}{\sigma}$ and σ is the standard deviation of the regression error term.

Sources of the Data

A CV survey was carried out among rural and urban households in the SPSB. As argued earlier, the SPSB was chosen as a case study because it uniquely represents a catchment area under water stress due to its diversity of water users potentially in conflict. To collect the necessary factual data, a semi-structured questionnaire was developed and tested. The questionnaire was designed to capture four types of data: location of the household by municipality and ward, its socio-economic characteristics, water availability and use. Households were also asked whether or not they were willing to pay, first for an improved availability of and more reliable access to water, then for improved quality of water (and the amount they were willing to pay).

The target population for the study was defined as households in the SPSB using water for domestic purposes. This population is distributed in 43 wards of 5 municipalities, namely Greater Groblersdal (GG), Greater Tubatse (GT), Highlands (HL), Makuduthamaga (MK), and Thaba Chweu (TC) (Figure 10.1).

Using the South Africa Explorer software (Jhagoroo et al, 2000) and the 1996 census data (South African Census, 1996), the total population in the sub-basin was estimated at 249,066 (47,892 households). This population was calculated by overlapping maps of the sub-basin from the DWAF with those from SA Explorer. The latter indicate the administrative units (wards and municipalities) to which

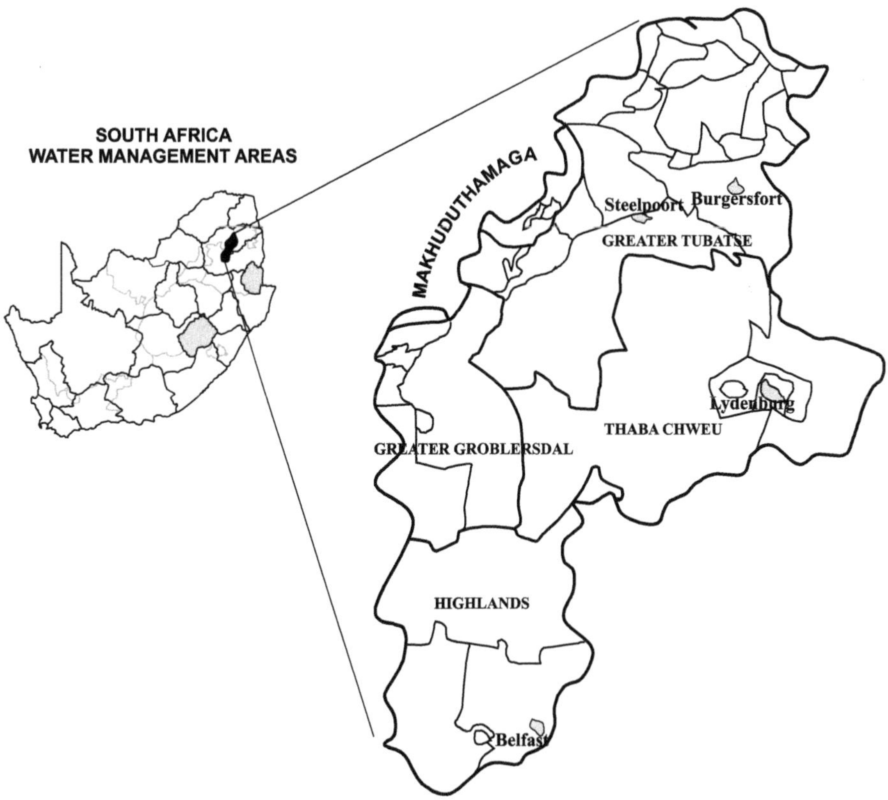

Source: chapter authors

Figure 10.1 *The Steelpoort sub-basin: municipalities, administrative wards and main urban centres*

census data refer. This procedure ensured that only households falling within the sub-basin borders were included in the target population (with a certain degree of precision).

Following one of the key questions of the study, i.e. investigation of the different domestic water uses and consumption patterns observable in rural and urban areas, a stratified sample of rural and urban households was chosen from the target population. The survey aimed at providing a description based on factual data of the current differences in terms of water supply and distribution in the two areas. The CVM was then adopted to elicit the WTP of rural and urban households for improved availability and quality of water, and the relation between their respective WTP and their current situation.

It was assumed that 'urban areas' were the wards including the four major centres in the sub-basin, namely Lydenburg (two wards), Belfast, Steelpoort and Burgersfort. According to this assumption, out of the 47,892 households in Steelpoort, 41,079 are in rural areas while 6,813 are in urban areas. The number of respondents to be selected from each stratum was found by multiplying the uniform sampling fraction (0.0055) by the size of the population in each stratum. This resulted in 226 rural households and 37 urban households. Because of the limited number of urban households resulting from this technique, the survey team decided to increase this sub-sample to 60 households. Rural households to be interviewed were also increased to 270. During the survey it was also possible to interview an additional group of 45 rural households. This brought the total sample size to 375 households (315 rural and 60 urban).

Seven rural wards were identified as being representative of their respective municipalities. Two wards were from GT (a third was added during the survey), with one from each of GG, HL, MK and TC. Five other wards in the sub-basin, mainly towns of certain relevance, were selected as urban areas.

To obtain the list of the households, a multistage selection was implemented consisting of the following steps. In rural areas:

1. a list of the villages in each rural ward was obtained from the ward councillor;
2. two villages were randomly selected in each ward;
3. a list of households in the village was obtained from the chief of the selected villages; and
4. a random method was adopted to identify the households to be interviewed. In urban areas, the household list for each town was drawn up by randomly selecting households from the telephone directory and from contacts with the councillors in each ward.

Once the survey was conducted, we realized that the characteristics of the 45 households interviewed in TC (3) were those of an urban area, and not rural as we thought when designing the sample. Furthermore, one of the rural questionnaires was not properly completed and was discarded. The final composition of our sample was therefore 374 households, of which 269 were rural and 105 urban. After these adjustments, the urban stratum represented 28 per cent of the sample, whereas in the mother population it accounts for 14.3 per cent.

Discussion of the Empirical Results

Socio-economic Characteristics of the Sample

An average household income of 1632 rand per month characterizes the surveyed sample (Table 10.1), with a relatively high level of dispersion. This result is consistent with the data from the 2001 Census (South African Census, 2001), in which the average household income for the Steelpoort sub-basin is 1787 rand per month.

The average income of the rural households surveyed was around one-third of that of the urban households (one-fourth in terms of per capita income). These considerations regarding income distribution among the surveyed households are important for analysing the uses of and willingness to pay for water, as income is an important determinant.

About 34 per cent of the heads of the surveyed households declared themselves to be unemployed (almost 40 per cent in rural areas), compared with a figure of 80 per cent real unemployment (declared unemployed + not economically active) in the sub-basin (South African Census, 2001). A further 26.5 per cent of the interviewed heads of households receive some form of state pension, while 20 per cent of the household heads are employed either in the public or in the private sectors. Another 9.4 per cent are self-employed or running small scale businesses.

As was observed for income, the level of education attainment was also very different in rural and urban areas (Table 10.2). Most rural household heads declared themselves as having no education or up to primary school level education. On the other hand, over 50 per cent of urban household heads have attained either secondary school or higher educational qualifications. These differences in educational attainment could explain the divergence in income levels between rural and urban households.

Table 10.1 *Monthly income per household and per capita in Steelpoort*

Location	*Mean household monthly income (rand)*	*St. dev. monthly income (rand)*	*Mean monthly income per capita (rand)*	*St. dev. monthly income per capita (rand)*
Rural	1109.7	1095.1	188.7	241.6
Urban	2972.5	4410.8	784.5	1539.1
Steelpoort	1632.7	2643.6	356.3	880.8

Source: chapter author

Table 10.2 *Level of education of the household head (%)*

Location	*None*	*Primary*	*Secondary*	*Diploma*	*Degree*	*Total*
Rural	43.5	30.5	23.4	1.9	0.7	100
Urban	14.3	18.1	50.5	10.5	6.7	100
Total	35.3	27	31	4.3	2.4	100

Source: chapter author

Patterns of Water Use

Four different sources of domestic water were identified in the SPSB: private tap, collective tap, river water and vending water. Private tap water users were mainly found in urban areas, with collective tap, river water and vending water users mostly in rural areas. Urban households using private taps were paying an average of 5.38 rand/m^3, while the rural households which had them were paying an average of 11 rand/m^3. Urban households were paying an equivalent of 1.28 rand/m^3 for collective tap water, while rural households were either not paying for collective tap water or they were getting water from the river.

The distribution of water sources for the whole sample indicates that most households get their water supply from collective taps (Figure 10.2). Private tap water represents the second most important source, followed by river water, while use of water supplied by vendors is marginal. These results are consistent with the South African Census (2001) data, which give the following distribution of sources in the sub-basin: tap in dwelling or inside yard 29.7 per cent, collective tap 35 per cent, river + stream 24.8 per cent, vendor 0.9 per cent and other 9.6 per cent.

Important geographical differences were observable when data were analysed by municipality (Figure 10.3). Households surveyed in TC live in the urban area

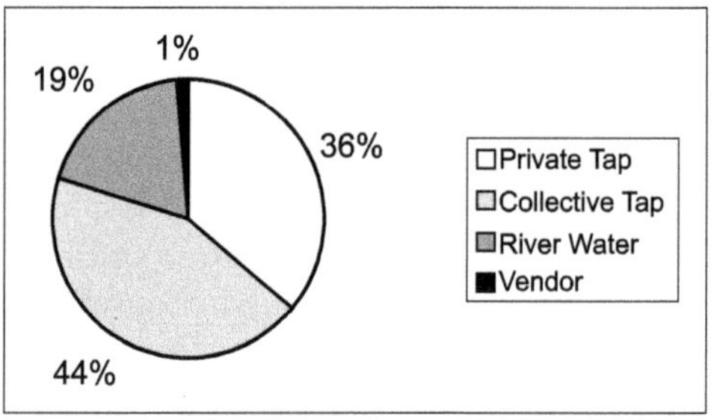

Source: adapted from Jhagaroo et al (2000)

Figure 10.2 *Sources of water in the Steelpoort sub-basin (2003)*

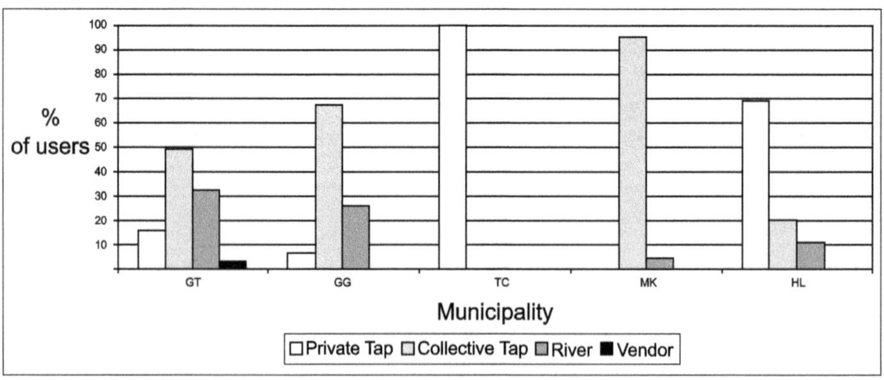

Source: adapted from Jhagaroo et al (2000)

Figure 10.3 *Water source distribution per municipality in Steelpoort*

Table 10.3 *Monthly water consumption per household and per capita*

Location	Household water consumption (m^3/month)	Per capita water consumption (m^3/month)
Rural	4.8	0.8
Urban	14.8	4.6
Steelpoort	7.6	1.8

Source: chapter authors

of Lydenburg and their source of water is exclusively private tap. Conversely, households in the rural MK, in the areas once occupied by Bantustans, rely almost exclusively on collective tap water. Households interviewed in GT and GG live either in peri-urban or in rural areas and their source of water is mainly collective taps or river water. GT households having private taps live in the urban areas of Steelpoort and Burgersfort. Finally, households surveyed in HL have good access to private tap water; they live either in the urban area of Belfast or in the relatively rich rural area represented by ward 3, where large commercial farms are located.

A clear dichotomy is observable between rural and urban water consumption patterns: urban households consume three times more water than rural ones; the per capita consumption in urban areas is five times higher than that in rural areas (Table 10.3).

A wide range of variation exists among urban areas, whereas water consumption in rural areas seems to be more uniform. A household typology was identified for both rural and urban units (Table 10.4). From the surveyed sample, three types of urban households and two types of rural ones were determined. 'Urban 1' is formed by households living in TC (1), GT (1) and GT (3). This group has the highest level of income, employment and literacy. 'Urban 2' is formed by households living in TC (2) and TC (3) and has an intermediate level of education, low per capita income and a very high unemployment rate. 'Urban 3' is formed by households living in HL (1) and it is characterized by the lowest urban per capita income and educational level. The high percentage of pensioners in this group keeps the figure of unemployment at a reasonable level. Households living in former Bansustans (GT (4, 17 and 29), GG (27), and MK (26)) form 'Rural 1'. This group has rates of employment, literacy and per capita income among the lowest of the whole surveyed sample. Households living in HL (2) constitute 'Rural 2'. This group has a per capita income comparable with that of the Urban

Table 10.4 *Monthly per capita income, unemployment rate and education attained by the head of the household in urban and rural groups*

Group	Monthly per capita income (rand)	Unemployment rate (%)	Education Attainment*
Urban 1	1733.1	0	2.6
Urban 2	303.4	40	1.5
Urban 3	128.9	20	0.8
Rural 1	164.7	40	0.9
Rural 2	311.4	10	0.7

Note: * 0 = none, 1 = primary, 2 = secondary, 3 = diploma, 4 = degree
Source: chapter authors

Table 10.5 *Per capita monthly water consumption (m^3) by source of water in urban and rural groups*

Group	*Private Tap*	*Collective Tap*	*River Water*	*Vendor*	*Total*
Urban 1	9.2	5.4	–	–	9.1
Urban 2	2.5		–	–	2.5
Urban 3	1.3	0.4	2.2	–	1.3
Urban	**4.8**	**1.4**	**2.2**	**–**	**4.6**
Rural 1	0.4	0.8	0.7	0.4	0.8
Rural 2	0.8	0.8	0.2	–	0.8
Rural	**0.7**	**0.8**	**0.7**	**0.4**	**0.8**

Source: chapter authors

2 group and is mainly composed of farm workers (30 per cent) and families receiving state pensions (30 per cent).

Table 10.5 indicates that the Urban 1 group has far higher water consumption than the other two urban groups. The source of water clearly heavily influences the consumption pattern. In fact, in urban areas households with private taps consume an average of three times more water than collective tap users and twice as much as river water users. On the other hand, in rural areas the water source does not seem to significantly influence water consumption.

Water consumption clearly correlates with some of the observed socio-economic variables. Figure 10.4 shows the correlation between per capita income and per capita water consumption.

Another factor influencing water consumption is the frequency of water availability. This variable shows a strong correlation with the source of water, as indicated in Table 10.6. Almost all households with private tap water have access to the resource all day every day, whereas most collective tap and river water users cannot afford more than one or two trips per day to fetch water. The distance to the closest source and the time needed to collect water were important limiting

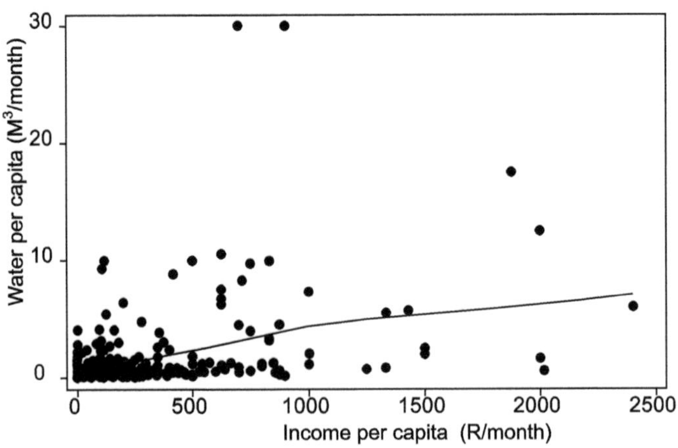

Note: bandwidth = .8
Source: chapter authors

Figure 10.4 *Water consumption and income in Steelpoort*

Table 10.6 *Frequency of using/fetching water by source (%)*

Frequency	*Private*	*Collective*	*River Water*	*Vendor*	*Total*
Less than once a day	3.7	8.0	0	50.0	5.1
Once a day	1.5	27.6	11.3	50.0	15.1
2x a day	1.5	31.3	35.2	0	21.0
3x or more a day	93.3	33.1	53.6	0	58.6

Source: chapter authors

Table 10.7 *Perception of water quality by source (%)*

Water quality	*Private tap*	*Collective tap*	*River Water*	*Vendor*	*Total*
Very poor	4.4	6.1	41.4	0	12.1
Poor	5.2	26.4	47.1	20.0	22.5
Just OK	3.7	10.4	4.3	0	6.7
Good	11.1	8.6	7.1	0	9.1
Very good	75.6	48.5	0	80.0	49.6

Source: chapter authors

factors. On the other hand, very few households in rural areas fetch water less frequently than once a day.

Water quality perception is also heavily influenced by the source of water. Table 10.7 indicates that 87 per cent of households with private tap consider the quality of water good or very good, while only 57.1 per cent of collective tap users and 7.1 per cent of households fetching water from the river rank the quality of the resource good or very good. These proportions are reversed for households considering the quality of their water bad. In fact 88.5 per cent of households fetching water from the river consider the quality of the resource poor or very poor, while 32.5 per cent of collective tap users and only 9.6 per cent of private tap users rank the quality of their water poor or very poor.

Willingness to Pay for Improved Water Quality and Quantity

Of the total sampled households, 61.9 per cent were willing to pay for improved availability of water, while only 40.6 per cent were willing to pay for improved quality of water.

The latent variable for the households' binary decision to pay (or not to pay) for improved availability of water has a different distribution for rural and urban households. Of the interviewed rural households, 82 per cent were willing to pay for a more regular and reliable source of water, compared with only 11 per cent of urban households, while 48.3 per cent of rural households were willing to pay for improved quality compared with only 21 per cent in urban areas. This confirms the importance of connecting rural households to regular supplies of clean water, while urban households place higher priority on quality as they are already connected to regular supplies. It is clear that among rural households there is very high demand for both quantity and quality attributes.

The application of the two-step model to the collected data revealed both the *incidence* (probability) and the *intensity* (quantification) of WTP for improved quantity and quality of available water. The logistic analysis (first step) providing the incidence of WTP can be found in Appendix 2. The results of the second step are illustrated in the following sections.

Willingness to pay for improved availability of water

The maximized probability from the logistic regression has a positive and significant influence on WTP for improved availability of water. A bootstrap of the standard error of the coefficient revealed, however, that the maximized probability was biased.[4] This bias is not peculiar, since we are dealing with a predicted value from a probability model designed as far as possible to attach high probability values to extreme values (0 and 1). The resulting distribution of the predicted probabilities has a peak at each extreme value.

The availability of water is significant and negative. This suggests a negative relationship between the availability of water, which is a measure of the reliability of water supply from the source, and willingness to pay for more water. It appears that the more reliable the water supply is to a household, the less the household is willing to pay for more regular water supply. Private tap and collective tap water users have a more regular supply of water and this has a negative impact on the mean willingness to pay. The tap water dummy in both regressions also confirms this.

The per capita monthly consumption of water is significant and positive. This result is consistent with the theoretical expectations that, at low levels of water consumption, the more the positively valued good is consumed, the more

Table 10.8 *WTP regression for improved availability of water*

Wtpq = WTP amount for quantity*	*TOBIT*		
	Coefficient	*t*	*P > t*
Constant	10.27	0.74	0.46
Pqty = (predicted probability from logistic regression)	66.17	8.48	0.00
Availability of water (frequency)	–9.11	–4.84	0.00
Income (square root of)	0.10	0.97	0.33
Water per capita	4.59	2.19	0.03
Tap = dummy for tap water	–24.41	–5.12	0.00
Age	–0.03	–0.18	0.86
_Se (Ancillary parameter)	32.11		
Number of observations	354#		
F (7, 215)			
Prob > F			
R-squared			
Root MSE			
Pseudo R^2	0.08		
LR chi2 (6)	187.21		
Prob > chi2	0.00		
Log likelihood	–1123.62		

Note: # Observations summary: 131 left-censored observations at wtpq ≤ 0, 223 uncensored observations
* *The stated WTP amount (rand per household per month)*
Level of significance: 10%
Source: chapter authors

a household is willing to pay a marginally higher amount to improve water availability. It can be noted, though, that some rural households using collective taps or river water, and therefore willing to pay for an improved and more reliable water quantity, have a relatively high water consumption. Income is also positive, but insignificant. Bootstrapping the standard error rejects the assumption of inconsistency in the income coefficient. The age of the respondent was found to be negative but insignificant.

Willingness to pay for improved quality of water

The maximized probability from the logistic regression has a negative but significant influence on the WTP for improved quality. This result would at first glance seem peculiar. However, this perception disappears as soon as the distribution of willingness to pay responses for the logistic model is interposed with the households' location, source of water and income. In fact, the large majority of bids for improved quality come from rural households with access to river or collective taps, whereas only 14 per cent of private tap users, mainly in urban areas, were willing to pay for improved water quality. Because of the income effect, the individual urban household's WTP is higher than rural household's. This fact affects the coefficient of the maximized probability from the logistic regression in the WTP function for water quality because urban households willing to pay individually more are few, and rural households willing to pay individually less are numerous. In line with the above interpretation, income has a positive and significant marginal impact on households' willingness to pay for quality. The

Table 10.9 *WTP regression for improved quality of water*

Wtpl = WTP amount for quality*	*TOBIT*		
	Coefficient	*t*	*P > t*
Constant	59.85	6.07	0.00
Pqlty = (predicted probability from logistic regression)	–28.50	–3.63	0.00
Income	0.001	2.74	0.01
Water = amount of water used per month (log of)	2.09	1.78	0.08
Wuaql = water user's quality ranking (0 = poor,..., 5 = very good)	–18.49	–8.68	0.00
_Se (Ancillary parameter)	18.32		
Number of observations	364#		
F (6, 142)			
Prob > F			
R-squared			
Root MSE			
Pseudo R2	0.13		
LR chi^2 (4)	212.51		
Prob > chi2	0.00		
Log likelihood	–693.23		

Note: # Observations summary: 215 left-censored observations at wtpl ≤ 0, 149 uncensored observations
* *The stated WTP amount (rand per household per month)*
Level of significance: 10%
Source: chapter authors

household's current consumption of water is also a positive and significant factor explaining the willingness to pay for improved quality of water. A household with abundant water would be more concerned with improving quality of water than if it had access to less water.

Water users' appreciation of quality is negative and significant. This means that as an individual's appreciation of quality increases, their marginal contribution to WTP declines. The quality of water was highly correlated with the dummy for tap water. The dummy variable for tap water was found to be negative but insignificant, and was subsequently eliminated from the regression.

Expected WTP and discussion

Equation 16 was used to estimate the predicted values (Table 10.10) for those households who were willing to pay a positive WTP amount.

Rural households consistently placed more value on improving availability and quality of water. Urban households were only marginally willing to pay for additional quantity and improved quality since, as indicated in the regression results, they are typically tap water users with almost guaranteed access to water.

It should be noted that WTP* for improved water availability as a proportion of the household's monthly income corresponds to 2.7 per cent and 0.5 per cent in rural and urban areas respectively. WTP* as a proportion of the actual monthly cost of water (estimated opportunity cost for river water users) corresponds to 107.5 per cent and 2.9 per cent in rural and urban areas respectively.

An immediate implication of the results is that rural households value improved availability of water more than the urban households. The welfare implication is therefore that improving water quantity and quality has a higher marginal benefit to rural households[5]. There is sufficient evidence that rural households would actually support some of the investment that would be necessary to improve availability and quality of water.

The creation or improvement of infrastructure and services to improve water services and distribution would significantly improve the welfare of both the rural poor and the middle-income urban households. Such infrastructure would increase access to clean water and improve the living standards of rural households. Equity would also be enhanced as rural households enjoyed the same level of water services as urban households.

Improving water provision in rural areas could negatively affect the revenue of the water-vending sector. But in Steelpoort, vending water users account for

Table 10.10 *Predicted willingness to pay for improved quantity and quality of water*

Location	*Median WTP for Quantity*	*Mean WTP for Quantity*	*Median WTP for Quality*	*Mean WTP for Quality*
Rural	22.3	21.6	4.4	6.3
Urban	0.8	3.0	0.5	4.5
Steelpoort	17.8	16.4	1.8	5.8

Note: WTP expressed in rand per household per month
Source: chapter authors

only 1 per cent of the sample, compared with the 19 per cent of river water users who stand to benefit from infrastructure development. The socio-economic gain would therefore be positive in the case of improved water provision.

Public investment in water facilities is the key to unlocking development in water services in rural areas. Where the main source of water is the river, for example, the government may help by providing collective taps. The results also show that households with collective taps wish to upgrade to private taps, a development that, along with the provision of safe convenient water, may allow households to improve other sanitation requirements, by allowing flush toilets, for example. The predicted WTP for both quantity and quality indicate that provision of collective taps or private taps where they previously did not exist would improve household welfare with the possibility of recouping some of the investment through charges.

Where public investment is unattainable, enhancing the participation of private water providers such as vendors would be an alternative to be explored. As indicated above, the success of the vending sector in Steelpoort is limited, probably because of its informal nature. It may be necessary for government to intervene to introduce standards, or at least provide a regulatory framework to enhance the success of the sector.

Conclusion

This chapter set out to estimate WTP for improved availability and quality of water in the Steelpoort river basin. Two broad conclusions can be drawn.

First, rural households stand to benefit from improved availability of water more than urban households. The local government responsible for providing water would be maximizing social welfare if investments were carried out to extend allocation of private tap water to those currently using collective tap water. In addition where budgetary considerations limit such investments, extending collective tap water to those using river water and vending water would be *Pareto* efficient.[6]

Second, there are cost-recovery avenues that may provide budgetary relief. At present there are many households with access to private taps and collective taps at zero cost, or at a cost not corresponding to the quantity used. This anomaly should alert the relevant authorities to inefficient billing and tariff collection by their agents. The household valuation of water would moreover to some extent provide an argument for investing in water resource improvement since some of the investment could eventually be recouped in the form of tariffs.

Notes

1 The household model is adapted from Freeman (1993).

2 This WTP exercise is applicable when only one profile change is proposed to respondents. When the number of profiles is two or more, other analytical frameworks such as McFadden's (1973) random utility model are appropriate. The value attached to each attribute of the resource under valuation is endogenous in the random utility model.

3 About 38 per cent of the sampled households were not willing to pay for more quantity, about 59 per cent were not willing to pay for improved quality.
4 A bootstrap procedure is a re-sampling technique from a sample from either an unknown or known population. In the case where the population is unknown, an empirical cumulative distribution function can be estimated after re-sampling with replacement from the sample.
5 Urban households with revealed WTP*>0 for quantity were willing to pay a higher amount averaging 24.45 rand per month compared with rural mean WTP for quantity of 22.03 rand per month. Similarly, urban households with WTP*>0 for quality were willing to pay a higher amount averaging 21.52 rand per month compared with rural mean WTP for quality of 9.62 rand per month. The predicted WTP for urban households was lower because of the high number of zero respondents who were included in the Tobit regressions. The total sample average WTP for households willing to pay a positive amount was 22.14 rand/month for quantity and 11.35 rand/month for quality. The predicted WTP for urban households was lower because of the high number of zero respondents who were included in the Tobit regression.
6 There is scope for improving the availability of water for collective tap users and river water users without having to raise the tariffs for private tap water users.

References

Amemiya, T. (1973) 'Regression analysis when the dependent variable is truncated normal', *Econometrica*, vol 41, no 6, pp997–1016

Anand, P. (2000) 'Decisions vs. willingness to pay in social choice', *Environmental Values*, vol 9, no 4, pp419–430

Blomquist, G. C. and Whitehead, J. C. (1998) 'Resource quality information and validity of willingness to pay in contingent valuation', *Resource and Energy Economics*, vol 20, pp179–196

Deffar, G. (1998) 'Economic valuation of environmental goods in Ethiopia: A contingent valuation study of the Abiyata-Shalla Lakes national parks', Studies in Environmental Economics and Development, no 6, Environmental Economics Unit, Göteberg University, Sweden

Dublin Statement (1992) *The Dublin Statement on Water and Sustainable Development*, Principle No 4, adopted at the International Conference on Water and the Environment (ICWE), Dublin, January

Freeman, A. M. (1993) *The Measurement of Environmental and Resource Values: Theory and Methods*, Resources for the Future, Washington, DC

Goldblatt, M. (1998) 'Assessing the Effective demand for improved water supplies in informal settlements: A willingness to pay survey in Vlakfontein and Finetown, Johannesburg', *Geoforum*, vol 30, pp27–41

Greene, W. H. (2003) *Econometric Analysis* (5th edition), Prentice-Hall, Upper Saddle River, NJ

Halpern, J., Gomez-Lobo, A. and Foster, V. (1999) 'Designing direct subsidies for water and sanitation services Panama: A case study', The World Bank Policy Research Working Paper Series, no 2344, World Bank, Washington, DC

Hokby, S. & Soderqvist, T. (2001) 'Elasticities of demand and willingness to pay for environmental services in Sweden', paper presented at the 11th Annual Conference of the European Association of Environmental and Resource Economists, Southampton, UK

Jhagoroo, R., Hepburn, T. and Pillay, L. (2000) *SA Explorer Version 1.0*, Municipal Demarcation Board, South Africa

Judge, G. G., Hill, R. C., Griffiths W., Lutkepohl, H. and Lee, T. (1987) *Introduction to the Theory and Practice of Econometrics* (2nd ed), John Wiley, New York

Kuriyama, K. (1998). 'Measuring the value of the ecosystem in the Kushiro wetland: An empirical study of choice experiments', Forest Economics and Policy Working Paper no 9802

McFadden, D. (1973) 'Conditional logit analysis of qualitative choice behaviour', University of California at Berkeley, US

Niklitschek, M. and Leon, J. (1996) 'Combining intended demand and yes/no responses in the estimation of contingent valuation models', *Journal of Environmental Economics and Management*, vol 31, no 3, pp387–402

NWA (1998) *The Republic of South Africa National Water Act (36 of 1998)*, Pretoria

Ryan, M. & Wordsworth, S. (2000) 'Sensitivity analysis of willingness to pay estimates to the level of attributes in discreet choice experiments', *Scottish Journal of Political Economy*, vol 47, no 5, pp504–524

South African Census (1996) *Population Census 1996*, available at www.statssa.gov.za/publications/populationstats.asp

South African Census (2001) *Population Census 2001*, available at www.statssa.gov.za/publications/populationstats.asp

Thompson, H., Stimie, C. M., Richters, E. and Perret, S. (2001) 'Policies, legislation and organizations related to water in South Africa, with special reference to the Olifants river basin', Working Paper 18 (South Africa Working Paper no 7), IWMI, Colombo, Sri Lanka

Whittington, D., Briscoe, J., Mu, X., and Barron, W. (1990) 'Estimating the willingness to pay for water services in developing countries: A case study of the use of contingent valuation surveys in southern Haiti', *Economic Development and Cultural Change*, vol 38, no 2, pp293–311

WSA (1997) *The Republic of South Africa Water Services Act (108 of 1997)*, Pretoria

Appendix 1: Probability Functions for the CVM Model

Since we are assuming a logistic distribution function,

$$\left.\begin{aligned} f(.) &= \frac{e^{Zi}}{\left(1+e^{Zi}\right)^2} = \Lambda(Zi)\left[1-\Lambda(Zi)\right] \\ \frac{\partial f(.)}{\partial X} &= \Lambda(Zi)\left[1-\Lambda(Zi)\right]\beta \end{aligned}\right\} \tag{A.1}$$

And since each observation is a single draw from a Bernoulli distribution (binary with one draw) and all observations are independent, the probabilities can be modelled as:

$$P(wj = ij) = \prod_{ij=0}\left[1 - F\left(\beta' X_j\right)\right]\prod_{ij=1} F(\beta' X) \tag{A.2}$$

Hence the likelihood (and log likelihood) functions to be maximized are:

$$\left.\begin{aligned} L &= \prod_{j=1}^{n}\left[F\left(\beta' X_j\right)\right]^{wj}\left[1 - F\left(\beta' X_j\right)\right]^{1-wj} \\ \text{Log}L &= \sum_{j=1}^{\text{n}}\left[wj \log F(\beta' X)\right] + (1 - wj)\log\left[1 - F(\beta' X)\right] \end{aligned}\right\} \tag{A.3}$$

Once the probability of a respondent is observed, the underlying value of the WTP is a function of the maximized probability:

$$WTP_i^* = \max\left(wi^*, 0\right) \tag{A.4}$$

Equation (A.4) is estimated using the standard censored Tobit model. It is assumed that the error term in equation (A.4) is normally distributed. Although the underlying population is assumed normal, however, survey data are rarely normally distributed. A bootstrap procedure is employed to assess the consistency of the results.

Appendix 2: Estimating Probability of WTP

The decision whether or not to pay for improved availability of water is mainly influenced by the source of water (Table 10A.1). The odds of WTP for improved availability increase for collective tap water users and river water users but decline for private tap water users. As demonstrated in the text, private tap users typically have easier and more reliable access to water, hence are unlikely to be in favour of paying for a better supply.

Location of a household is another significant explanatory variable in the quantity logistic regression. Urban respondents are less likely to be willing to pay for improved quantity of water. This is consistent with the finding of the descriptive analysis that urban water users have higher per capita water consumption (about three times the sample mean).

The level of education is also a significant factor explaining the binary choice to pay or not to pay for more water. Respondents with qualifications below tertiary school are more likely to be willing to pay for more water because most of them live in rural areas and have access only to collective taps or river water. In contrast, tertiary qualifications holders live mainly in urban areas, have access to regular private tap water and would be less likely to be willing to pay for improved access to water.

The odds of WTP for respondents with primary education are higher than for those with secondary qualifications. This is because only 27 per cent of the respondents with primary qualifications have access to more regular water from private taps, compared with 53 per cent of the respondents with secondary education. Respondents with no education would be less likely to pay for more water because they have low or no income.

Water quality perception is the main determinant for the decision whether or not to pay for improved quality of water. The source of water is also an important determinant of the binary choice to pay or not for improved quality of water.

Private tap water users typically have better quality water and so the odds of them choosing to pay for improved quality are lower. Collective tap and river water users are inclined to want better water quality, due to their negative perception of the current resource quality.

Income is another statistically significant explanatory variable for those willing to pay for improved quality. Income has a positive influence on the probability of a household being willing to pay for improved quality of water, although the marginal odds for income are only 0.02.

Table 10A.1 *Probability models for willingness to pay for quantity and quality of water*

	Logistic for willingness to pay for quantity Wtpqt (1 = Yes, 0 = No)				Logistic for willingness to pay for quality Wtpql (1 = Yes, 0 = No)			
Explanatory Variables	*Coefficient*	*Std. Err.*	*Z*	*P > |z|*	*Coefficient*	*Std. Err.*	*Z*	*P > |z|*
Location: 1 = urban, 0 = rural	–1.43	0.46	–3.12	0.002				
Source of water dummies								
private tap#	–9.57				–3.01			
collective tap	5.08	0.56	9.07	0.000	1.48	0.34	4.41	0.000
river	4.49	0.62	7.25	0.000	1.52	0.58	2.64	0.008
Education dummies								
none	0.22	0.46	0.49	0.627				
primary	1.02	0.56	1.81	0.071				
secondary	0.78	0.47	1.67	0.095				
tertiary#	–2.02							
Income					0.02	0.01	2.48	0.013
Water Quality (1 = Poor, 0 = otherwise)					3.29	0.37	8.88	0.000
Constant	–2.44	0.45	–5.43	0.000	–3.13	0.42	–7.53	0.000
Number of observations	373.00				372.00			
Wald chi2 (6)	156.53				131.05			
Prob > chi2	0.00				0.00			
Log pseudo-likelihood	–81.00				–148.07			
Pseudo R2	0.67				0.41			

Note: Level of significance: 10%
Calculated after the regression as the negative of the sum of the coefficients of the other categories
Source: chapter authors

11

Water Resources and Food Security: Simulations for Policy Dialogue in Tanzania

Sindi Kasambala, Abdul B. Kamara and David Nyange

Introduction

Studies linking food security and trends in water availability are receiving increasing attention in developing countries; the issue has become one of concern to policy analysts and development practitioners alike. Central to that concern is the increasing food grain demand resulting from population increase, coupled with diminishing water resources for food production. There are concerns that in sub-Saharan Africa, high population growth and inadequate agricultural production may combine with increasing water scarcity to pose serious constraints to future economic development (Webb and Iskandarani, 1998). Against a background of drought and uneven spatial distribution of water within Southern Africa, water stress is likely to impede economic development and increase the likelihood of major food access problems and malnutrition in the inevitable drought years (Rosegrant and Ringler, 1999).

Water is important for food production not only because of its direct effects on yields and cultivated area, but also because reliable water supplies induce farmers to invest in other essential crop inputs, such as improved germplasm, fertilizers and capacity building for better resource management (Rosegrant, 1997).

Water is likely to be a major constraint to the achievement of food security in many developing countries in the future. Food demand growth caused by increasing population and shifting consumption patterns will necessitate future food production increases. But sometimes arable land is limited, placing the burden for these increases on technology-driven yield improvements. Growing urban and industrial demands on existing water supplies and the need for improved water quality further complicate the situation.

Based on the 2002 population census and projections, the population of Tanzania is currently 34.5 million and will reach 59.8 million by 2025. This increase in population is certain to increase pressure on water resources for food production and domestic and industrial use. Turton and Warner (2002) argue that, in a situation where water resources are relatively finite within any given country, doubling of that country's population will halve the volume of water available per capita.[1]

Water Resources in Tanzania

Tanzania faces a water stress situation in many parts of the country. 'Water stress' here is taken to refer to non-availability rather than under-utilization of the available potential water resources. Tanzania is rather a dry country, besides lying close to the equator. In recent years (period ranging from 1978 to 2001), the country received an average of 992mm of rainfall per year. About one-third is arid or semi-arid, with annual rainfall below 500mm. Another two-thirds of the country is upland with rainfall over 1000mm (United Republic of Tanzania, 2002).

Currently, Tanzania has 89km^3 of annual renewable water resources, which is equivalent to 2700m^3 of water per person per year. Based on population statistics, in 2025 the annual renewable water resources per capita will be 1500m^3 (i.e. 45 per cent less than today) (ECA, 2001, cited in United Republic of Tanzania, 2002). Thus if people do not manage their water resources consumption, then the country will face a water stress situation, characterized by the UN as a water supply of 1000–1600m^3 per person per year (Meinzen and Rosegrant, 1997). Given the projected value of available water per capita per year for 2025 (1500m^3), it is clear that Tanzania might face water scarcity in the near future if the water resources are not well managed now.

Tanzania is one of the countries bordering international water bodies, in the form of lakes covering about 60,000km^2, which clearly represent an ample freshwater resource. Among the international water bodies are the three great East African lakes (Victoria, Tanganyika and Nyasa). Lake Tanganyika, which is Africa's deepest and longest freshwater lake, is also the world's second deepest lake. Lake Victoria, which is shared by Tanzania, Kenya and Uganda, is the world's second largest freshwater lake and drains into the Nile.

For the purposes of resources planning and administration, the country is divided into nine basins: Lake Victoria; Lake Nyasa; Lake Tanganyika; Lake Rukwa; Pangani River; Rufiji River; Wami/Ruvu River; internal drainage; and Ruvuma/Southern Coast Rivers. In this chapter PODIUM simulations were carried out at both national and basin levels. The analysis at basin level is justified by the existing spatial diversity in water resources and agricultural production systems in the country.

The Tanzanian Water Policy of 2002 provides a guideline for water resources management. The policy stipulates the need for prioritization in water use so as to ensure equitable and sustainable socio-economic activities in all sectors of the economy (eg agriculture, domestic use and industrial use). The policy also aims to ensure sustainable management of the environment, increasing productivity and mitigating conflicts (United Republic of Tanzania, 2002). The Tanzanian Water Policy aims at providing 25 litres of potable water per capita per day through outlets located no more than 400 meters from the furthest homestead and each serving 250 people.

This chapter presents the results of an application of the PODIUM policy dialogue model to Tanzania. The objective of this analysis is to assess the water situation in the country as it relates to food security at various levels. It concludes by presenting various policy options on how national level challenges of food security could be met through a combination of domestic production (under various scenarios of water constraints) and trade. National level food requirements are appraised from relevant macro-economic variables such as present population, growth rates, daily dietary intake and proportion of cereals and animal products in diet. Options such as domestic production via irrigated or rainfed agriculture are examined using various parameters such as water requirements, yields (irrigated and rainfed) and potential irrigable land.

The PODIUM Policy Dialogue Model

The PODIUM policy dialogue model was developed as part of the Vision 2025 exercise, a joint effort by the International Water Management Institute (IWMI) in Colombo, Sri Lanka and the International Food Policy Research Institute (IFPRI) in Washington DC. PODIUM is a user-friendly decision-support tool for testing policy options that aim at creating a balance between water allocations for food production on the one hand and satisfying domestic, industrial and environmental needs on the other. PODIUM gives a complete picture of a country's future food and water trends (IWMI, 2000).

PODIUM provides analyses of 'what if' questions relating to assumptions about macro-economic variables such as population growth, per capita calorie intake, changing proportion of grain–meat ratio in dietary requirements, contribution from rainfed and irrigated agriculture, import–export options and potentially utilizable water resources from other sources or models (Kamara and Sally, 2004). In the analysis, a number of questions are addressed. Given current population growth rate, available water resources and existing production technologies, will Tanzania be able to feed itself in 2025? Taking into account technological factors (such as high yielding varieties and more efficient irrigation systems), is irrigation land expansion sufficient to meet the growing food grain demand? How would urbanization and increase in industrial water use put pressure on water demand and compete with water use in the agricultural sector?

The basic structure of the model consists of three main components: the cereal demand drivers, the cereal production drivers and water balance indicators. The cereal demand drivers relate to total population and projected growth rates, food and feed requirements for humans and livestock, and import/export balances. The cereal production drivers relate to irrigated and non-irrigated areas, crop yields and cropping intensities. The cereal balance and proportions of cereals grown through irrigated and rainfed agriculture will then determine the water requirement for irrigation. The water diversions for irrigation, domestic and industrial uses constitute the water requirements of the country that must be met from primary diversions (diversion for the first time) and from recycling of water that has been diverted at least once. The model structure and key variables in the analysis are shown in Figure 11.1.

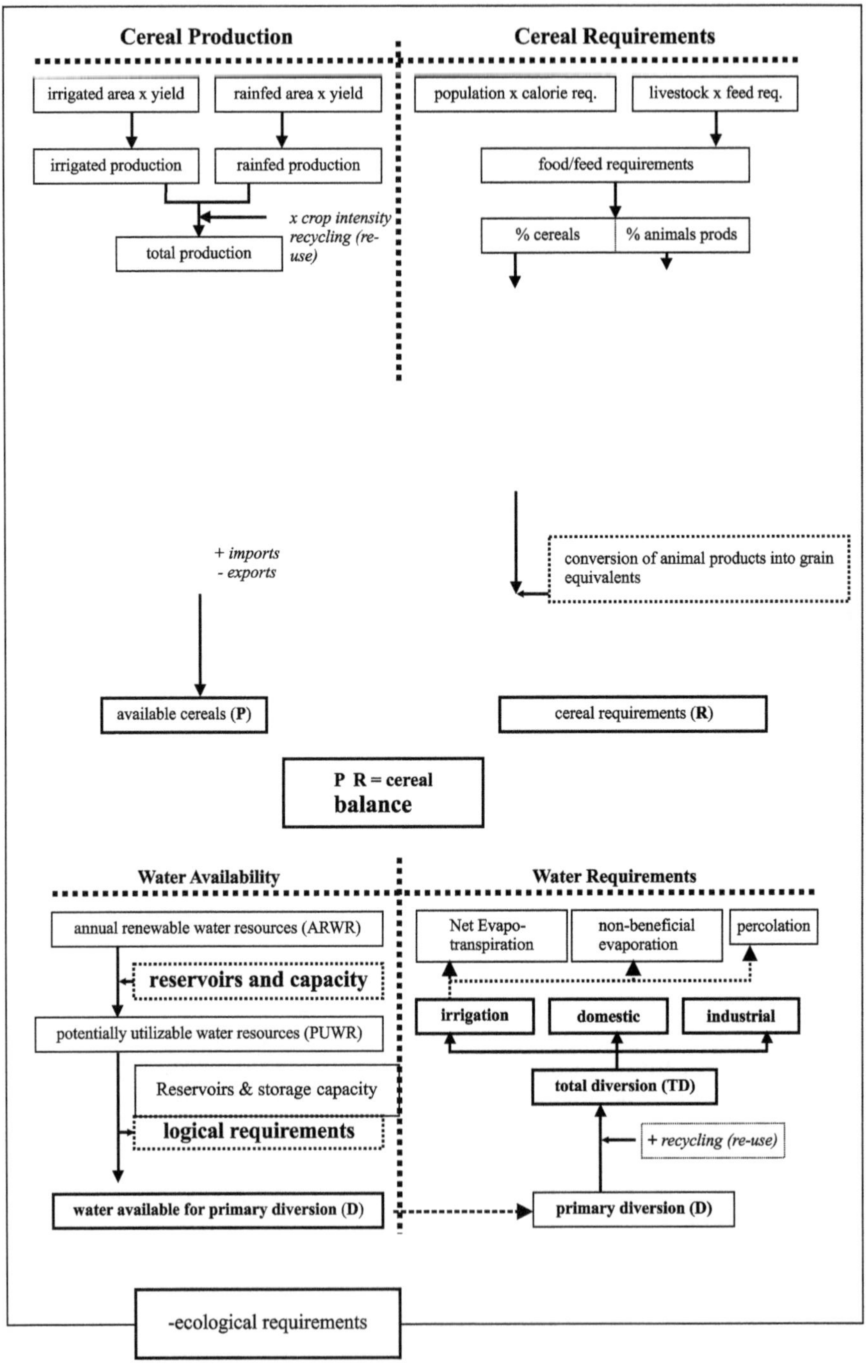

Source: Kamara and Sally (2004)

Figure 11.1 *PODIUM conceptual framework*

Water Resources and Definitions of Key Indicators in the PODIUM Framework

Annual renewable water resources (ARWR) are the sum of all the run-off generated within a country, plus any inflows into the country less outflows to neighbouring countries. Only a certain proportion of the ARWR will actually be available for utilization in the country, this depending on physical and economic storage facilities, including reservoirs and aquifers. This proportion is referred to as the potentially utilizable water resources (PUWR).

One of the key indicators in PODIUM that determines the national-level water situation is the 'degree of development', which refers to primary diversions expressed as a percentage of PUWR. The degree of development is what determines whether or not current strategies for water resources development and management are sustainable. This assessment is expressed in terms of physical and economic water scarcity.

In PODIUM, a country is said to be physically water scarce if the degree of development is greater than the proportion of the PUWR available for diversion after the ecological reserve has been met. In the model, the threshold value for degree of development is 60 per cent, but this may vary from country to country. In Tanzania, the water policy only emphasizes the necessity to consider environmental water use, without providing for a specific amount to be set aside for this (United Republic of Tanzania, 2002).

The second important indicator in PODIUM is the growth in total diversions (TD), which indicates whether or not a country is economically water scarce. A country is considered economically water scarce if it becomes necessary to increase water supplies by the development of additional storage, conveyance and regulation systems equivalent to more than 25 per cent of 1995 levels (Seckler et al, 1998). Economic water scarcity therefore reflects a country's inability to make the necessary investments in water development and associated infrastructure.

Data and Model Estimation

Data on yields of various crops of both rainfed and irrigated agriculture by river basin was derived from the national agricultural survey. Water resources data was obtained from the Tanzanian meteorological agency, relevant river basin authorities and IWMI publications. Production data, daily calorific requirement and per capita water demand were gathered from various secondary sources.

The computations in the PODIUM model involve three major steps. The national cereal requirement of a country, based on population, daily calorie intake and composition of diets, is estimated first. In the second step cereal production, based on expected yields and areas under both irrigated and rainfed conditions, is estimated. Finally the water demand for the projected food production is estimated. In order to work out the drivers of the PODIUM model, the three steps above have to be implemented, based on the Tanzania situation. The human and animal populations of the country must be recorded, along with daily per capita calorie intake and the agricultural systems used and crops grown under rainfed and irrigated conditions. Different agricultural means and different crops have different water requirements. Production and yield under both rainfed and irrigated agriculture are some of the drivers obtained. The population of an area and its cereal requirement were explored based on daily per capita calorie intake. Water availability and demand sectors were also studied and entered in the model. Then available water was analysed to see whether or not it was able to support the grain production and other uses.

Simulations and Scenario-testing

Urbanization

The rate of urban migration in Tanzania is high. The urban population is growing at 6 per cent per annum, far above the national population growth rate of 2.8 per cent. Furthermore, the consumption habits of the urban population, with their preference for cereals such as maize, rice and wheat, are quite different from those in rural areas, which may explain increased grain consumption, mostly of cereals. It is also projected that in Tanzania, as in the rest of the developing world, per capita urban income growth will be translated into soaring demand for animal and horticultural products (IFPRI, 2000). Income elasticity of demand suggests a slow increase in demand for maize, cassava and sweet potatoes, with per capita income increase demand for wheat, rice, potatoes, fruits, vegetables and animal products rising quickly. Increase in rural–urban migration also has effects on water resources. Increase in the urban population will require more water supplies for domestic consumption. PODIUM simulations results in Table 11.1 show a decrease in growth in diversion while the degree of development in Tanzania is slowly increasing.

The slow growth in per capita water availability despite increase in water diverted for urban use is partly explained by water lost before it reaches consumers. A recent study by Japan International Cooperation Agency (JICA) estimates that over one-third of water supply to Dar es Salaam is lost through leakages.

It is also noted that around 80 per cent of the water supplied to urban areas ends up as wastewater (United Republic of Tanzania, 2002).

Urban agriculture, families engaging in vegetables and livestock production, is another phenomenon associated with rapid urbanization. While the urban poor are more engaged in vegetable production, livestock keeping is more practised by

Table 11.1 *Urban population increase with water resources development and food security*

Population (%)	*Growth[a] (%)*	*DOD[b](%)*	*Food security[c] (million tonnes)*
22	50.79	3.48	0.45
35	49.02	3.60	0.45
50	46.91	3.75	0.45
60	45.50	3.80	0.45

Notes: [a] Growth in total water diversions to agriculture, domestic and industrial uses (in %)
[b] Degree of development (% of potentially utilizable water resources)
[c] National level food security: surplus (+) and deficit (–) in million tonnes
Source: PODIUM model results, 2002/03

Table 11.2 *Estimated feed requirement for poultry, dairy and pigs*

	Feed required (thousand tonnes per year)		
	1990	*1995*	*2000*
Poultry	148	172	206
Dairy	76	87	142
Pig	26	31	38

Source: MALC (2000)

middle class citizens to supplement their incomes. Urban agriculture increases water demand while livestock keeping creates demand for feed.

The urban population in Tanzania is estimated to be 22 per cent of the total population (Population Reference Bureau, 2002). In terms of livestock population Tanzania ranks third in Africa. While most indigenous breeds are grazed on natural pastures, improved breeds mostly raised in the highlands, urban and peri-urban areas are fed on cereal feeds. In the 1998 and 1999 seasons, the number of improved cattle was 451,142 while the indigenous cattle numbered 15,943,827. The revival of the dairy industry is likely to contribute to the increase in numbers of improved dairy cattle. As the population becomes more health conscious, consumption of red meat is likely to decrease while that of white meat (such as poultry and pork) is projected to grow. Such a shift in consumption patterns is likely to further increase the demand for livestock feed and hence cereal demand (Table 11.2).

From 1990 to 2000 cereal demand for livestock feed was increasing (Table 11.2). Population scenarios were tested in the model with relation to food requirements and water resources development, with assumptions that there will be an increase in rural–urban migration, industrialization and urban agriculture (Figure 11.2).

Industrialization

The recent rapid growth in the industrial sector has been attributed to privatization of state industries and attainment of macro-economic stability, which in turn have attracted foreign direct investment (FDI) in the sector. The industrial sector grew by 3.6 per cent in 1999 and attained a growth rate of 4.8 per cent in 2000. It is projected to continue to grow in line with the overall economy, which is currently growing at more than 6 per cent. It is further projected that by 2025 the industrial sector's contribution to the country's GDP will be three times the current figure. Tanzania's Development Vision 2025 envisages transforming the economy to that of a semi-industrialized country (United Republic of Tanzania, 2002). The growth of the sector implies that adequate and reliable water supply will be required. Another sector that has contributed to increase in water demand is the mining sector, which requires large quantities of water during processing. In 2000 the mining sector grew by 13.9 per cent.

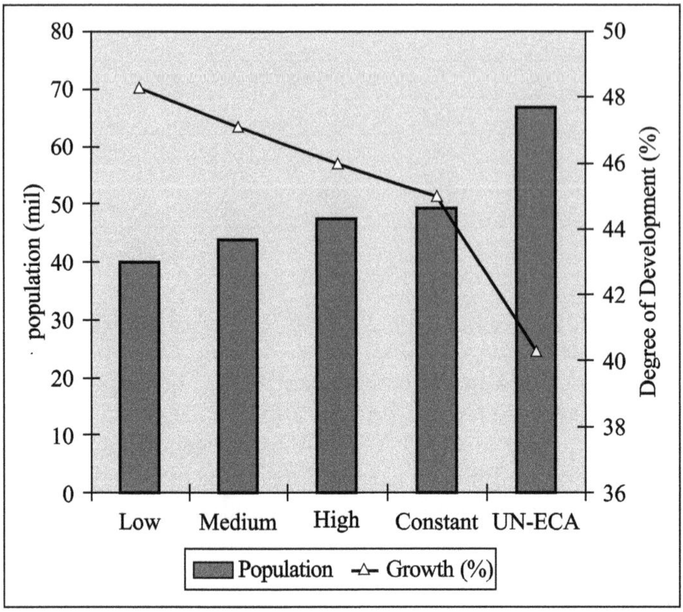

Source: PODIUM simulation results (2002/2003)

Figure 11.2 *Population growth and water resources development scenarios*

Orientation of government expenditure towards provision of domestic clean water and other social services

Under this scenario, due to the increase in urbanization, the current water supply systems are not adequate to meet the clean water needs of domestic uses. In Table 11.3, growth in diversions is seen to be decreasing while the percentage of the population with access to clean water is increasing. This is perhaps due to competition for water between different sectors.

Table 11.3 *Increase of population with access to clean water: impacts on water resources development and food security*

Population (%)	*Growth[a] (%)*	*DOD[b] (%)*	*Food security[c] (million tonnes)*
50	53.96	3.25	0.45
75	50.79	3.48	0.45
80	50.26	3.51	0.45
85	49.6	3.56	0.45
100	47.8	3.69	0.45

Notes: [a] Growth in total water diversions to agriculture, domestic and industrial uses (in %)
[b] Degree of development (% of potentially utilizable water resources)
[c] National level food security: surplus (+) and deficit (–) in million tonnes
Source: PODIUM model results (2002/03)

Seventy per cent of the urban population in Tanzania has access to reliable water supply services (United Republic of Tanzania, 2000), but the increase in water demand in urban areas is not proportional to the rate of expansion of the water supply. This is due to the high rate of increase in industrial activities, and significant water losses through leakage, wastage and illegal connections. Wastewater quantity in urban areas is estimated to be 80 per cent of water supplied. United Republic of Tanzania budget figures show that the budget for the water sector increased by 21 per cent and 43 per cent in 2001/02 and 2002/03 respectively (estimates calculated from United Republic of Tanzania, 2000). The number of people with piped water in rural areas increased from 24.5 per cent in 1991/92 to 28.3 per cent in 2000/01; in urban areas, there was an increase from 72.7 per cent in 1991/92 to 75.6 per cent in 2000/01 (United Republic of Tanzania, 2001). Therefore, while the budget is increasing, there are still a lot of inefficiencies in the sector. Part of the explanation lies in the fact that management of the water during diversions is very poor, with much of it lost on the way. The challenge to policy-makers is to ensure the increase in public investment in water projects commensurate with population growth and also that water systems are maintained to reduce leakages.

Irrigated agriculture and water resources

Both water and agricultural policies envisage expansion of irrigated land. Most irrigation practised in Tanzania is traditional with application efficiencies of 10–15 per cent (United Republic of Tanzania, 2002) and the country has potential for growth in terms of irrigated area. With the irrigation potentials and other opportunities well utilized, 'crop per drop' can probably increase (Table 11.4), even without increase in the water diversions to the agricultural activities. As has been suggested by the National Irrigation Master Plan (NIMP), improved traditional schemes need low capital investment compared with modern irrigation schemes. Modern irrigation schemes in Tanzania bring high irrigation efficiencies (65–80 per cent above traditional schemes) but cover only a small area of 40,000ha (equivalent to 20 per cent of total irrigated land which is 200,000ha). Development of modern irrigation leads to increased efficiency in water use and usually modern irrigation schemes are coupled with the introduction of high yielding varieties (HYV); both of these factors are considered in PODIUM simulations. However, results suggest that expansion of irrigated area does not do much to reduce food deficit unless irrigation development is coupled with the use of high yielding varieties (Table 11.4).

Tanzania has a potential 1 million ha for irrigation. Around 200,000ha are currently under irrigation, but only 20 per cent of this is modern irrigation. Growth in diversions and degree of development in water resources are, however, increasing.

Table 11.4 *Yield increase with the current irrigated area: water and food situation (2025)*

Irrigated area	*Increase in rice yields*	*Water resources development and food security indicators*		
Area (ha)	*kg/ha*	*Food security*[a]	*Growth (%)*[b]	*DOD*[c] *(%)*
200,000	2.84	0.45	50.79	3.47
200,000	3.0	0.45	50.79	3.47
200,000	3.8	0.46	50.79	3.47
200,000	4.5	0.47	50.79	3.47
200,000	5.5	0.47	50.79	3.47

Notes: [a] National level food security: surplus (+) and deficit (–) in million tonnes
[b] Growth in total water diversions to agriculture, domestic and industrial uses (in %)
[c] Degree of development (% of potentially utilizable water resources)
Source: PODIUM model results (2002/03)

Table 11.5 *Irrigated area increase: water and food situation (2025)*

Increase in net irrigated area	*Water resources development and food security indicators*		
Area (ha)	*Food security*[a]	*Growth (%)*[b]	*DOD*[c] *(%)*
180,000	0.45	50.79	3.47
200,000	0.45	50.79	3.47
250,000	0.45	50.79	3.47
500,000	0.45	50.79	3.47
850,000	0.45	50.79	3.47
1,000,000	0.45	50.79	3.47

Notes: [a] National level food security: surplus (+) and deficit (–) in million tonnes
[b] Growth in total water diversions to agriculture, domestic and industrial uses (in %)
[c] Degree of development (% of potentially utilizable water resources)
Source: PODIUM model results (2002/03)

Summary and Conclusions

Using national and basin level data, this chapter has examined the relationship between food security and availability of water resources under various scenarios of demand for food, agricultural production and water use efficiency in Tanzania. PODIUM simulation results indicate that in the medium term, Tanzania could increase food production through farmland expansion even without improvement in irrigation efficiency. This observation seems plausible since only 200,000ha of land is currently under irrigation in Tanzania, a mere 20 per cent of the irrigable potential. In the long term, improvement in irrigation efficiency is necessary since increased food production through land expansion will not match increasing food demand from rapid urbanization coupled with growth in income. This is supported by the fact that of the total irrigated land, 25,000ha is under modern large-scale irrigation while the rest (87 per cent) is under the traditional system. Traditional irrigation offers room for improvement in water-use efficiency. Moreover, irrigation land expansion and improvement in water-use efficiency must be coupled with the use of high yielding varieties in order to cope with increasing food grain demand.

Based on PODIUM simulations it is evident that with the prevailing low levels of farm productivity arising from under-utilization of irrigation potential land and inefficient traditional irrigation technologies the country's food production will be unable to keep up with predicted population growth rates. There is thus a need to expedite implementation of the national irrigation plan, which envisages increase in irrigated land while modernizing land under traditional irrigation.

Acknowledgements

Funding for this study was provided by the IWMI; this chapter benefited greatly from comments from Upali Amarasinghe of the IWMI. The authors are also indebted to staff of the Soil and Water Management Research Group at Sokoine University of Agriculture for facilitating the project and various people who made data available for the study.

Note

1 *Editor's note:* This may appear to be stating the obvious, but it all depends on the interpretation of the water available per capita. If it's only domestic water, that's obvious. If it's the overall amount of water available/allocable, as per capita, that's a questionable background assumption for further calculations. Demographic growth can just be used as a proxy for estimating increase in water consumption since domestic uses represent only a minority portion of overall consumption, which includes major users such as industrial and mining uses, and agricultural uses (notably for exports).

References

ECA (United Nations Economic Commission for Africa) (2000) *Transboundary River/Lake Basin Water Development in Africa: Prospects, Problems, and Achievements*, ECA, Addis Ababa, Ethiopia

IWMI (2000) 'PODIUM, The Policy Dialogue Model Brochure', available at www. cgiar.org.iimi/tools/(accessed in December 2002)

Kamara, A. B. and Sally, H. (2004) 'Water management options for food security in South Africa: Scenarios, simulations and policy implication', *Development Southern Africa*, vol 21, no 2, pp366–384

Keenja, C. (2001) 'Food security in Tanzania: The way forward', paper presented at the eighth Sokoine Memorial Lecture, SUA Morogoro, Tanzania

Meinzen, R. S. and Rosegrant, M. W. (1997) 'Managing water supply and demand in Southern Africa', in L. Haddad (ed) *Achieving Food Security in Southern Africa: New Challenges, New Opportunities*, IFPRI, Washington, DC

MALC (2000) *Basic Statistics 1999–2000: Agriculture*, Ministry of Agriculture, Land and Cooperatives, Dar Es Salaam, Tanzania

Population Reference Bureau (2002) 'Demographic and population data, New York', available at www.prb.org/pdf/WorldPopulation DS02_Eng.pdf (accessed in November 2002)

Rosegrant, M. W. (1997). *Water Resources in the Twenty-First Century: Challenges and Implications for Action*, Vision 2020 discussion paper no 20, International Food Policy Research Institute, Washington, DC

Rosegrant, M. W. and Ringler, C. (1999) 'Impact on food security and rural development of reallocating water from agriculture', discussion paper no 47, Environment and Production Technology Division, IFPRI, Washington, DC

Seckler, D., Amarasinghe, U., Molden, D., de Rhadika, S. and Barker, A. (1998) *World Water Demand and Supply, 1990 to 2025: Scenarios and Issues*, Research Report no19, International Water Management Institute, Colombo, Sri Lanka

Turton, A. R. and Warner, J. F. (2002) *Exploring the Population/Water Resources Nexus in the Developing World*, The Woodrow Wilson Institute, Washington, DC

United Republic of Tanzania (2000) *Poverty Reduction Strategy Paper*, Government printers, Dar es Salaam, Tanzania

United Republic of Tanzania (2001) *Household Budget Survey, Final Report, vol 32*, Government printers, Dar es Salaam, Tanzania

United Republic of Tanzania (2002) *National Water Policy*, Ministry of Water And Livestock Development, Dar es Salaam, Tanzania.

Webb, P. and Iskandarani, M. (1998) 'Water insecurity and the poor: Issues and research needs', discussion papers on development policy, Centre for Development Research, University of Bonn, Bonn, Germany

IFPRI (2000) *Agriculture in Tanzania since 1986: Follower or Leader of Growth*, International Food Policy Research Institute, Washington, DC

Appendix: Summary of PODIUM Model Results for Base Year (1995) and Projections for 2025

Table 11A.1 *Rainfed agriculture*

Rainfed crop production	*1995*	*2025*	*Annual growth*
Total Rainfed grain production (bn US$)	0.853	1.646	2.21%
Total Rainfed non-grain production (bn US$)	1.129	4.289	4.55%
Total Rainfed crop production (bn US$)	1.983	5.934	3.72%
Grain production – percentage of crop production	**43%**	**28%**	
Total cereal Rainfed area (m ha)	3.11	4.17	0.98%
Total grain Rainfed area (m ha)	3.42	5.14	1.37%
Total crop Rainfed area (m ha)	4.52	5.59	0.71%
Rainfed cereal production (m tonne)	4.15	7.49	1.99%
Rainfed grain production (m tonne)	4.30	8.16	2.15%
Cereal Rainfed area – percentage of total	69%	75%	
Grain Rainfed area – percentage of total	76%	92%	

Table 11A.2 *Food grain balance*

	1995	*2025*	*Growth Rate*
Total grain requirements (m tonnes)	4.02	7.06	1.89%
Total grain production (m tonnes)	4.47	8.56	2.19%
Grain production surplus/deficit (m tonnes)	0.45	1.50	
Production surplus/deficit (% of consumption)	11%	21%	
Total grain requirements (bn US$)	0.82	1.45	1.89%
Total grain production (bn US$)	0.89	1.73	2.25%
Grain production surplus/deficit (bn US$)	0.07	0.28	
Production surplus/deficit (% of consumption)	8%	19%	
Total crop requirement (B US$)	3.5	7.1	2.34%
Total crop production (bn US$)	2.3	6.7	3.63%
Crop production surplus/deficit (bn US$)	−1.3	–0.4	
Production surplus/deficit (% of consumption)	−35%	–6%	

Table 11A.3 *Water availability variables*

TANZANIA	*1995*	*2025*	*Annual Growth*
Utilizable surface water (natural) (km^3)	65.6	65.6	0.00%
Interbasin transfer: import (km^3)	0.0	0.0	
Interbasin transfer: export (km^3)	0.0	0.0	
Water for environment and navigation (km^3)	0.0	0.0	
Net utilizable surface water (km^3)	65.6	65.6	0.00%
Reported groundwater resources (km^3)	16.4	16.4	0.00%
Total live storage (km^3)	0.0	0.0	
Evaporation rate from reservoirs	15%	15%	0.00%

Table 11A.4 *Reference evapotranspiration (ETo) and rainfall*

		1995			*2025*		
Month	*Days*	*ETo (mm/day)*	*P_{75} Rainfall (mm/month)*	*Effective Rainfall (mm/month)*	*ETo (mm/day)*	*P_{75} Rainfall (mm/month)*	*Effective Rainfall (mm/month)*
Jan	31	4.32	94.08	85.9	4.32	94.08	85.9
Feb	28	4.48	93.92	85.7	4.48	93.92	85.7
Mar	31	4.30	110.97	97.7	4.30	110.97	97.7
Apr	30	3.84	104.66	93.3	3.84	104.66	93.3
May	31	3.75	30.34	15.2	3.75	30.34	15.2
Jun	30	3.83	3.32	1.7	3.83	3.32	1.7
Jul	31	4.05	1.44	0.7	4.05	1.44	0.7
Aug	31	4.58	2.39	1.2	4.58	2.39	1.2
Sep	30	5.19	5.05	2.5	5.19	5.05	2.5
Oct	31	5.50	11.24	5.6	5.50	11.24	5.6
Nov	30	5.05	43.81	21.9	5.05	43.81	21.9
Dec	31	4.41	93.64	85.5	4.41	93.64	85.5
Total			595	497		595	497

Table 11A.5 *Water use balance for 1995, base year*

	1995 (base year)		*Utilizable Return Flow*		*Non-utilizable Outflow*	
	Total Diversions	*Evaporation*	*Ground-water*	*Surface*	*Sea and deficit/ surplus (d/s)*	*Swamps*
Irrigation	2.03	0.96	0.58	0.06	0.11	0.32
Domestic	0.52	0.10	0.22	0.02	0.21	0
Industrial	0.03	0.01	0.01	0.00	0.01	0
Total	2.58	1.07	0.81	0.09	0.32	0.32
Total %			54%	6%	22%	21%

Table 11A.6 *Water use balance for 2025, projection year*

2025 (Target year)		*Utilizable return flow*			*Non-utilizable return flow*	
	Total diversions	*Evaporation*	*Ground-water*	*Surface*	*Sea and d/s*	*Swamps*
Irrigation	3.08	1.84	0.67	0.07	0.12	0.37
Domestic	2.54	0.51	1.10	0.12	0.81	
Industrial	0.00	0.00	0.00	0.00	0.00	
Total	5.62	2.34	1.77	0.20	0.94	0.37
Total %			54%	6%	29%	11%

Table11A.7 *Water diversion for 1995 and 2025, base year and projection year respectively*

Water use (km³)	*1995*	*2025*	*Growth (%)*
Total diversions (irrigation)	2.03	3.08	1.40
Total diversions (industry)	0.03	0.00	−100.00
Total diversions (domestic)	0.52	2.54	5.43
Evaporation from reservoirs	0.00	0.00	0.00
Total diversions (surface plus groundwater)	2.58	5.62	2.63
Groundwater use	1.12	2.67	2.94
Water for environment & navigation	0.00	0.00	

Table 11A.8 *Water availability for 1995 and 2025 base year and projection year respectively*

Water availability (km³)	*1995*	*2025*	*Growth(%)*
Total surface water resources	65.6	65.6	0.00
Surface return flow (regeneration)	0.09	0.20	2.62
Groundwater resources	16.4	16.4	0.00
Groundwater recharge	0.81	1.77	2.62

Table 11A.9 *Water balance for 1995 and 2025*

Water balance	*1995 (%)*	*2025 (%)*	*Growth (%)*
% of surface resources diverted	2.2	4.5	2.37
% of groundwater resources diverted	6.5	14.7	2.75
% of water resources diverted	3.1	6.7	2.59
% of total diversions evaporated	41.5	41.7	0.02

Table 11A.10 *Water surplus or deficit for 2025*

Water surplus or deficit (km^3)	*1995*	*2025*	*Growth (%)*
Surface water	64.23	62.85	–0.07%
Groundwater	16.09	15.50	–0.13%
Total	80.33	78.35	–0.08%

12

How More Regulated Dam Release Can Improve the Supply from Groundwater and Surface Water in the Tadla Irrigation Scheme in Morocco

Thomas Petitguyot and Thierry Rieu

Introduction

From an international perspective, groundwater tapping in irrigated agriculture has been exponentially growing for about 40 years, thanks to the generalization of mechanized techniques for drilling and pumping (Shah et al, 2003). Furthermore, this evolution has been achieved in a lot of collective irrigation schemes where irrigation has been traditionally practised from surface water (Chohin-Kuper et al, 2002). This development, known as 'conjunctive use' as both surface water and groundwater are withdrawn by users, could be seen as a consequence of a reduction in the availability of surface resource or of the degradation of water services and/or the increase in water tariffs. This chapter looks at ways of improving water management within this type of irrigated scheme in the context of its becoming more and more common all over the world.

Irrigation systems with a conjunctive use are often seen as tapping two separate water resources: a centrally managed water service with a declining water supply on the one hand, on the other a free water resource without any regulation process due to the high degree of asymmetry in information between users and water managers. The situation can easily be compared to a non point source pollution (Xepapadeas, 1991) as the regulation agency is unable to determine the effort each user is making to avoid water wastage. Facing both these difficulties, we propose to improve the groundwater management when considering the tight hydraulic links between resources in conjunctive use systems. In such systems, groundwater is generally used as a complement to surface water, and water demand for both resources are mutually dependent. Moreover, in our case study in Tadla (Morocco), infiltration of surface water in irrigated schemes is responsible for a major part of aquifer replenishing. We therefore propose to reconsider the design of water surface management as an indirect tool to improve conjunctive use of water resources.

The situation of a conjunctive use of groundwater and canal water is well illustrated by the Tadla Office in the Oum-Er-Rbia basin in Morocco, and using this example we explore the extent to which new management of surface water resources can improve the comprehensive water management. Groundwater is practically subject to free access for farmers equipped with a pumping facility (i.e. 30 per cent at least). The delivery of canal water is reputed to be efficient and distributed in an equitable way to irrigators through water turns, but the global amount of this supply falls far short of crop requirements (40 to 70 per cent of the design allowance depending on the drought intensity, Petitguyot, 2003). In this context we focus on the design of release rules from the dam, on an annual basis, and look for further information that might help to indicate tapping intensity.

Case Study

Our case study is located in the Oum-Er-Rbia catchment area in Morocco, where there is a wide agricultural plain about 200km south of Casablanca extending over 3,600 km². Its agricultural output is closely related to a wide-scale irrigation scheme (100,000 ha), whose equipment was set up in the early 1920s (Préfol, 1986). Wheat, sugar and dairy production in this area represent a major part of the national agricultural output.

During the first period, only surface water was used for irrigation: water was collected and stored in two dams, designed for both inter and intra-annual storage. Two distinct management institutions were involved in providing water to farmers:

1. a resource manager[1] (or 'dam manager'), who operated all hydraulic work in the river basin and was in charge of water allocation among users; and
2. a water service manager[2] (or 'scheme manager'), who received water from the former, and was in charge of the irrigation service and distributing water among the irrigators. Authorized by the state and with a mission of agricultural development, this manager's function was to provide farmers with the best conditions for crop production and give them incentives to produce according to the national agricultural policy.

However, many changes have occurred since the 1980s:

- Chronic droughts drastically reduced resource availability in dams.
- The share of the scheme in resource allocation was reduced due to increasing urban demand and the development of competition with other irrigation schemes.
- Liberalization of agricultural production in 1996 led to the evolution of new cropping patterns.

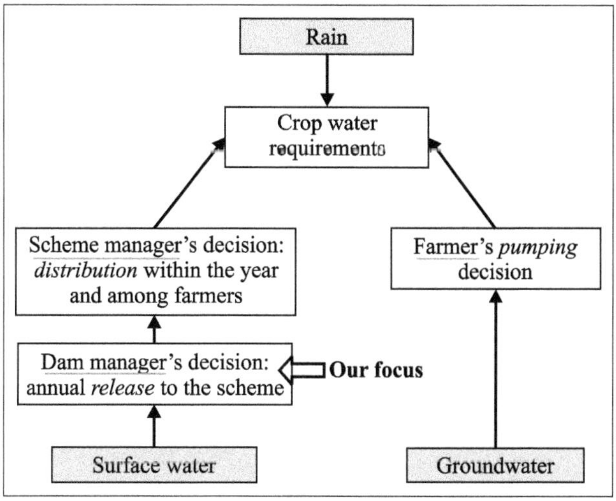

Source: chapter authors

Figure 12.1 *Water resources and allocation decisions*

These conditions increased the demand for groundwater use, and many farmers have since invested in pumping facilities, which were sometimes highly subsidized by the Moroccan government. Although the exact number of wells is unknown, it is roughly estimated at around 10,000 units within the irrigation scheme. The great majority of these are individually operated by farmers, and no regulation of abstraction has been implemented by any manager so far.

As a result, water at the farm level comes from three different sources, as shown in Figure 12.1. The figure also illustrates the way each resource is governed by different stakeholders' decisions.

To complete this brief presentation, other details about groundwater status which have been used in designing our model are as follows:

- Both the geometry of the aquifer and the dynamics of groundwater are not known fully, so it is not possible to derive the withdrawals in the aquifer from its piezometric drop.
- Local public managers share a common assumption about groundwater that for the main part it comes from seepage lost from irrigation as practices rely on the traditional micro-basin system.
- As water in this shallow aquifer is of poor quality, the water table is perceived as belonging to the agricultural sector.

The main body of this chapter, as already mentioned, focuses on the dam manager's annual release decisions (in our case study, the Agence de Bassin Hydraulique (ABH)).

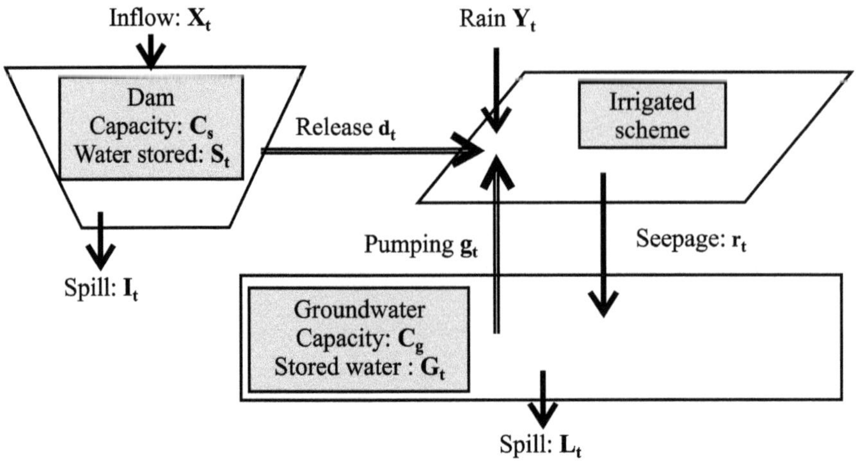

Source: chapter authors

Figure 12.2 *Physical model and variables definition*

Model and Method

Modelling the Hydrological System

Our model represents two connected reservoirs whose water is solely used by agriculture within a unique irrigation scheme, as depicted in Figure 12.2. Index t is for time, discrete in the model. The time unit is one year.

At the beginning of the agricultural campaign, the dam managers inform irrigators about the volume to be available for the coming year. This is made before the rainy season, and without knowledge of inflow quantity.

Each reservoir is of finite capacity: when the stored water reaches this capacity and exceeds it, water spills and is then unavailable for any agricultural use. The dam is subject to stochastic inflows, controlled release and spills. The dynamic of the water stock is specified by the recursive mass balance equation:

$$S_{t+1} = S_t + X_t - d_t - I_t \tag{1}$$

We consider that inflows take place during the rainy season and no water is released before. Then overflow is:

$$\begin{aligned} I_t &= 0 && \text{if } S_t + X_t \leq C_s \\ I_t &= S_t + X_t - C_s && \text{otherwise.} \end{aligned} \tag{2}$$

This specification overestimates the spilled volume as irrigation usually starts before the end of the rainy season.

Like the dam, the groundwater reservoir is subject to seepage and pumping and may also spill:

$$G_{t+1} = G_t + r_t - g_t - L_t \tag{3}$$

We consider that seepage and pumping take place simultaneously, and overflow is:

$$\begin{aligned} L_t &= 0 && \text{if } G_t + r_t - g_t \leq C_g \\ L_t &= G_t + r_t - g_t - C_g && \text{otherwise.} \end{aligned} \tag{4}$$

Like Knapp and Olson (1995), here we make the strong assumption that the aquifer could be seen as a bathtub with an infinite transmissivity, which means that all farmers face the same water table level. This level may be directly derived from groundwater stock and is expressed in terms of variation of the piezometric level h_t:

$$h_t = \mu - v.G_t \tag{5}$$

where μ and v are parameters related to the real reservoir (see Appendix). h_t is constant within a period, and changes occur at the end of the year, as a result of the second equation.

The two reservoirs (the dam and the aquifer) are indirectly connected in the sense that while no water runs directly from one to the other, inflows to the groundwater reservoir consist only of seepage flows r_t, defined as a fixed part τ of the total irrigation volume v_t:

$$r_t = \tau.v_t \tag{6}$$

τ is linked to irrigation technology and exogenously determined.

Irrigation in the scheme uses water resources from three sources – water abstraction from the two reservoirs and rain:

$$v_t = d_t + g_t + Y_t \tag{7}$$

Inflows to the reservoirs and rain are independent and stochastic variables. For simplicity, we neglect rain variability and its contribution to groundwater recharge. Its constant value is, however, taken into account through the calculation of crop requirements. For dam inflows (annual values), we assume all X_t variables are mutually independent, identically distributed over years and accord with Gaussian law.

Management Model

Abstraction

Of the three water resources, farmers may command only the pumped volume. Pumping decisions are taken in the knowledge of dam release and rain, so that groundwater acts as a complement to crop requirements. Within the scheme, the release is distributed to farms according to their irrigated areas, regardless of their particular crop requirements or pumping capacity.[3]

Pumping is modelled as an endogenous variable of the model, on which no direct command exists for the resources manager. The farmers' programme is to maximize their profit by choosing the level of pumping that equalizes the marginal benefit for water and the marginal cost, provided their pumping facility enables them to reach the water table. This individual decision is taken with knowledge about the water level h in the aquifer and about the volume of surface water d_i they are granted (a part of the release). In this attempt, we consider N types of farms that differ in their size a_i, water demand per hectare $P_i(v)$, pumping capacity (i.e. maximum pumping depth H_i) and pumping cost $c_i(g_i,h)$. All these functions were calculated based on our fieldwork; water demand was derived from linear programming (Petitguyot, 2003).

If g_i^* is a solution of

$$\frac{\partial P_i(d_i + g_i)}{\partial g_i} = \frac{\partial c_i(g_i, h)}{\partial g_i} \tag{8}$$

for farm i, with $d_i = \frac{d_t}{N}$ (all farms are the same size and receive the same part of the release), then

$$g_t = \sum_{i=1}^{N} (\varphi_i . a_i . g_i^*) \tag{9}$$

where

$$\begin{aligned} \varphi_i &= 1 \quad \text{if } H_i \geq h \\ \varphi_i &= 0 \quad \text{otherwise} . \end{aligned} \tag{10}$$

This is illustrated in Figure 12.3.

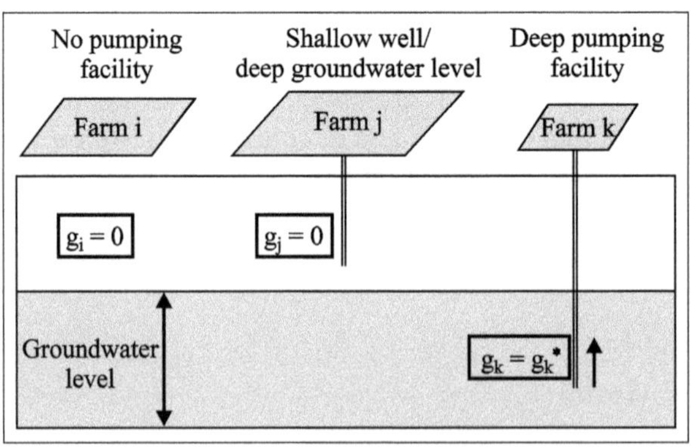

Source: chapter authors

Figure 12.3 *Groundwater withdrawal according to pumping facilities*

Farms' characteristics are given in the model; they do not vary either over time or according to water table evolution. This means farms cannot adapt themselves to their available water resource.

Release

The decision variable in the model is the volume annually released from the dam. Each year, before the beginning of the agricultural season, the dam managers forecast the volume to be released to the irrigated scheme. For this decision, the managers may use information about the distribution of dam inflows X_t, water levels in reservoirs S_t and G_t, and the farm demand from groundwater (g_t knowing d_t and G_t).

We use a very simple rule that has been widely developed in the literature on reservoir theory and the linear decision rule (LDR, initiated by Revelle et al, 1969). According to the original rule, the release is a linear function of the reservoir stock:

$$d_t = S_t - b \tag{11}$$

where b is a decision parameter for the manager. We refine this rule by introducing an additional decision parameter (a):

$$d_t = a.S_t + b \tag{12}$$

The managers' decision is then restricted to the choice of a vector (a, b) constant over time, according to their objectives. As we do not restrict a priori the value of parameters, we must extensively define release thus:

$$\begin{aligned} &d_t = a.S_t + b \quad \text{if } a.S_t + b \leq S_t + X_t \Leftrightarrow X_t \geq (a-1).S_t + b \\ &d_t = S_t + X_t \quad \text{otherwise.} \end{aligned} \tag{13}$$

In practical terms, when the first rule is applied, of stock S_t at the beginning of the year, one part $[a.S_t]$ will be delivered to the scheme, and the remainder $[(1-a).S_t)]$ will be kept for the next year. And of inflow X_t to be recoded in the season, one part will be delivered to the scheme $[b]$, and the remainder $[X_t - b]$ will be stored for the future.

Thus the first parameter a is linked to the managers' interest for inter-annual regulation of releases: when a is large, most of the water stored at the beginning of the season will be released in the coming year and little $[(1-a).S_t)]$ will be left as insurance for the future. In this sense, we choose a in $[0,1]$.

The second parameter, when positive, may be seen as a fixed volume delivered whatever the stock is, expecting that rain in the coming year will enable its release (we recall that, of course, rain is independent of the water). b may be negative, as the managers may decide to use inflow to increase the inter-annual stock.

In the literature on reservoirs and LDRs, the issue is to choose the parameters that either minimize dam capacity (Revelle et al, 1969) or optimize other reservoir performances. These objective values are generally associated with other constraints:

- The release must be lower than a given value (to prevent channel erosion), but higher than basic agricultural need.
- The storage must keep a sufficient safety margin to enable flood control, but cannot be lower than a level needed for recreational, environmental or aesthetic purposes.

The model does not incorporate such constraints, essentially as we are considering a reservoir solely devoted to irrigation.

Simulation

Combining equations 1–4, we form two dynamic equations for reservoirs' combined evolution:

$$S_{t+1} = f_S\left(S_t, X_t\right) \tag{14}$$

and

$$G_{t+1} = f_G\left(G_t, d_t\right) = f_G'\left(G_t, S_t, X_t\right) \tag{15}$$

As X_t are independent and identically distributed over time, the vector of variables (S_t, G_t) at time T only depends on the value of this vector at time $t - 1$; this may be characterized as a single-step Markov process. This is a basic property on which the probability theory applied to the reservoir was built (Moran, 1959). In the literature, until recently (Raheem and Khann, 2002), most of the papers focused on the design of the dam and the related probability of emptiness or overflow; release was generally seen as a constant. Our focus is on managing water releases from the dam, which is the duty of the dam manager. The dam manager's decision should be adapted to the actual level of water stored. Jarvis (1964) proposed a release rule depending on the level of storage, a precursor of the LDR.[4]

We chose several discrete values for the vector (a,b). For each of these we simulated the evolution of our system and drew out of this simulation statistics on released and pumped volumes over years (S_t, G_t). To rank the vectors we tested, we used indicators of water availability in the reservoirs and at farm level. The optimal vector related to an indicator was the one that led to the best results in this regard.

Within the range of real values for the depth of the aquifer, we could calculate the marginal cost of pumping, which is always lower than the marginal value of water. Farmers will withdraw the volume they need subject to the technical capacity of their pumping facilities: either nil when the water is deeper than the facility, or equal to the difference between surface allocation and the maximum crop requirements. On the basis of fieldwork (Petitguyot, 2003), we assume this maximum requirement to be 20,000m^3/ha/year.

In order to obtain results consistent with our case study, we include the fact that d_t is shared between the scheme and other areas. As this share is a constant percentage, it may be added to the model without any theoretical incidence. α = 75% (official rule in use in Tadla).

Results and Issues

In this study, we compare two different scenarios for the dam managers, depending on the information they use. In scenario 1, only surface water use is considered in the irrigation scheme. This means they should choose a release rule that optimizes indicators based on the available stock in the dam (i.e. the probabilities of overflow or emptiness) and on the release (average release and variability). These indicators are used as measures of system performance.

However, as explained above, groundwater stock, which includes the amount of water that can be pumped out, is closely related to surface supply. Moreover, it is well known that in the scheme, agricultural production heavily depends on both surface and groundwater for irrigation. Therefore it might be more suitable to optimize the decision rule according to this global irrigation volume and not only according to the release. This will define scenario 2.

Performances Only Related to Dam Release

To define scenario 1, we suppose that the water resource managers only consider indicators relating to surface water, ignoring the fact that a large part of the water resources dedicated to agriculture comes from groundwater. As a consequence, performance criteria and managers' objectives are only linked to the available supply in the dam and to its release.

The amount of stored water in the dam varies with the LDR's parameters as shown in Figure 12.4.

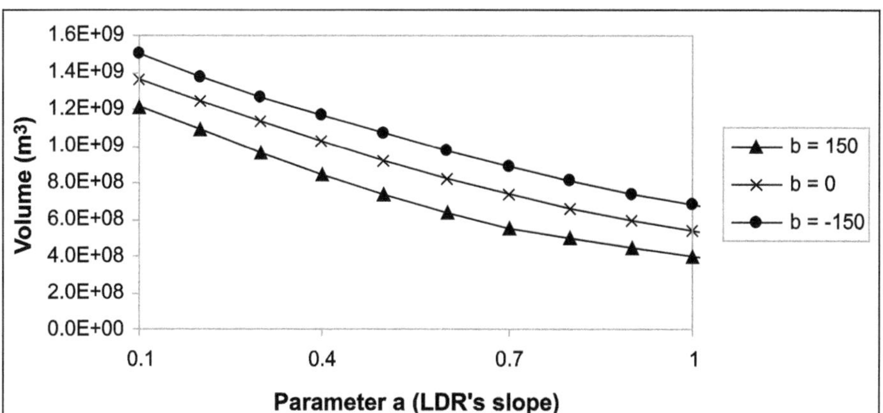

Source: chapter authors

Figure 12.4 *Average dam stock (S_t)*

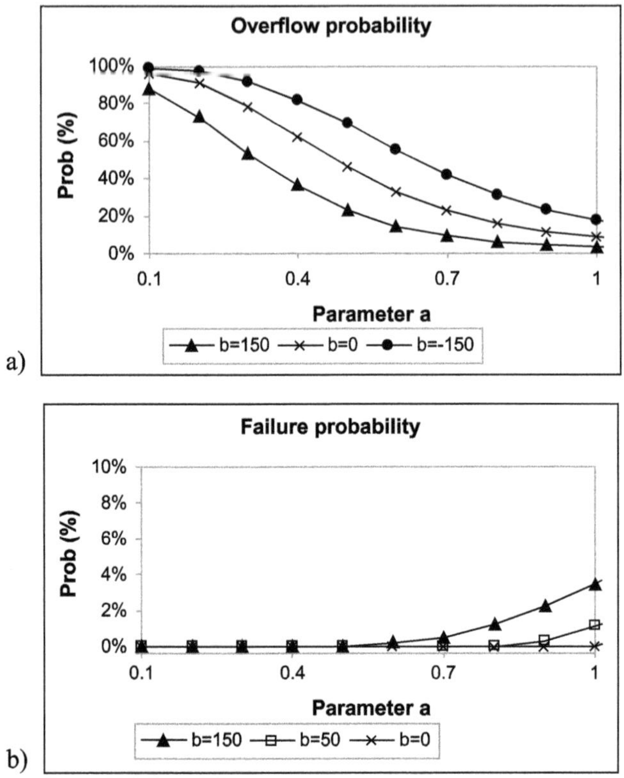

Source: chapter authors

Figure 12.5 *Probability of overflow (a) and failure (b) of the dam*

The average stock decreases when either *a* or *b* increases. Closely related to this first result is the probability of overflow and failure of the water service, i.e. when the commitment based on the release rule cannot be fulfilled because the water level in the dam is too low.

When *a* or *b* increases, the average stock decreases, as the dam overflows during some periods.[5] Obviously failure probability is nil when $b < 0$. A positive *b* might be seen as a guarantee to the irrigators: when the stock is low, it is a fixed volume that could be released, given the expectation of forthcoming inflows. In dry years this stock-independent volume may exceed inflows, in which case the difference must be taken out of the undelivered stock (i.e. $(1 - a).S$). The dam managers then face the risk of being unable to deliver the release they committed to. This risk increases with the guarantee *b* and with the LDR slope *a*.[6]

As a result of overflow, the average release is influenced by the choice of LDR (Figure 12.6).

When *a* and *b* increase, the probability of overflow decreases and the average release increases. However, this release has an upper bound, linked to the average inflow: whatever the decision rule is, dam outflow may never exceed its inflow, on average and for a long period.

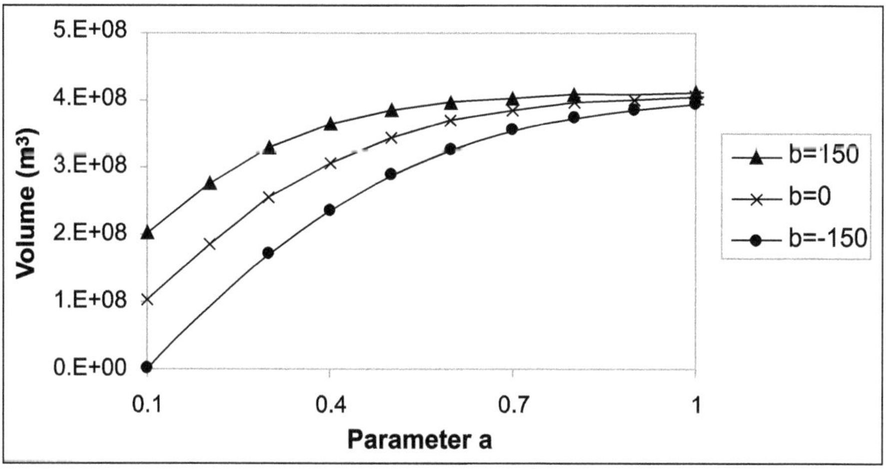

Source: chapter authors

Figure 12.6 *Average release (d_t)*

According to these first results, an optimization might be carried out taking into account average release and probability of failure. For example, if failure is never accepted, maximizing the release leads to (a = 1, b = 0). In this case, all the stock is released, and the release in year T is equal to inflow in year T – 1. The dam is used only for intra-annual flow regulation and will not absorb inter-annual variations. However, this inter-annual buffer property is highly relevant in arid countries with very variable inflows, and should be taken into account. In our model, the buffer is higher when the average stock is high, because annual inflow has a smaller influence on annual stock and annual release.

In order to analyze this buffer role, we considered the inter-annual variability of the release. As already explained, this depends on average stock and decreases when (a, b) increase. Another more operational criterion linked to the dam's buffer role is the potential volume that can be released with a given probability α (Figure 12.7).

As Figure 12.7 shows, (a,b) can be chosen to maximize this guaranteed level. However, this performance criteria leads to a different choice of LDR than when considering only the average release from the dam. In our case, the second indicator leads to a pair of equations (a = 0.45, b = 150) that represent a trade-off between average release (smaller than with [a = 1, b = 0]) and variability (also smaller is the second case). For many reasons, the dam managers may prefer to increase the steadiness of the release, even if this requires a (limited) decrease in the average release.

Performance Relating to Both Dam Release and Pumped Groundwater

In a second scenario, we took into account the dam managers having information about water availability and the piezometric level of groundwater and using this to increase their performance.

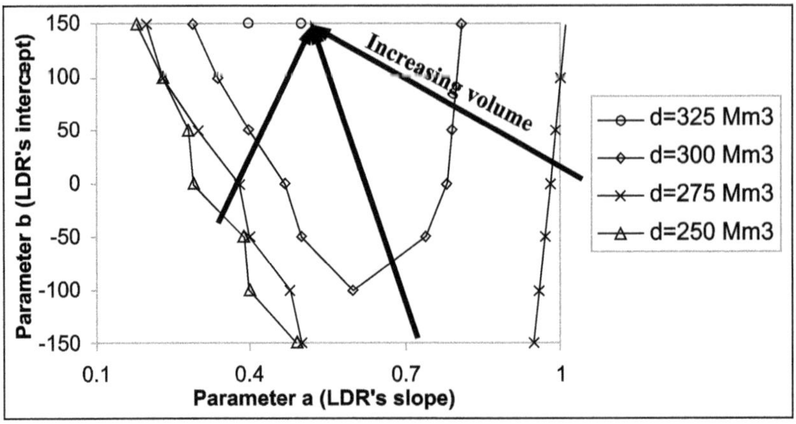

Source: chapter authors

Figure 12.7 *Guaranteed release (d_t) with 80% probability*

In our model, the average aquifer level was always found to be far from its physical limit, and its fluctuations remained small compared to the average; probabilities of failure and overflow are therefore always close to 0. Average irrigation volume follows the same evolution[7] as average release, increasing with *a* and *b* (Figure 12.8).

Thus considering the average irrigation volume instead of just one release would not change their rule as maximum is still reached for maximum (*a,b*). However, sensibility of volume variability (Figure 12.9) and α-guaranteed volume (Figure 12.10) changes even more when we take the whole water resource into consideration.

Above a given range for *a*, volume variability is relatively insensitive to the choice of parameters. Moreover, the guaranteed volume increases when (*a,b*) increases, ie when average volumes increase.

This may be explained by the 'second buffer' role played by groundwater (this effect was first analysed by Tsur and Graham-Tomasi, 1991): when the release is

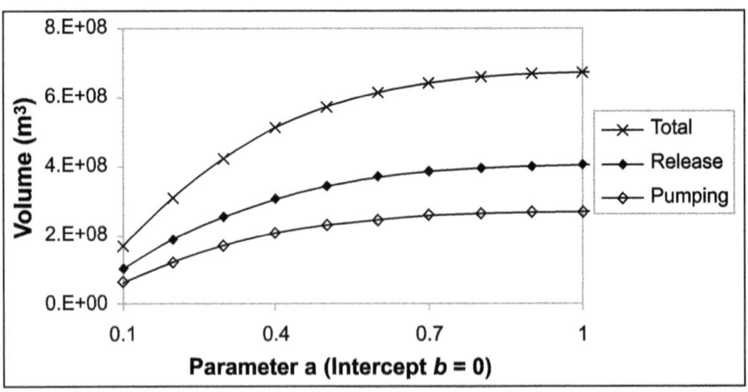

Source: chapter authors

Figure 12.8 *Average volumes from the 2 reservoirs and their sum (d_t, g_t, and $s_t + g_t$)*

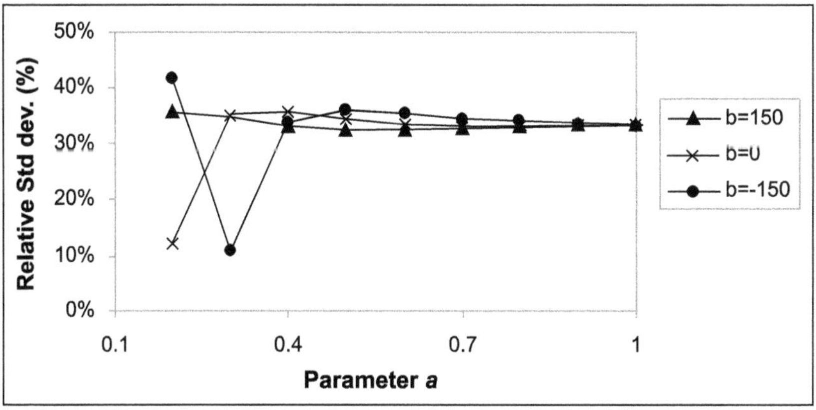

Source: chapter authors

Figure 12.9 *Relative standard deviation of irrigation volume* ($s_t + g_t$)

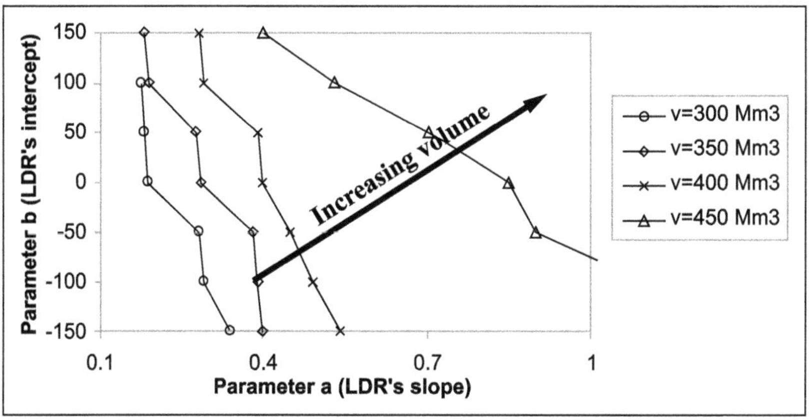

Source: chapter authors

Figure 12.10 *Guaranteed volume* ($s_t + g_t$) *with 80% probability*

low, farmers tap the water table more and its level decreases; on the other hand, groundwater stock replenishes when farmers pump less. Thanks to this buffer, the increase in release fluctuations (when *a* increases) is not transmitted to total irrigation volume, which remains relatively stable. Nevertheless, in the present case study this complementation remains uncompleted and the irrigation volume remains very variable. This may be explained by the fact that, as we will see (Table 12.1), not all farmers have access to the ground buffer, and the water table is often too deep for some pumping facilities to reach it.

This result is very attractive for managers, as they have the option to give smaller weight to inter-annual performance of the dam and protect themselves from the related risk with groundwater. Thus they may increase the variability of dam release, decrease overflow probability and reach a greater average release.

Table 12.1 *Percentage of equipped land units*

Type of pumping facility	*No well*	*Well*	*Tube well*
ORMVAT's survey (2002)	68%	30%	2%
Our model	50%	40%	10%

Source: chapter authors

Equity Concerns

Although our model was not originally designed to tackle this issue, it provides an interesting perspective on water resource distribution among farmers. As already explained, surface water is delivered according to irrigated area without any discrepancy among farmers, whereas groundwater is accessed according to the pumping facilities that farmers have been able to buy. In our case study, about half of the irrigated area has access to groundwater (Table 12.1). The figures used in the model are slightly different from those of the scheme manager, Office Régional de Mise en Valeur Agricole du Tadla (ORMVAT), as the 2002 survey only accounted for facilities declared by farmers themselves and are thus underestimated.

This inequality in pumping facility creates huge differences in terms of available irrigation water (Figure 12.11), as farmers with deep wells may use more than three times the volume others get when they only rely on the scheme's delivery.

As shown in Figure 12.11, these types of access to water do not have the same sensibility to LDR parameters. Above a certain range for *a*, farmers with deep wells are relatively insensitive to the choice of LDR: their access to groundwater enables them to fulfil their crop requirements whatever the release. For other farmers, however, irrigation volume increase with *a*, and increases quicker when they have access to groundwater. This difference is even greater when we consider volume variability (Figure 12.12).

Farmers without pumping facilities are far more affected by release variability, as they cannot rely on the groundwater insurance; farmers with deep wells may

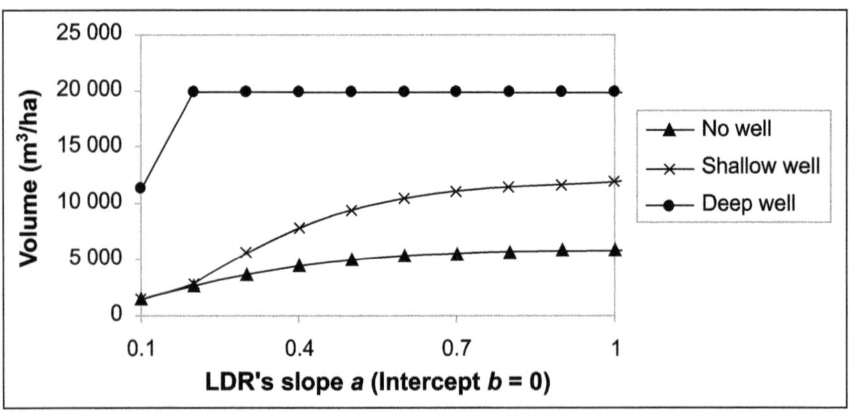

Source: chapter authors

Figure 12.11 *Average irrigation volume (s_t + g_t) according to pumping facility*

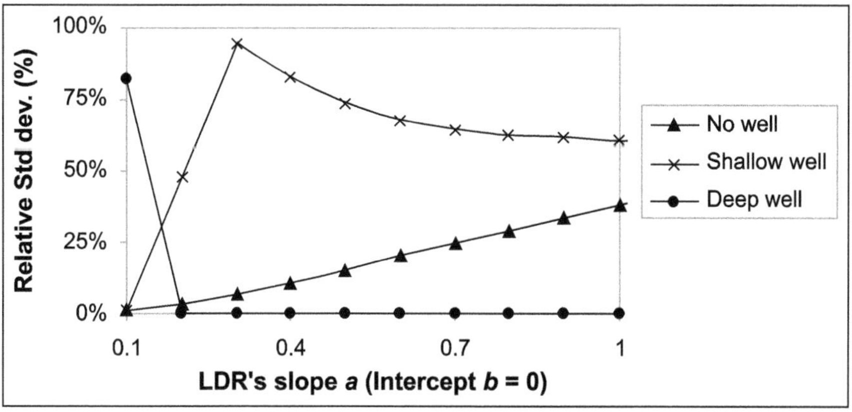

Source: chapter authors

Figure 12.12 *Relative standard deviation of irrigation volume ($s_t + g_t$) according to pumping type*

not be affected by surface water variability at all and are indifferent to the release rule, as they are always able to complement water supplies with groundwater. These farmers face no risk.

The intermediary case represented by shallow wells is interesting: when *a* increases, the release variability increases. For small values of *a*, the pumped volume is very low, and variability of irrigation volume is mostly influenced by release. For higher values of *a*, the share of pumped volume increases, balancing the release variability and reducing total volume variability.

As a result, when managers choose an optimal rule considering total volume performance, they might also consider the distributive effects of their policy. Transferring the buffer role to groundwater would enable them to improve the dam's performances. However, groundwater acts as an insurance only for farmers with an adequate pumping facility. Those farmers (who may be quite numerous) without any access to groundwater have a much lower income and are very sensitive to variations in surface water release. Choosing a rule based on global irrigation volume or single release is equivalent to supporting farmers who can or cannot afford to use a pumping facility. Then the regulators, willing to provide the most favourable conditions for fragile farms have very few incentives to integrate groundwater use into the dam management.

Conclusion

In the context of irrigation schemes where water supply depends both on surface water and groundwater, and where a large number of users both have individual access to water through wells and are benefiting from a collective water service, it has been demonstrated that the global efficiency of the system can be improved when the water service provider takes into account information about the groundwater availability and not only information relating to the surface irrigation system. This result might seem obvious, but it should be noted that this comprehensive approach to water resources is far from being implemented in a lot of countries and situations.

More precisely, we have shown that when managers consider the global use of water from different sources, they may deliver a more variable release from the dam, as this variability is mostly reduced by groundwater, which acts as a second buffer in the system. Thus they may choose a lower average level in their reservoir and reduce its overflow probability.

Nevertheless, some equality concerns would often thereby arise as the majority of farmers do not have any access to groundwater and would not be offered a high quality service from this new kind of water management. As this equality issue is likely to counterbalance the benefits from a more comprehensive approach in dam management, it would be of interest to address how another distribution rule for surface water allocation among farmers within the scheme may lead to a better compromise.

Finally, ORMVAT is the service provider and must balance its budget. As nearly all its income derives from water sales to farmers, with a proportional price that cannot vary year on year, release variability is automatically translated into income fluctuation. This variability is a serious concern for such managers (Loubier, 2003), and may explain why the water managers have so far preferred the most stable release option.

Notes

1 Agence de Bassin Hydraulique de l'Oum Er Rbia, (Oum Er Rbia River Basin Agency).
2 Office Régional de Mise en Valeur Agricole du Tadla (Regional Office for Agricultural Development of Tadla).
3 Discussing this distribution rule is outside our present focus but might be usefully carried out as a extension to this study.
4 This theory uses one basic property of Markov chains, 'ergodism' (Kottegoda and Rozzo, 1997): provided the process respects specified conditions, the distribution of its related variables converges to a stationary state, independent of the initial conditions. Ergodic conditions require non-periodicity and irreducibility of the variables, what is usually realized. Rather than an inter-temporal optimization whose result is strongly influenced by initial conditions, we decided to optimize the LDR's parameters according to their effect on this stationary distribution.
5 Note that in our model, overflow is structurally overestimated, as we consider release to only take place after inflows.
6 As the undelivered storage $(1 - a).S$ decreases with a.
7 If the aquifer never spills, its outflow (pumping) equals inflow (seepage), and we may easily show that on average $g = \frac{\tau}{1-\tau}.d$ and $v = \frac{\tau}{1-\tau}.d$.

References

Chohin-Kuper, A., Rieu, T. and Montginoul, M. (2002) 'Les outils économiques pour la gestion de la demande en eau pour la Méditerranée', Forum: *Avancées de la Gestion de la Demande en Eau dans la Région Méditerranéenne*, Fiuggi, 3–5 October

Dinar, A. and Mody, J. (2002) 'Irrigation management policies: Allocation and pricing principles and implementation experiences', paper for Irrigation Water Policies: Micro and Macro Considerations, World Bank/AMAECO seminar, Agadir, Morocco, 15–17 June

Faÿsse, N. (2003) 'Allocating irrigation water: The impact of strategic interactions on the efficiency of rules', *European Review of Agricultural Economics*, vol 30, no 3, pp305–332

Jarvis, C. L. (1964) 'An application of Moran's theory of dams to the Ord River Project, Western Australia', *Journal of Hydrology*, vol 2, pp232–247

Johansson, R. C., Tsur, Y., Roe, T. L., Doukkali, R. and Dinar, A. (2002) 'Pricing irrigation water: A review of theory and practice', *Water Policy*, vol 4, pp173–199

Jourdain, D. (2003) 'Impact des politiques visant à réduire la consommation brute en eau des systèmes irrigués: Le cas des puits gérés par des collectifs de producteurs au Mexique', PhD thesis, Department of Economics, University of Montpellier, France

Knapp, K. C. and Olson, L. J. (1995) 'The economics of conjunctive groundwater management with stochastic surface supplies', *Journal of Environmental Economics and Management*, vol 28, no 3, pp340–356

Kottegoda, N. T. and Rozzo, R. (1997) *Statistics, Probability, and Reliability for Civil and Environmental Engineers*, McGraw-Hill, New York

Loubier, S. (2003) 'Gestion durable des équipements d'hydraulique agricole: Conséquences sur la tarification et les politiques publiques en hydraulique agricole', PhD thesis, Department of Economics, University of Montpellier, France

Moran, P. A. P. (1959) *Theory of Storage*, Methuen, London

Petitguyot, T. (2003) 'Agriculture irriguée et utilisations durables des ressources en eau souterraines et de surface: Une exploration micro-économique dans la Plaine du Tadla, Maroc', Master's thesis, Université Paris X, France

Petitguyot, T., Rieu, T., Chohin-Kuper, A. and Doukkali, R. (2004) 'Modernisation de l'agriculture irriguée et durabilité des ressources souterraines dans le périmètre du Tadla au Maroc', Wademed Seminar, Rabat, Morocco

Préfol, P. (1986) *Prodige de l'Irrigation au Maroc. Le Développement Exemplaire du Tadla, 1936–1985*, Nouvelles Editions Latines, Paris

Raheem, E. and Khan, S. H. (2002) 'Combining probability of emptiness and mean first overflow time of a dam to determine its capacity', *Journal of Spatial Hydrology*, vol 2, no 2

Revelle, C., Joeres, E. and Kirby, W. (1969) 'Linear decision rule in reservoir management and design: 1) Development of stochastic model', *Water Resources Research*, vol 5, no 4, pp767–777

Shah, T., Deb Roy, A., Qureshi, A. S. and Wang, J. (2003) 'Sustaining Asia's groundwater boom: An overview of issues and evidence', *Natural Resources Forum*, vol 27, no 2, pp130–141

Tsur, Y. and Graham-Tomasi, T. (1991) 'The buffer value of groundwater with stochastic surface water supplies', *Journal of Environmental Economics and Management*, vol 21, no 3, pp201–224

Xepapadeas, A. P. (1991) 'Environmental policy under imperfect information: incentives and moral hazard', *Journal of Environmental Economics and Management*, vol 21, pp113–126

Appendix: Parameter Values for Tadla Irrigation Scheme

Inflow to the dam $E(X_t) = 550\ Mm^3$
$\sigma(X_t) = 220\ Mm^3$

Dam Capacity = $1.5\ Gm^3$

Groundwater reservoir Capacity = $7\ Gm^3$
$\mu = 200$
$\nu = 2.82\ 10^{-8}$

Irrigation scheme Irrigated area = 70,000ha
Infiltration: $\tau = 40\%$
Crop requirements = 20,000 m^3/ha

Table 12A.1 *Farm characteristics*

Type	*1*	*2*	*3*	*4*
Land (ha)	14,000	14,000	28,000	7,000
Facility pumping depth (m)	0	30	80	175

Source: chapter authors

13

Impact of Institutional Changes within Small-scale Groundwater Irrigated Systems: A Case Study in Mexico

Damien Jourdain

Introduction

Exploitation of aquifers has traditionally been regarded as a typical example of management of a common property resource where collective ownership induces inefficient use due to strategic interactions between users (Provencher and Burt, 1994; Roemer, 1989). Indeed, an increasing number of regions have unsustainable groundwater balance due to overdraft (Shah et al, 2000). This chapter takes groundwater extraction rights owned and administered collectively by groups of farmers[1] and considers the potential impact of changes in groups' operational rules on the resource use.

In arid and semiarid regions of Mexico, farming systems based on irrigation have been developing since the 1960s. Much of the water used for irrigation is tapped from aquifers; however, groundwater reserves are now being rapidly depleted. This situation is especially critical in the state of Guanajuato where the water table is decreasing in most aquifers by 2m/year due to overdraft (CEAG, 2000). In this state, 60 per cent of irrigated production systems rely on groundwater extracted with electric pumps (CEAG, 1999). Seventy per cent of wells in the region are owned and managed by groups of farmers (Jourdain, 2004), the size of the groups varying between 5 and 30 participants, and the management of wells is decentralized (i.e. farmers' groups craft their own 'institutions', in the sense used in Ostrom, 1992, and Crawford and Ostrom, 1995). Water and cost allocation rules are decided by farmers' groups and the institutional settings vary between wells (Gillet and Ollivier, 2002).

In this context, two different types of externalities should be considered: externalities between wells sharing an aquifer and externalities between farmers sharing an extraction right for a given well. For the first type of externalities, Gisser

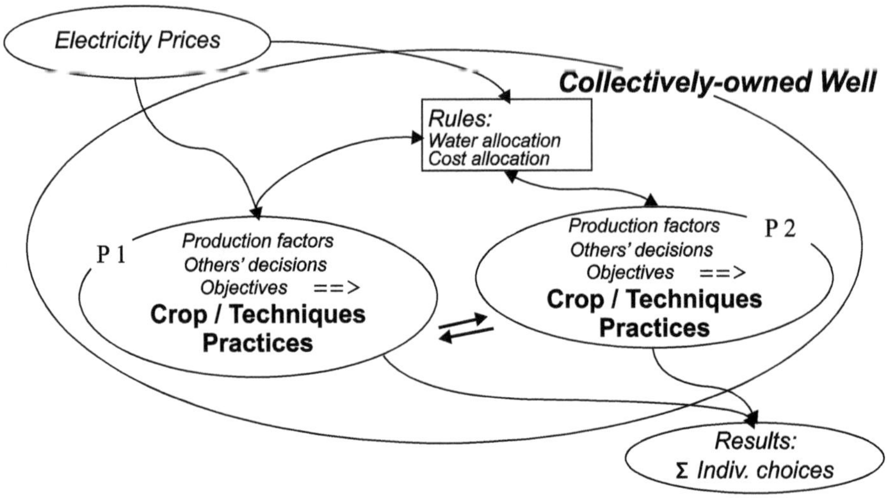

Source: chapter author

Figure 13.1 *Schematic view of interactions between farmers sharing a well*

and Sánchez (1980) and, more recently, Rubio and Casino (2003), using realistic numeric examples, have shown that the differences between open access, common property and optimal control equilibriums are small when the storage capacity of the aquifer is relatively large and can be ignored for practical considerations. Therefore, we will only consider strategic interactions between farmers using a common extraction right and ignore interactions between wells (Figure 13.1).

Some rules, particularly those relating to energy costs allocation, may induce strategic behaviour and hence inefficiencies in the use of an extraction water right. Here we quantify the impact of strategic behaviours on water consumption with collective wells. For a particular set of operational rules, we compare cooperative and non-cooperative equilibriums. A large gap between the two will mean that this institutional environment does not favour individual water-savings and that institutional change could bring behavioural changes.

Model

The nature of strategic interactions between producers depends on groups' operational rules. We focus on rules relating to the allocation of extraction costs. In the study area, electricity is the only energy source used to extract groundwater. In most cases, extraction costs are allocated among members according to irrigated surface. We suppose this rule is chosen because it is easy to implement and enforcement costs are very low. However, with higher extraction costs, eg when water tables are lower, a new allocation rule emerges whereby extraction costs are allocated according to an indicator of consumption.

Modelling Choices

Mathematical programming offers an adequate environment for modelling production decisions in agriculture (Boussard and Daudin, 1988; Hazell and Norton, 1986), because it allows (a) integrating existing or under development production functions, (b) clearly illustrating farmers' constraints and (c) taking into account interactions between the various productive activities of the production system. It is generally used to represent individual choices.

In the context of collective ownership and non-cooperative behaviour, each farmer will individually maximize his own utility function arising from water diversions given his constraints and the actions of other farmers. Farmers are linked together through the extraction cost resulting from the interaction of all of them, and the final solution is a Nash equilibrium. Cost-allocation rules also affect behaviours (Figure 13.2). The equilibrium is the result of simultaneous optimization problems under individual and global constraints. The decision variables are crop and technical choices resulting in water consumption.

Simultaneous and inter-related optimization problems can be formulated as a mixed complementarity problem (Ferris and Pang, 1997; Rutherford, 2002). In this type of formulation, we substitute the first-order optimality conditions (or Karush–Kuhn–Tucker (KKT) conditions) to the typical problem of optimization (Figure 13.3).

In Figure 13.3, L represents the Lagrangian of the optimization problem and the coefficients μ represent the dual variables associated to the set of G constraints. First-order conditions are written as three equations sets. The first of these cancels the first derivative of the Lagrangian with respect to decision variables. The second set corresponds to the constraints themselves (in the form of inequalities). The third set corresponds to the associated complementary slackness constraints. In the following we present this formulation in greater detail.

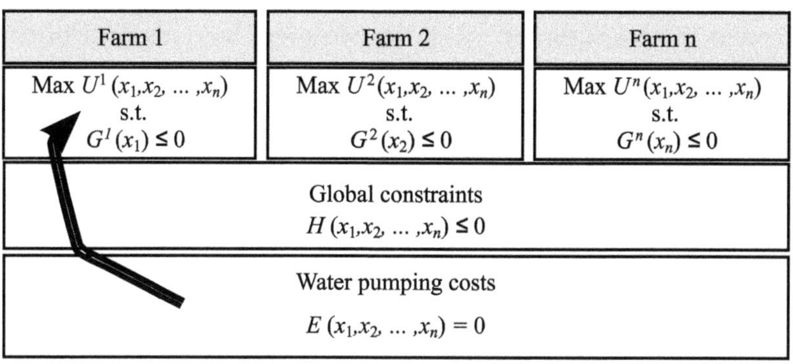

Source: chapter author

Figure 13.2 *Non-cooperative equilibrium model*

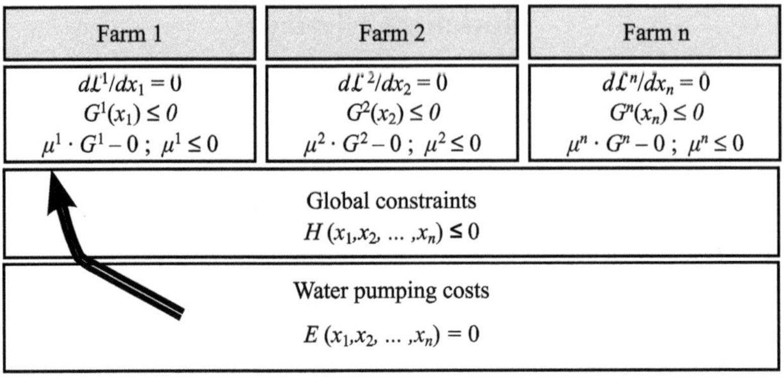

Source: Inspired by Ventosa et al (2000b)

Figure 13.3 *Non-cooperative equilibrium model (formulated as a mixed complementarity problem)*

Farms Model

First, we consider that the farmer is isolated from the group to which he belongs and consider other producers' consumptions as constant.

Constraints

Farmers are limited in their choices by resources and the availability of production factors. The limited resources used in our model are land, water, labour and money. Considering the importance of migrations in the area, we separate constraints linked to migration phenomena (i.e. labour shortages, financial constraints and risk management). Risk considerations are integrated in farmers' objective functions.

Land constraints

A farmer f has a given quantity of irrigated land, $area_f$. Thus land constraints for each farm can be defined by

$$Land_f : \sum_{ai} X_{f,ai} \leq area_f$$

where $X_{f,ai}$ is the area of crop ai grown by farmer f. Equations are given names – here '$Land_f$' – since their dual values will be expressly used in the mixed complementarity problem formulation.

Water constraints

There are physical limits to water extraction. For each period t, farmers decide collectively on an irrigation scheduling acceptable for the whole group. Thus the water constraint for the group is given by

$$WatAv_t : \sum_{f, ai} X_{f, ai} \cdot wat_{f, ai, t} \leq dis \cdot leng_t$$

where $wat_{f, ai, t}$, is water applied and dis is the pump discharge (supposed constant), and $leng_t$ is the period length.

Constraints related to migration issues

Migration is of particular importance in the studied region since Guanajuato is Mexico's first 'migrant exporting state' (Consejo Nacional de Población, 2000). Very often, several household members migrate to the US, while a limited number of family members remain to manage agricultural production. This influences labour balance, money flows and, as a result, farmers' risk considerations.

Labour constraints

For each period t, total workload cannot exceed the household working capacity (including hired workers). Furrow irrigation work requires a great deal of specialized work to obtain good application efficiency (monitoring, furrows maintenance, etc). Many farmers carry out irrigation work alone, but renting the services of specialized day labourers is also possible.

For activities handled by family members, we distinguished two different labour requirement coefficients: $wirri_{ai, t}$ for the irrigation work and $work_{ai, t}$ for others. Similarly, two labour types are defined: ($hlfirr_{f, t}$ the family labour able to complete irrigation works; $hlf_{f, t}$ for the others). Labour able to complete irrigation work can also complete other works but the reverse is not true. Therefore we used a transfer variable $TRANS_{f, t}$ between those two resources.

First, land is distributed between the one irrigated by family members and the one irrigated by outside labour (equation $REG1_{f, ai}$):

$$XF_{f, ai} + XR_{f, ai} = X_{f, ai}$$

where $XF_{f, ai}$ is the area irrigated by family members, and $XR_{f, ai}$ is the area irrigated by outside labour.

For the irrigation work completed by family labour, we have a first labour constraint, at each period (equation $REG2_{f, t}$):

$$\sum_{ai} \left(XF_{f, ai} \cdot wirri_{ai, t} \right) + TRANS_{f, t} \leq hlfirr_{f, t} .$$

For the irrigation work realized with outside labour, we will have (equation $REG3_{f,t}$):

$$\sum_{ai} \left(XR_{f,ai} \cdot wregad_{ai,t} \right) \leq REGAD_{f,t}$$

where $REGAD_{f,t}$ is outside labour employed to irrigate land and $wregad_{ai,t}$ outside labour needed to complete irrigation work.

We have assumed here that farmers are able to impose their technical choices on employees they contract, even if that is against the employee's interest. Thus, we suppose farmers spend some time ($TSURV_{f,t}$) to monitor their employees' activities (equation $MONIT_{f,t}$):

$$TSURV_{f,t} = survRate \cdot SURV_{f,t}$$

where *survRate* is expressing a time ratio to monitor employees.

For activities other than irrigation, we have a second labour constraint (equation $WORKC_{f,t}$):

$$\sum_{f,ai} X_{f,ai} \cdot work_{ai,t} - TRANS_{f,t} + WOFF_{f,t} - DAYL_{f,t} \leq hlf_{f,t}$$

where $WOFF_{f,t}$ are family members off-farm activities, and $DAYL_{f,t}$ is contracted labour force for non-irrigation works.

Financial constraints

For each period, and for each state of nature, we will have (equation $TREAS_{f,t,n}$):

$$\begin{aligned} & FB_{f,t-1,n} - famCons_{f,t} - IWATCOST_{f,t} \\ & + LOAN_{f,t,n} - (1 + ctrans) \cdot sal \cdot DAYL_{f,t} + (1 - xtrans) \cdot sal \cdot WOFF_{f,t} \\ & - pregad \cdot REGAD_{f,t} + remit_{f,t} + \sum_{ai} \left(X_{f,ai} \cdot \left(sales_{ai,t,n} - costs_{ai,t} \right) \right) = FB_{f,t,n} \end{aligned}$$

where $FB_{f,t,n}$ is the cash balance, $LOAN_{f,t,n}$ are seasonal loans contracted, *sal* is average salary for non-specialized work, *pregad* is the cost of contracted labour for irrigation work, $remit_{f,t}$ are money transfers from migrants, $sales_{ai,t,n}$ are proceeds from the sale of crop products, $costs_{ai,t}$ are production costs (other than water and labour), and *ctrans* are transaction costs on the labour market. $IWATCOST_{f,t}$ is defined below.

Individual costs depend on the allocation of the total cost of water extraction between members. With a uniform electricity price, for each period *t*, total energy costs are (equation $CWAT_{f,t}$):

$$TOTALCOST_t = p_e \cdot \sum_{ai} X_{ai,f} \cdot wat_{ai,t}$$

Total cost is then allocated between members. When allocation is done according to irrigated area (equation $SIR_{f,t}$):

$$IWATCOST_{f,t} = \frac{TOTALCOST_t}{\sum_{ai} X_{f,ai} \cdot paying_{ai,t}}$$

where $paying_{ai,t}$ is a binary parameter equal to one when activity *ai* requires water during period *t* and to zero otherwise.

Objective Function and Risk

Farmers are assumed to be risk-averse. In this chapter, we adopt a modified version of the MOTAD model (Hazell, 1971), as proposed by Donaldson et al (1995):

$$E(U(w)) = E(w) - \rho \cdot \sum_n (NEGDEV_n + POSDEV_n)$$

where $NEGDEV_n$ and $POSDEV_n$ are the absolute size of the deviation in income from its mean.

Average incomes are calculated according to the equation $RU_{f,n}$:

$$\begin{aligned}&\sum_{ai,t} X_{f,ai} \cdot (sales_{ai,t,n} - costs_{ai,t})\\&+\sum_t \big(WOFF_{f,t} \cdot sal \cdot (1 - ctrans) - REGAD_{f,t} \cdot pregad\\&- DAYL_{f,t} \cdot sal \cdot (1 + ctrans) - IWATCOST_{f,t}\big)\\&-\sum_t (LOANS_{f,t,n} \cdot intRate) = REV_{f,n}\end{aligned}$$

Then deviations from means are calculated (equation $DN_{f,n}$):

$$REV_{f,n} - \frac{1}{n} \cdot \sum_n REV_{f,n} + NEGDEV_{f,n} - POSDEV_{f,n} \geq 0$$

The objective function for a farmer becomes:

$$\frac{1}{n} \cdot \sum_n REV_{f,n} - \rho_f \cdot \sum_n (NEGDEV_{f,n} - POSDEV_{f,n})$$

Cooperative Equilibrium

This situation corresponds to a well where producers maximize a common objective function. To represent this, we aggregate the models from individual farmers described above. The objective function of the group is then the sum of individual utility functions:

$$Z = \sum_f \left(\frac{1}{n} \cdot \sum_n REV_{f,n} - \rho_f \sum_n \left(NEGDEV_{f,n} + POSDEV_{f,n} \right) \right)$$

Non-cooperative Equilibrium

To obtain simultaneous optima, the basic idea is to replace the individual maximization problems by their first-order conditions. The equilibrium is the solution of a system of equations where farmers' utility functions do not appear directly. The different farmers' optimization problems are systemized side by side while maintaining explicit the interactions between them (Figure 13.3).

As previously discussed, KKT conditions also include individual models constraints. The optimality conditions can be expressed as sets of equations. The first set cancels the gradient of the Lagrangian function with respect to the decision variables. The second set is formed by the complementary slackness conditions associated with the inequality constraints.

Here we present only the first derivative of the Lagrangian with respect to the decision variable $X_{f, ai}$. (The other equations, not presented here, follow the same logic.) The derivative of the Lagrangian with respect to the decision variable $X_{f, ai}$ gives us the first equation $CoX_{f, ai}$.

$$\begin{aligned} & lLAND_f + lR1_{f,ai} + \sum_t \left(p_e \cdot wat_{f,ai,t} \cdot lCWAT_t \right) \\ & + \sum_t \left(wat_{ai,t} \cdot lWatAv_t - lSIR_{f,t} \cdot paying_{ai,t} \right) \\ & + \sum_t \left(work_{ai,t} \cdot lWW_{f,t} - lSIR_{f,t} \cdot paying_{ai,t} \right) \\ & + \sum_{t,n} \left(lB_{f,t,n} \cdot \left(costs_{ai,t} - sales_{ai,t,n} \right) \right) \\ & + \sum_{t,n} \left(lRU_{f,t,n} \cdot \left(costs_{ai,t} - sales_{ai,t,n} \right) \right) = 0 \end{aligned}$$

We then find the usual result of equalization of the costs and marginal profits of the variables of surface. In this context, each resource, in particular water, is evaluated at its real cost. Thus in the cost of water the dual cost of the constraint of water scheduling ($lWatAv_t$) and the real cost of water ($p.lCWAT_t$) are integrated. The dual value $lCWAT_t$ is a value without unit which derives from the extraction costs allocation function.

Table 13.1 *General characteristics of simulated wells*

	W1	*W2*	*W3*	*W4*	*W5*
Well flow	34l/s	34l/s	28l/s	34l/s	40l/s
Aquifer depth	50m	50m	65m	50m	50m
Farm areas (hectares)	5/5/ 12/12	5/5/ 12/12	5/5/ 12/12	8.5/8.5/ 8.5/8.5	5/5/ 12/12
Allowed crops	Cereals Fodder crops Chick peas	Cereals Chick peas	Cereals Chick peas	Cereals Chick peas	Cereals Chick peas

Source: chapter author

Dual equations are then added to the constraints of the cooperative model. Lastly, the complementary slackness equations are clearly specified using the algorithm syntax.

The model was implemented in GAMS (Brooke et al, 1998). Two specific commercial solvers are available for large-scale complementarity problems, MILES (Rutherford, 1997) and PATH (Ferris and Munson, 2000, 2003); these are based on a generalization of the classic Newton method. Simulations presented here were carried out using the PATH solver.

The model represents the joint decisions of 4 producers irrigating 34 hectares of land. In order to represent the diversity of the situations met, we recorded the characteristics of representative wells (Table 13.1). Using the same generic model, these parameters allowed us to study the differentiated behaviour of the main types of wells and producers encountered in the region (Jourdain, 2004).

The reference situation corresponds to groups not having particular internal rules, i.e. no internal quotas and where extraction costs are allocated among members according to cultivated area. Electricity price as of January 2002, calculated in constant 1994 Pesos was used. We also made the assumption of a non-cooperative equilibrium. Allowed crops and techniques may vary depending on the wells studied. Risk aversion and certain prices were adjusted in order to approach the productive behaviours observed within the groups.

Results

Potential Gains of Cooperation

In terms of water consumption

Collective consumption

When considering groups' annual consumption, it is noted that the collective wells always consume less in the cooperative mode than in the non-cooperative mode (Figure 13.4). However, at current electricity prices, consumption differences between the two modes are relatively small (between 0 and 12 per cent of additional consumption annually).

These differences also vary with the groups' configurations. When water is scarce (W3), the possibilities of over-consumption being weak, the consumption differences between the two modes of operation are smaller. Similarly, for the wells

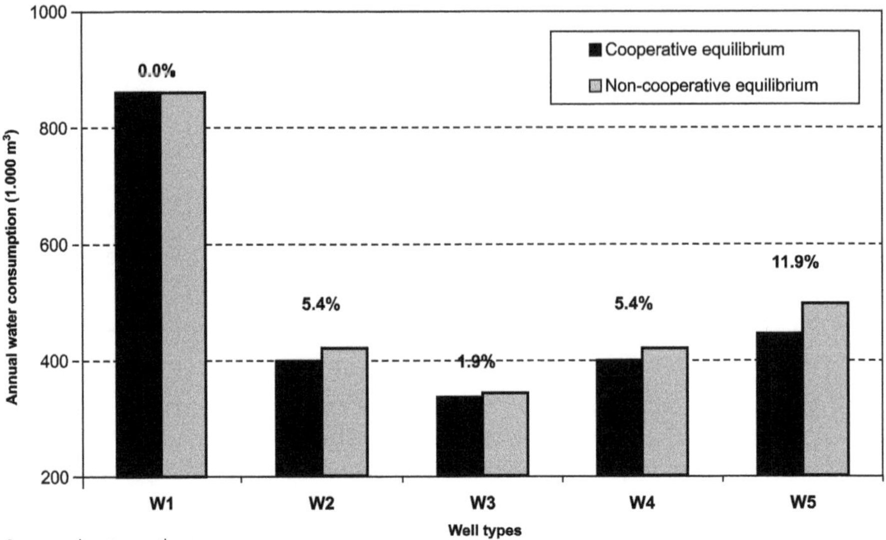

Source: chapter author

Figure 13.4 *Individual consumptions for cooperative and non-cooperative equilibriums (Group W2)*

with equally distributed land (W4), over-consumption is relatively less important. Conversely, when water is not a real constraint and land is distributed relatively unequally, the differences in consumption can reach 12 per cent between the two modes (Figure 13.5).

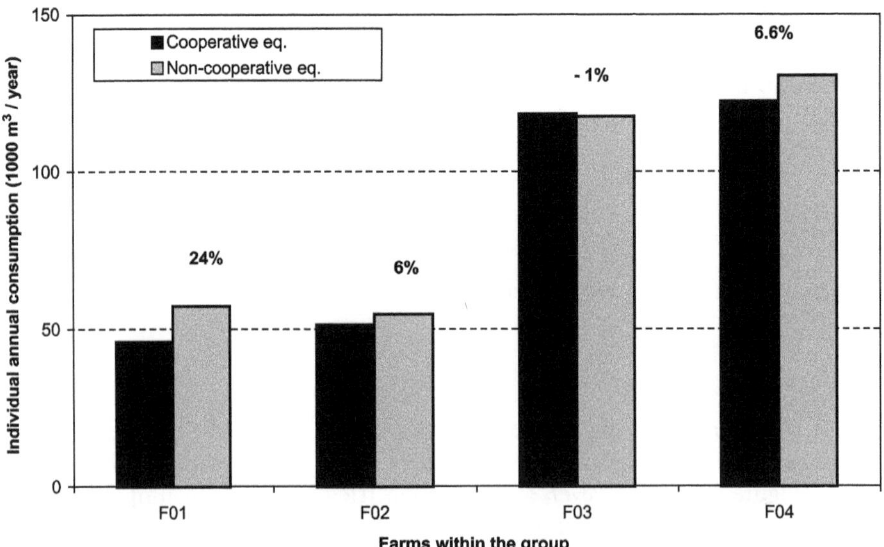

Source: chapter author

Figure 13.5 *Well water consumption for cooperative and non-cooperative equilibriums*

INDIVIDUAL CONSUMPTIONS

Individual consumption, in general though not systematically, is higher under a non-cooperative equilibrium (Figure 13.4). However, a relatively small difference between the two equilibriums for a group may hide great disparities between members. Figure 13.6 show the effects of cooperation for the different farms of group W2. While the overall difference in water consumption is only 5.4 per cent, farm F01 has a difference in consumption of 24 per cent. Hence cooperation has a greater impact on water allocation between members within a group than on aggregate water demand.

Moreover, for non-cooperative wells, farmers' shares of the well irrigated land influence their consumption levels. With a greater share of land, farmers tend to have similar water demand under cooperative and non-cooperative equilibriums. Indeed, the closer the farmer's share is to 100 per cent, the closer the energy cost (per unit of land) is to the one he would face if he was alone to manage the well. Conversely, the closer the share to 0 per cent, the cheaper the effective energy cost for the farmer.

In terms of income

GROUP INCOME

With regard to the agricultural income generated by the group, with the exception of the 1a group, non-cooperation between members induces income losses (Figure 13.6).

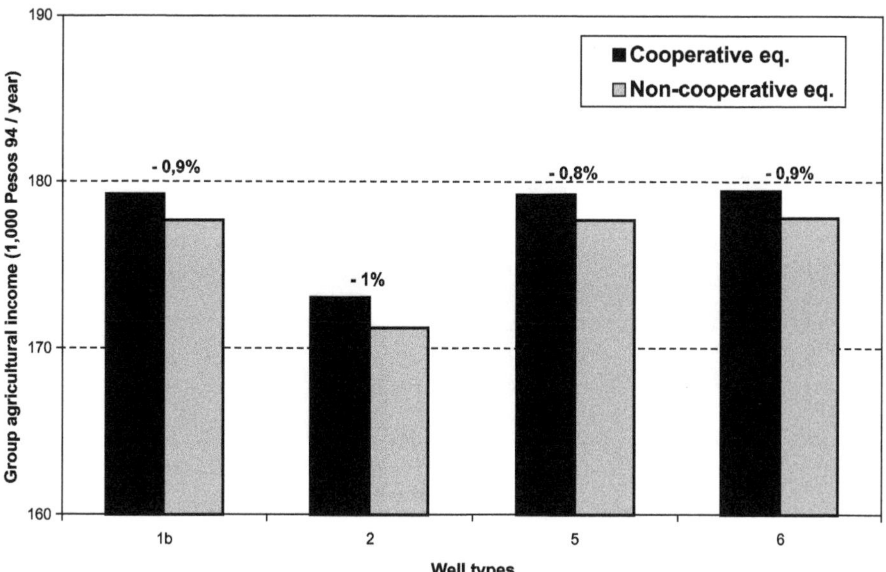

Source: chapter author

Figure 13.6 *Group agricultural income for cooperative and non-cooperative equilibriums*

As in the case of water consumption, income differences between the two modes are very low; non-cooperation results in an income loss of about 1 per cent and thus does not have a strong influence on group income.

Distribution of the cost of non-cooperation

Although the income impact is small for the group, it is interesting to look at its impact on individual incomes – there is a much stronger impact for at least some members of the group (Figure 13.7, indicating incomes per area unit).

It can also be seen that while some producers do gain through non-cooperative behaviour, others lose out.

However, at initial electricity prices, individual income differences are not very important.

Cooperative behaviour can generate some water savings. However, at current electricity prices, its impact remains limited. More important, cooperation between members has an impact on water allocation and individual revenues.

Changes of operational rules without electricity price readjustment are not likely to significantly modify consumption behaviours. Thus, institutional changes alone may not be of significant interest for policy-makers who are especially concerned with global water consumption. Nevertheless, at this stage it is imperative to check the possible interactions between institutional settings and electricity pricing.

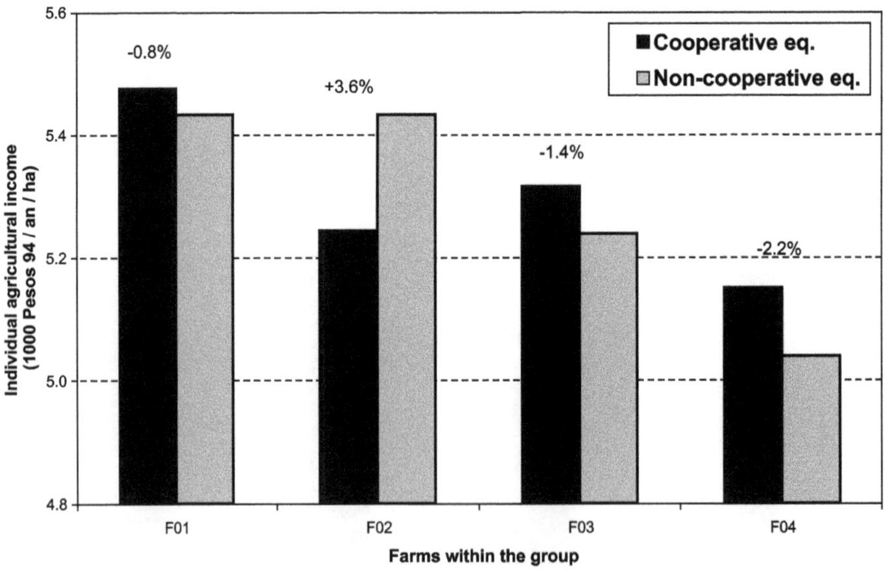

Source: chapter author

Figure 13.7 *Farm agricultural income for cooperative and non-cooperative equilibriums (Group W2)*

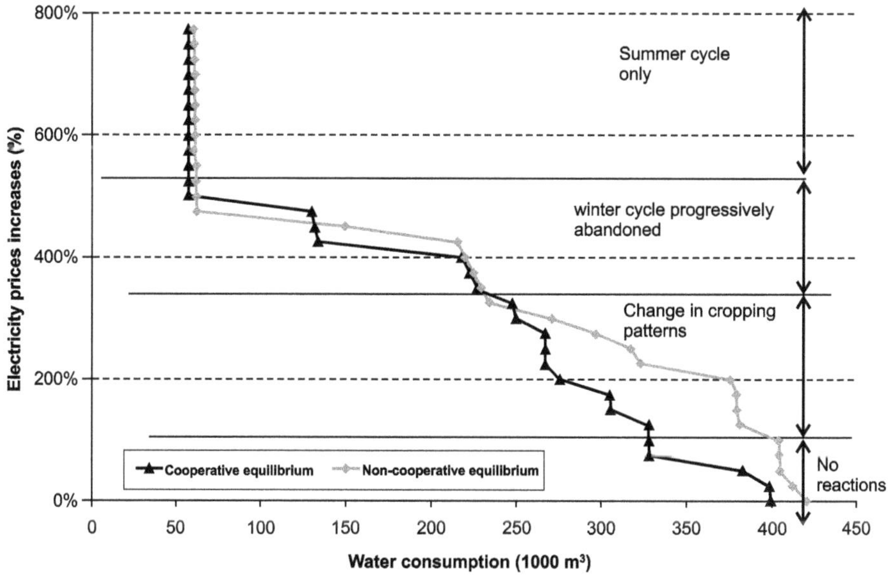

Source: chapter author

Figure 13.8 *Group W2 aggregate demand*

Interactions Between Institutions and Pricing

Successive electricity price changes from 0 to 1200 per cent of the initial price[2] were simulated to obtain water demand curves (Figure 13.8). These are compared for cooperative and non-cooperative equilibriums.

Four major phases

Water demand curves can be divided into four phases (Figure 13.8). The first corresponds to a consumption plateau: farmers are not reactive to price changes. The second phase corresponds to an adjustment in cropping patterns and techniques, allowing farmers to reduce their water consumption without reducing their irrigated area. The third phase corresponds to a rapid reduction of the irrigated area; farmers can no longer adjust their crops and practices and are forced to reduce the area they irrigate. The fourth phase, where the producers are no longer reactive to price changes, corresponds to the total abandonment of winter crops.

Strong interactions between the non-cooperation and electricity pricing

In a general way, non-cooperative behaviours have several effects on water demand. First, non-cooperation delays changes of crops and practices: farmers are more inelastic to electricity prices. Second, when price increases are more important, the producers abandon the irrigated crops more quickly. When not cooperating, producers must support higher individual energy costs, causing them to abandon the winter cycle more quickly. Third, non-cooperation also has effects on the distribution of the water between small- and large-scale farmers. For example, for the W2 group, the difference in total consumption between the two equilibriums is about 10 per cent on average, but this breaks down to 20 per cent for the small producers and 5 per cent for the large ones. It follows that any institutional work is likely to have more impact on the distribution of water between group members than on the group consumption.

Conclusion

This study focused on the impact of changes of extraction cost allocation rules on water demand for small-scale groundwater irrigated systems in the State of Guanajuato in central Mexico. These systems are owned and managed collectively by farmers' groups.

Two types of equilibrium were modelled and compared. The first is characterized by socially optimal behaviour: each producer struggles for a common objective without concerns about later income transfers. The second is characterized by non-cooperative behaviour. Modelling of non-cooperative groups rests on mixed complementarity problem (MCP) algorithms. This tool, already used in the literature on oligopolistic markets (Lise et al, 2003; Ventosa et al, 2000a) was adapted to our water management problem with water viewed as a regulated common property. We have shown this approach's usefulness: we were able to describe and analyse in a detailed way the strategic interactions between the producers beyond stylized analytical models since producers' strategies include the irrigated area, crop selection and irrigation practices.

Finally, from a methodological point of view, to complete this analysis it would be interesting to study and compare the same problem of allocation of costs from a cooperative game theory perspective (Tijs and Koster, 1998; Tijs and Driessen, 1986).

In terms of results, we first showed that the impact of changes in the cost allocation rules is highly dependant on well characteristics, eg extraction cost structure, and water constraints. In particular, in the Mexican environment, we showed that changing cost allocation rules within characteristic collective wells, at current electricity prices, may not bring about a significant reduction in water consumption. Cost allocation rule changes would have an impact on resource allocation between users rather than on the total consumption of the group. Therefore, in the current economic environment, the new cost allocation rules proposed to reduce water use inefficiencies may not have the beneficial effects that were a priori expected.

Conversely, an increase in the costs of extraction, via electricity pricing for example, has a strongly differentiated impact according to the cost allocation rules. When extraction costs are shared according to irrigated areas, wells are more inelastic to electricity prices: even a significant rise in extraction cost does not induce water saving behaviours or penalize farmers in terms of revenues. Our study presents a complementary explanation for the relative inelasticity of irrigated systems to that of water pricing studies for mainly large irrigation infrastructure (de Fraiture and Perry, 2002; Perry, 2001).

We would conclude that it is not advisable for decision-makers to choose either a strictly institutional approach, eg by implementing changes to cost-allocation rules within collective wells, or a strictly economic approach, eg via pricing, as to do so would *not* bring the expected results (i.e. reduction of water demand) and would be likely to have important side-effects (i.e. reduction of farmers' revenues). Instead, we have demonstrated the need for a combination of the two types of instruments to reduce the water consumption of these irrigated systems. For this reason, it will be more beneficial to devote further study to the impacts of combinations of instruments, rather than the impacts of isolated instruments (see, for example, the work of Bullock and Salhofer, 1998).

Notes

1 In the remainder of this chapter, the term 'collective well' is used for wells that are owned and managed collectively.

2 This price range may appear disproportionate, but it is similar to price ranges met in the literature

References

Boussard, J. M., Daudin, J. J. (1988) *La Programmation Linéaire dans les Modèles de Production.* Masson-INRA, Paris

Brooke, A., Kendrick, D., Meeraus, A. and Raman, R. (1998) *GAMS: A User's Guide*, GAMS Development Corporation, Washington, DC

Bullock, D. S. and Salhofer, K. (1998). 'Measuring the social costs of suboptimal combinations of policy instruments: A general framework and an example', *Agricultural Economics*, vol 18, pp249–259

CEAG (1999) *Plan Estatal Hidraulico de Guanajuato, 2000–2025, Fase I: Diagnostico de la Situacion Hidraulica del Estado de Guanajuato; Fase II: Estrategia Estatal en Materia de Agua*, Comisión Estatal del Agua de Guanajuato, Guanajuato, Mexico

CEAG (2000) *Estudios Hidrogeológicos y Modelos Matemáticos de los Acuíferos del Estado de Guanajuato*, CD-ROM, Comisión Estatal del Agua de Guanajuato, Guanajuato, Mexico

Consejo Nacional de Población (2000) *Migración México-Estados Unidos: Presente y Futuro*, Consejo Nacional de Población, Mexico

Crawford, S. E. S. and Ostrom, E. (1995) 'A grammar of institutions', *American Political Science*, vol 89, pp582–600

de Fraiture, C. and Perry, C. (2002) 'Why is irrigation water demand inelastic at low price ranges?' in proceedings, Irrigation Water Policies: Micro and Macro Considerations, International Conference, Agadir, Morocco, 15–17 June 2002

Donaldson, A. B., Flichman, G. and Webster, J. P. G. (1995) 'Integrating agronomic and economic models for policy analysis at the farm level: The impact of CAP reform in two European regions', *Agricultural systems*, vol 48, pp163–178

Ferris, M. C. and Pang, J. S. (1997) 'Engineering and economic applications of complementarity problems', *Siam Review*, vol 39, pp669–713

Ferris, M. C. and Munson, T. S. (2000) 'Complementarity problems in GAMS and the PATH solver', *Journal of Economic Dynamics and Control*, vol 24, pp 165–188

Ferris, M. C. and Munson, T. S. (2003) *PATH 4.6*, GAMS Development Corporation, Washington, DC

Gillet, V. and Ollivier, I. (2002) 'Evolution des règles de gestion de l'eau superficielle et souterraine et impact du transfert', MSc thesis (Gestion Sociale de l'Eau), CNEARC, Montpellier, France

Gisser, M. and Sánchez, D. A. (1980) 'Competition versus optimal control in groundwater pumping', *Water Resources Research*, vol 16, pp638–642

Hazell, P. B. R. (1971) 'A linear alternative to quadratic and semivariance programming for farm planning under uncertainty', *American Journal of Agricultural Economics*, vol 53, pp53–62

Hazell, P. B. R. and Norton, R. D. (1986) *Mathematical Programming for Economic Analysis in Agriculture*. Macmillan Publishing Company, New York

Jourdain, D. (2004) 'Impact des politiques visant à réduire la consommation brute en eau des systèmes irrigués: Le cas des puits gérés par des collectifs de producteurs au Mexique', doctoral thesis, faculty of social sciences, University of Montpellier, Montpellier, France

Lise, W., Kemfert, C. and Tol, R. S. J. (2003) 'Strategic action in the liberalised German electricity market', European Association of Environmental and Resource Economists 12th Annual Conference, Bilbao, Spain, 28–30 June 2003

Ostrom, E. (1992) *Crafting Institutions of Self-governing Irrigation Systems*. Institute for Contemporary Studies, San Francisco, US

Perry, C. (2001) *Charging for Irrigation Water: The Issues and Options, with a Case Study from Iran*, research report no 52, International Water Management Institute (IWMI), Colombo, Sri Lanka

Provencher, B. and Burt, O. (1994) 'A private property rights regime for the commons: The case for groundwater', *American Journal of Agricultural Economics*, vol 76, pp875–888

Roemer, J. E. (1989) 'A public ownership resolution of the tragedy of the commons', *Social Philosophy and Policy*, vol 6, pp74–92

Rubio, S. J. and Casino, B. (2003) 'Strategic behavior and efficiency in the common property extraction of groundwater', *Environmental and Resource Economics*, vol 26, pp73–87

Rutherford, T. F. (1997) *MILES: A Mixed Inequality and Non-Linear Equation Solver*, Economics Department, University of Colorado, Boulder, CO

Rutherford, T. F. (2002) 'Mixed complementarity programming: Applications in economics', Applied Math Colloquium, Economics Department, University of Colorado, Boulder, CO

Shah, T., Molden, D., Sakthivadivel, R. and Seckler, D. (2000) *The Global Groundwater Situation: Overview of Opportunities and Challenges*, IWMI, Colombo, Sri Lanka

Tijs, S. and Koster, M. (1998) 'General aggregation of demand and cost sharing methods', *Annals of Operations Research*, vol 84, pp137–164

Tijs, S. H. and Driessen, T. S. H. (1986) 'Game theory and cost allocation problems', *Management Science*, vol 32, pp1015–1028

Ventosa, M., García-Alcalde, A., Mencía, A., Rivier, M. and Ramos, A. (2000a) 'Modeling inflow uncertainty in electricity markets: A stochastic MCP approach', 6th PMAPS Conference, Madeira

Ventosa, M., Rivier, M., Ramos, A. and García-Alcalde, A. (2000b) 'An MCP approach for hydrothermal coordination in deregulated power markets', IEEE PES Summer Meeting, Institute of Electrical and Electronics Engineers–Power Engineering Society Summer Meeting, Seattle, WA, 16–20 July, pp2272–2277

14

Local Empowerment in Smallholder Irrigation Schemes: A Methodology for Participatory Diagnosis and Prospective Analysis

Sylvain Perret

Introduction

Over the past three decades, the world's irrigation sector has been increasingly exposed to decentralization and privatization. Many countries have embarked on a process to transfer the management of smallholder irrigation systems from government agencies to local management entities (Vermillion, 1997). This process of irrigation management transfer (IMT) includes state withdrawal, promotion of water users' participation, development of local management institutions, and transfer of ownership and management. South Africa has only recently initiated IMT in government smallholder irrigation schemes located in former homeland areas and most transfer operators are still unsure about how to design and implement the process.

At present, South Africa has an estimated 1.3 million ha of land under irrigation. Owing to history and past policies, different types of irrigation schemes have been developed (Perret, 2002a). Most smallholder irrigation schemes (SIS) were developed during the early Apartheid era. These cover approximately 47,000ha (Bembridge, 2000), and account for about 4 per cent of irrigated areas in SA. It is estimated that about 250,000 rural black people are dependent, at least partially, on such schemes for their livelihoods, and in spite of the relatively small contribution the schemes make to overall productivity, it is believed that they could play an important role in rural development, hence the rehabilitation and transfer policies. Furthermore, the National Water Act (NWA) of 1998 promotes the creation of Water Users' Associations (WUAs) (Perret, 2002a). It is envisaged that such local institutions will take over most irrigation management functions, i.e.

water allocation and distribution, maintenance, water charging systems, financial management, and so on. The situation, however, gives cause for concern as most SIS are currently moribund and have been inactive for many years (Bembridge, 2000). Several causes have been mentioned for this (IWMI, 2001): infrastructure deficiencies resulting from inappropriate design, management and maintenance; both beneficiaries and government-assigned extension officers lacking technical know-how and ability; absence of people involvement and participation; inadequate institutional structures; inappropriate land tenure arrangements; local political power games; a history of dependency and subsistence orientation; low land productivity; and high cash costs. Following the dismantlement of Apartheid, parastatal management agencies were liquidated and government gradually withdrew from its past functions in SIS (extension, marketing and financial support). With regard to rehabilitation and the IMT process, the above raises a series of questions at different levels: at the national and provincial government levels, rehabilitation policy and implementation and IMT procedure; at the WUA level, collective management of newly-transferred irrigation schemes and institutional arrangements; and at the farmers' level, farming and cropping systems management.

The objectives of the approach presented here is to contribute to investigating the sustainability of SIS in the context of IMT and to accompany and support decisions and actions undertaken by development operators by promoting collective solution seeking through scenario-testing. This chapter limits itself to a presentation of the approach and its principles, the model's conceptual framework and results from a case study.

Rationale, Material and Methods

Societal and Policy Rationale

In the Limpopo Province, most SIS display similar traits: a significant proportion of non-farming plot occupiers; a diversity of practices and performance among irrigation farmers (generally not very productive and subsistence-oriented); a simple conception of infrastructures (a gravity-fed system with dams, canals and furrows), these generally in a state of deterioration; a lack of support services; a weak agri-business environment; a lack of markets, water allocation and sharing; and water availability problems, especially in winter.

The Limpopo Province Department of Agriculture has embarked on a rehabilitation/revitalization programme, the first pilot phases of which started in 2001, with the final Revitalization of Smallholder Irrigation Schemes programme (RESIS) launched in 2004. It aims at addressing about 130 SIS over the next 5 years. As soon as the economic viability and overall sustainability of these schemes is established, the national Department of Water Affairs and Forestry (DWAF) will establish WUAs in the schemes.

The action-research approach that is described here has been developed and applied since 2001 in various case study schemes of Limpopo Province and the Eastern Cape Province of South Africa (Touchain and Perret, 2002; Perret, 2003;

Ntsonto, 2004; Perret et al, 2003). More particularly, it aims to support and inform the RESIS programme through case study analysis and the set up and transfer of adapted methodologies.

Principles, Theoretical Background

First, the approach acknowledges that there are costs involved in supplying water and water-related services to farmers, and that an objective of financial viability is pursued at the scheme level (involving partial or total cost recovery) (Perry, 2001; Le Gal et al, 2003). In an IMT context, this means:

- the management entity (WUA) provides irrigation water and related services to farmers;
- such services generate costs (capital, maintenance and operation costs, along with personnel-related costs); this forms the 'manager's cost function' (Le Gal et al, 2003);
- the management entity charges the farmers according to a system to be established; and
- farmers tap into their monetary resources (generated by irrigated or rainfed cropping systems, by off-farm income-earning systems) to pay these water service fees; this forms the 'farmers' income function' (le Gal et al, 2003).

Second, smallholders' agricultural and resource-management systems face a quickly changing economic, legal and social environment. For the necessary adaptations to occur, renewed approaches require facilitation of collective learning and negotiated agreement (Jiggins and Roling, 1997). Action-research strives to play this facilitation role. As defined by Liu (1994), it combines (a) the convergence of a will for change and a research intention, which entails a two-fold objective, i.e. problem solving and knowledge generation (with local and generic scope); (b) an ongoing long-term joint project between researchers, development operators and users; and (c) a common ethical framework negotiated and accepted by all stakeholders.

Argyris et al (1985) and Liu (1994) identify autonomization of beneficiaries, whereby local stakeholders and beneficiaries self-empower and acquire capacity, as a key objective of action-research. Oja and Smulyan (1989) highlight that action-research must trigger and organize a major and sustainable shift (i.e. long after the research is finished) in terms of the technical and organizational practices of communities. The difficult and essential point is to implement properly the participation of stakeholders, not only for data collection but also during recurrent, interactive workshops (Perret and Le Gal, 1999).

Third, SIS are not only constituted by individuals and assets, but also by knowledge, rules and information (Ostrom, 1992). Such information may be organized as and take different forms, such as databases, indicators, maps, worksheets, management boards, schedules and production forecasts. It may be used to monitor and assess the activities performed and to support decisions. These formalized representations are called management tools and form an information system (Moisdon, 1997). Owing to the increasing complexity and dynamics of organizations, and to the increasing uncertainty of their economic environment, management tools no longer seek optimal solutions and one-way prescriptions or recipes, but now favour information, learning processes, adaptability, discussion, collective awareness, and so forth (Ostrom, 1992). Developing information systems and management tools goes along with developing the organization itself and its strategy (Moisdon, 1997). From the information system, simulation tools may be developed to support and accompany the knowledge and exploration of reality. The objective then is not only to manage and monitor, but also to fuel discussion and make people participate and interact, challenge hasty judgements and support sound decisions, raise new questions, foresee issues and problems, and test solutions.

Developing the Model: Conceptual Framework

The approach as a whole is rooted in the above principles. Its implementation vehicle is a model, and a software, called Smile (Sustainable Management of Irrigated Land and Environments).The model's conceptual framework takes into consideration the economic and financial aspects of scheme's management and looks at certain technical indicators (eg water resource availability) in order to confirm that scenarios are realistic. Five input modules form the basis of the information system, as interfaces for data capture by the user (see Figure 14.1).

Each cost-generating item is listed in the 'cost' module. This module generates output variables that reckon the costs incurred by the scheme and its management (i.e. capital costs, maintenance costs, operation costs and personnel costs). Such information answers the question of how much it costs to operate the scheme in a sustainable manner, regardless of who is going to pay for it.

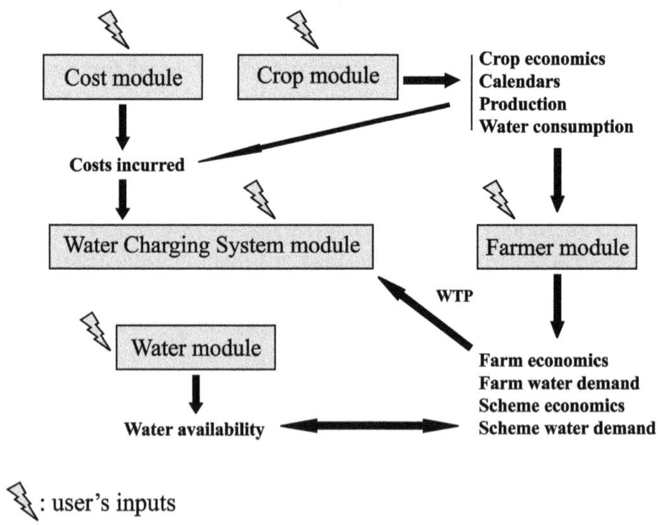

Note: the Smile software is available for free from the internet: www.smile-cirad.co.za
Source: chapter author

Figure 14.1 *Smile: structure*

In the 'crop' module, each irrigated crop is listed with its technical and economic features (eg management style, cropping calendar, water demand, yield and production costs). The module then generates micro-economic output variables (eg total revenue and gross margin) that allow comparative evaluation of crops in terms of costs, profitability, land productivity and water productivity.

The 'farmer' module captures the different farmer types and their cropping systems (combination of crops that have been documented in the 'crop' module), average farm size, percentage of scheme's size and willingness to pay (WTP) for irrigation water services. This module generates type-related output variables (eg aggregated income per type and crop calendar) and scheme-related output variables (eg number of farmers and aggregated water demand) when combined with the 'scheme' module.

The 'scheme' module lists the scheme's characteristics (eg size, rainfall and resource-availability patterns and tariff structure). This module is combined with the 'farmer' and 'cost' modules and generates output variables on water pricing, tariffs, cost recovery rate and contribution per type. This provides answers to questions of who should pay, and how much, for water services. It also generates social and equity-related indicators and resource-related indicators (eg total number of farmers, area per type, number of farmers per type, type income, scheme total gross margin, total water consumption and overall weekly water balance).

The initial inputs (real data) form the base scenario. Additional scenarios may be tested through the capture of hypothetical or prospective data, especially when the given scheme has not yet been rehabilitated or transferred (eg alternative crops and cropping systems, emerging farmer types, changes in the scheme's management patterns, options for a charging system, new infrastructure, etc).

The calculation principles of Smile have been discussed in previous publications (Touchain and Perret, 2002; Le Gal et al, 2003); here the focus is on the overall approach and the use of Smile as a vehicle fostering participation and capacity-building.

Participatory Diagnosis and Prospective Analysis: Implementation and Results

As shown in Figure 14.2, the approach implies three phases:

1. data collection at the household and scheme levels for a given scheme;
2. data processing and information-system development, which requires typologies of crops and farmers; and
3. running the model on a scenario-testing basis, evaluating the impact of certain measures or decisions, or certain farmers' strategies on agricultural and production features, land allocation, costs and cost recovery, and sustainability-related indicators. This supposes close interactions with experts and local stakeholders, who should validate findings and identify topics for further investigations and scenario-testing (Perret and Le Gal, 1999).

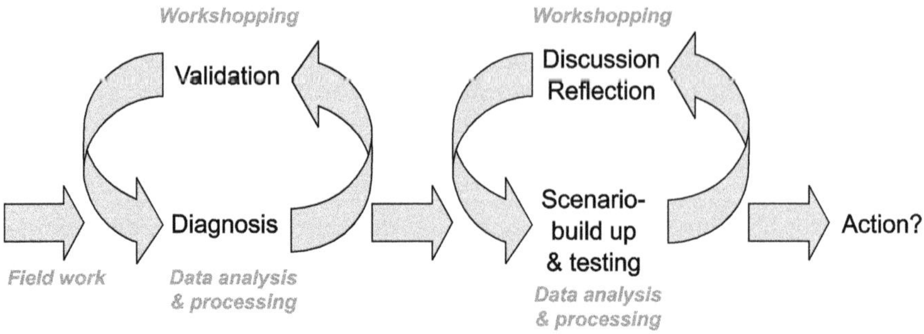

Source: chapter author

Figure 14.2 *The Smile approach: scheduling the action-research process*

Data Collection

The more accurate and reliable the data, the better the modelling and simulation development. The approach makes use of questionnaire-based, individual interviews of farmers (sampling proves necessary in large case study schemes), discussions with local experts, literature review and secondary data gathering. Engineers, agronomists, extension officers, economists, development operators, farmers and policy-makers are first involved on an individual basis, before experts and stakeholders are involved in an informal and flexible steering committee for the latter phases.

After data collection is completed, a database is created on the basis of information collected, in the form of spreadsheets. The basic cluster is the interviewed household, to which variables and information are attached. This includes information on households' characteristics, production features, resources and assets, inputs, costs and so forth. More qualitative information is also included such as issues and prospects from the perspective of the interviewees.

Other data might also be collected according to the area of focus. For example, contingent analysis has been carried out in certain schemes in order to evaluate farmers' WTP for water related services and then to investigate a possible water charging system (see Table 14.1).

Table 14.1 *Thabina farmers' WTP for water supply and related services under conditions of self-management of the scheme*

Not willing to pay	*Less than 100 rand/ha/yr*	*100–120 rand/ha/yr*	*120–240 rand/ha/yr*	*240–600 rand/ha/yr*	*More than 600 rand/ha/yr*	*Willingness to pay (average of all answers, in rand/ha/yr)*
5.0%	8.3%	3.3%	51.7%	21.7%	10.0%	235.00

Source: Perret et al (2003)

Data Analysis: Typologies

The first step of analysis consisted of a selection of the most representative and common crops grown in the scheme (each representing more than 1 per cent of the total area cropped by the sampled households). Each crop is described thoroughly, in the form of a crop management style (CMS), which includes crop yield, inputs, production costs and budget, water consumption (from bioclimatic standards – SAPWAT, using data from the closest climatic station, is used in South Africa; see www.sapwat.org.za), calendar, market price, type of product sold or consumed, and so forth (see Table 14.2 for an example). Each product is given a monetary value (even those self-consumed) according to the market prices for equivalent products.

The second step of the analysis consisted of the establishment of a farmer typology. Farmers were grouped into types, as one can neither address each farmer individually nor consider all farmers similar. Different farmers' strategies and practices coexist within a scheme, and grouping irrigation farmers into several types helps to represent this reality as shown by Lamacq (1997). The idea is to reflect the diversity of strategies and situations that exist within the scheme in an intelligible and manageable manner. The typology usually includes sources and level of income, number and type of crops grown and proportion of crop sold as key criteria (Perret, 1999) (see Figure 14.3 as an example).

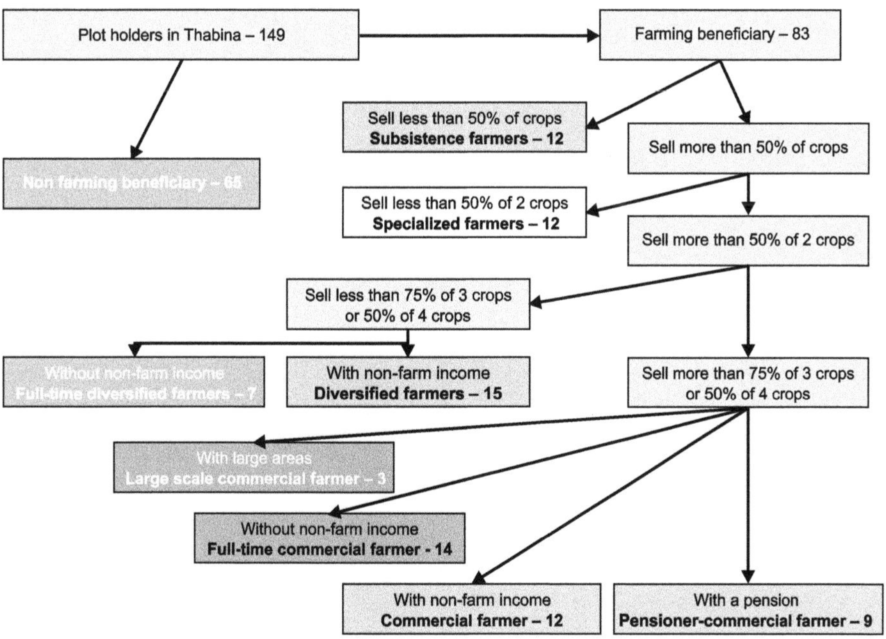

Source: Perret et al (2003)

Figure 14.3 *Example of a farmer typology: classification tree in the Thabina irrigation scheme, South Africa*

Table 14.2 *Examples of crop management styles in the Thabina irrigation scheme*

Crop management style	*Yield (unit)*	*Market price/unit (rand)*	*Production costs (rand)*	*Total revenue (rand)*	*Gross margin (rand)*	*Water consumption (m^3) ** *
Dry maize – Low yield *	6.6 bags	166	1470.00	1095.60	–374.40	5840
Dry maize – High yield *	18.20 bags	154	1640.00	2802.80	1162.80	5840
Green maize – Average *	1690 cobs	0.76	1170.00	1284.00	114.00	6520
Spinach – Extensive	560 bunches	3.80	585.00	2128.00	1543.00	3020
Spinach – Intensive	1010 bunches	4.00	1820.00	4040.00	2220.00	3020
Tomato – Extensive	130 boxes	22.90	1020.00	2977.00	1957.00	3910
Tomato – Intensive	510 boxes	23.50	3530.00	11985.00	8455.00	3910

Note: All figures are expressed per ha. All figures are averages obtained from different interviewees (except ** from SAPWAT). * Summer crops
Source: Perret et al (2003)

The categorization of types must also consider the different CMSs, as elementary clusters that are combined at farm level and contribute to determining farm profit, water consumption, etc. In practice, a given type incorporates all economic, hydraulic and agronomic features of the different crops it grows. Simplification and trade-offs are therefore part of typology construction, a necessity in the typological modelling of a complex reality.

A Diagnosis of the Current Situation as a First Outcome

The combination of typologies in Smile allows for calculations at both farm and scheme level. The simulation tool makes it possible to display results in a simple and comprehensible way for all stakeholders, through figures, graphs and tables. All this forms a base situation, a model that reflects the current state of affairs, captured in the Smile platform. The base situation is submitted to the management committee or farmers' association. It is especially important that the CMS and farmers' types are discussed and validated by the farmers. A first series of scenarios can also be presented to the farmers to familiarize them with the exercise.

Box 14.1 provides an example of such a base situation in the Thabina scheme, and Table 14.3 provides a summary of the situation from a farmer type viewpoint.

Scenario-testing and Prospective Analysis

Scenarios are based on questions in the form of 'What if...?' Smile allows for testing alternatives and the effect of such alternatives on a number of topics: costs and infrastructures, crops and cropping systems, farmers' strategies and practices, water resources, market features, etc.

After the diagnosis presentation and validation phase, farmers and management committee members were invited to think about alternatives, ideas and questions they would like to see tested. Specific workshops had to be held, first to raise ideas on scenarios, then to report back and discuss the outcomes of such scenarios. The definition of scenarios was done in close partnership with local stakeholders and experts.

Past and recent experiences involving Smile show that most scenarios consider the major issues facing the schemes, as perceived by farmers, and involve land redistribution options, the emergence of commercial farming, the set up of small-scale food plots, intensification and diversification of crop production at the farmers' level, water charging systems options, and rehabilitation options.

A number of recommendations, measures and decisions have been drawn from the simulations. Even more important is the fact that the approach has proved conducive to an in-depth, collective discussion and reflection on the schemes' prospects and collective investigation on the possible solutions to current issues.

Several important points must be highlighted here:

- Ideas, questions and scenarios as raised by farmers and local stakeholders should remain unchallenged by the research team; realistic or not, feasible or not, they have to be discussed and ultimately tested if this is requested.

Box 14.1 *Elements of diagnosis of the current situation in the Thabina irrigation scheme*

The whole operation of the Thabina irrigation scheme requires about 670,000 m^3 of irrigation water per year. The total operational costs (fixed and variable) incurred are 41,400 rand, which equates to 177 rand per irrigable hectare per year. In other words, supplying one m^3 at plot level costs 0.06 rand. The current water charging system in place covers 68 per cent of these costs. Active farmers make an average gross margin of 675 rand per hectare per year. Of these, commercially-oriented pensioners are the most efficient, with an average gross margin of 3092 rand per hectare per year, and also achieve the highest water productivity (0.53 rand of gross margin per m^3 used). These are the kind of outcomes that can be reported back to and discussed with the farmers.

In Thabina, simulations on the current situation showed that costs are not covered, and they can hardly be reduced as the bulk relate to capital and maintenance. The biggest issue is that the majority of non-farming plot occupiers have low ability or willingness to pay water fees. Low land productivity also strongly limits farmers' income and capacity to pay for water services. (A number of realistic alternative scenarios were also defined; these consider changes that are very likely to occur and/or are likely to considerably affect output indicators.)

The total annual water demand, including losses, amounts to 668,150m^3. On a weekly balance basis, and with solely the canal assuring supply, the total demand exceeds water supply from April to early July and from October to December. During these times of shortages (which have been acknowledged by the farmers), the pumps must be functioning (hence additional costs), adaptations and reductions in irrigation must be made, and conflicts sometimes occur.

The current costs incurred by management and operation for that is 41,400 rand, comprising 11,400 rand fixed operational costs and personnel and 30,000 rand variable operational costs (electricity and diesel for pumps). Assuming that all beneficiaries pay 120 rand per hectare per year as water and management fees, 68 per cent of those costs are covered in the current situation. The research team understood that some beneficiaries were not paying their fees and that some others (influential members of the management committee and/or commercial farmers) covered the costs of diesel. The situation is in any event worrying, since bills from the electricity provider Eskom remain unpaid, with the threat of power being cut off.

The most striking figures that come out are (a) the limited land use by all types (although some areas covered with marginal crops where not considered), and (b) the very low margin made at farm level. Farmers blame water shortages, the low insurance of water supply, the high production costs and the lack of finance for those shortcomings. The huge difference between types' water productivity (roughly calculated as gross margin per farm/water consumption per farm) is also worth noticing. Pensioner commercial farmers make 0.53 rand per m^3 used, while large commercial farmers make 0.11 rand and subsistence farmers only 0.01 rand. For comparison's sake, it is worth mentioning that the supply of 1m^3 costs 0.06 rand, based on the current operation and maintenance costs incurred.

Source: Perret et al (2003)

Table 14.3 *Traits and performances of farming styles and strategies in Thabina*

Farmer type	*Farm area (ha)*	*Cropped area (ha)*	*Gross margin per ha*	*Gross margin per farm*	*Gross margin per m^3*	*Water consumption (m^3 per farm)*	*Number of farmers*
Non-farmer plot occupiers	1.5	0	0	0	–	0	66
Subsistence farmers	1.0	0.67	56	37	0.01	6884	12
Specialized farmers	1.4	0.54	–443	–239	0	5904	12
Diversified farmers with income	1.7	1.0	1594	1594	0.15	10,315	15
Diversified full-time farmers	1.0	0.57	1260	718	0.14	4965	7
Full-time commercial farmers	1.2	0.5	2379	1189	0.34	3467	14
Pensioner commercial farmers	1.8	0.45	3092	1391	0.53	2611	9
Commercial farmers with income	1.6	1.15	215	248	0.02	10,471	12
Large commercial farmers	8	5.8	903	5240	0.11	45,782	3

Source: Perret et al (2003)

- All such questions have to be translated into quantified scenarios that can be simulated into the Smile platform (see examples in Box 14.3).
- The accuracy of the outcomes of simulations is not really the issue; the closer to reality the better and it is indeed important that local stakeholders see some coherence in the figures that are presented and discussed, yet the way action-research is conducted, with real participation, discussion and reflection is more important: this forms the core point of the approach.

Box 14.2 *Scenario definition for the Thabina irrigation scheme*

After the baseline situation was presented and discussed with the management committee (June 2003), the research team first introduced changes onto the baseline situation (or base scenario) to demonstrate the procedure to the farmers. The idea was to illustrate the concept of scenario, i.e. the possibility to test all sorts of changes, to foresee the impact of such changes on the situation in Thabina on a simulation basis. The farmers were then asked to formulate their own questions. Five types of questions were expressed (session in August 2003):

1. What would be the consequence of **upgrading the main canal** (i.e. by deepening it), with an augmented intake at the weir? This question shows how much the farmers are focusing on the issue. The underlying idea is to use the pumps as little as possible, hence reducing operational costs, while limiting water shortages and sharing issues within the scheme. This idea accompanies all the following ideas, which involve increased use of water (more users/more water-consuming crops).
2. What would be the consequences of **increased vegetable production** by most farmers? Faced with the current situation, the management committee acknowledges that most farming systems are not profitable. The underlying idea of the scenario is to investigate possible changes in terms of intensification and shift towards vegetable production, yet with pending questions in terms of water supply, farmers' skills and marketing.
3. What would be the consequences of **land reallocation**, from non-farming occupiers to willing new settlers? The proportion of unused land is unacceptable, especially given the long list of applicants. The idea is to investigate the impact of such reallocation. Negotiations have been initiated with the traditional authorities in order to make this process possible.
4. What would be an **appropriate water charging system** for Thabina? In the current situation, non paying beneficiaries and the low fees (10 rand per hectare per month) mean that only 68 per cent of the operational costs are covered.
5. What would be the **best farming model** in Thabina? The members of the management committee acknowledge the diversity of farming systems that take place, with diverse performance and efficiency. The majority of farming strategies are not profitable, while much land remains uncropped. The idea is to identify the best options and combinations, and to highlight their feasibility.

Source: Perret et al (2003)

Box 14.3 *Scenario testing for the Thabina irrigation scheme*

One important question that was raised by farmers in Thabina was related to the introduction of vegetable crops into the scheme. This has been translated into the scenario of all farmer types growing 'extensive spinach' (ie agronomically extensive, as opposed to intensive) on 0.25ha each (scenario 2a). A second related scenario considered the additional production of 'extensive tomato' on 0.5ha (scenario 2b). 'Extensive spinach' (Swiss chard) has been chosen because it seems easy to grow yet gives a reasonably good return. 'Extensive tomato' is already well known by the farmers (44 per cent grow it). The respective areas chosen reflect current practices and are reasonably realistic. Table 14.4 recaps the results obtained with scenarios 2a and 2b.

The results show that growing vegetables, even in an extensive way, leads to a better utilization of resources, ie water and land. The increase in the overall profit proportionally outweighs the increase in water consumption. This outcome is due to increased land utilization (more hectares are irrigated and cropped), while average margin per irrigated hectares increases only slightly.

Scenarios 2a and 2b refer to extensification, whereby increased use of resource leads to more production. This might, however, be a problem in terms of water availability. The total water demand is such that this scenario is feasible only with intensive use of pumping. The modest 120 rand that is currently paid cannot cover the increasing cost incurred by pumping.

If the hypothesis of a threefold inflow of water in the main canal (scenario also tested) is introduced, the water issue is solved (more water), as is the cost recovery issue (no need for pumping). Scenario 2b results in a slight water shortage limited to May, which can be overcome with some pumping and/or crop calendar shifts.

The results on extensive spinach and tomato also show that there's a need to intensify production systems. Shifting to other crops such as vegetables will prove fully efficient only if intensification takes place (a better use of the same resources leading to increased production).

Source: Perret et al (2003)

Table 14.4 *Comparing results from scenarios on vegetable production in Thabina*

	Base situation	*Scenario 2a*	*Scenario 2b*
Total water demand (m^3)	668,150	785,366	1,088,893
Total costs incurred (rand)	41,400	46,665	60,291
Percentage of costs covered	68%	60%	47%
Average cost per irrigable ha (rand)	177	199	258
Average gross margin per irrigated ha (rand)	675	690	785
Gross margin at scheme level (rand)	73,930	103,662	180,296

Note: Gross margin: before payment of any tax or water fee.
Source: Perret et al (2003)

Box 14.4 *Testing scenarios on the water charging system in Thabina*

Farmers already pay water fees in Thabina (10 rand per hectare per month). Such fees currently cover 68 per cent of operational and personal costs. Current costs amount to 174 rand per hectare per year; to cover these a fee of 14.50 rand per hectare per month would be required. The 10 rand fee was initially established by rule of thumb and does not fall far short of calculated requirements. However, even these do not cater for non-ordinary operation and maintenance costs. Recent vandalism has necessitated the replacement of wires, which in turn requires specific fund raising among farmers. Tables 14.4 and 14.5 show that any attempt towards crop diversification or land reallocation requires more pumping and hence increased fees (given the current state of the infrastructure). For instance, a scenario involving land reallocation to commercial farmers would require all to pay 17.00 rand per hectare per month. Scenario 2b (increased tomato and spinach cropping by all) would require 21.50 rand per hectare per month.

Regardless of the amount, the charging system remains the same for all scenarios, i.e. everyone has to pay, but it does not further charge irrigators that are openly commercially-oriented and use more water. Nor does it generate any incentive towards water saving. Most farmers indicated that they were ready to pay more if necessary (see Table 14.1), yet with some differences among them, depending on their farming profits and their commitment to the scheme. Furthermore, water demand greatly varies among types. Some committee members are willing to investigate alternative systems for charging, regardless of the amounts required. It has been suggested that farmers pay according to their consumption, to encourage reasonable use. From the simulation and base situation, it is established that 1m^3 costs 0.06 rand. The problem is to evaluate correctly each farmer's consumption: there is no measuring device in Thabina, and the SAPWAT standards are probably too far from reality to be used as references. Furthermore, establishing water fees from actual consumption would exempt non-farming plot occupiers, a disadvantage since there must be some form of incentive to either make them farm, leave, or in any case pay their share.

Therefore, if the objective is to combine (a) reasonable water use and fair pricing, and (b) incentive to farm, a mixed charging system could be established, with a fixed component applying to all beneficiaries, and a variable component, increasing with increased consumption, applying to irrigators. Based on this idea, further investigations were carried out using Smile. The main issue remained the lack of records or measurements on actual consumptions, hence the need to use cropping systems as references. During report back sessions, such a mixed system was presented to the farmers and to the DWAF as an example.

The system is as follows: all beneficiaries would pay a fixed component, per hectare owned, regardless of what is done with the land (eg 10 rand per hectare per month). Then farming beneficiaries would pay an additional variable component per hectare cropped (eg 40 rand per hectare per month). As a result, a non-farming plot occupier with 1ha would pay 10 × 12 rand = 120 rand per year. A farmer with 1ha, growing maize from September to December (4 months) on 0.5ha, and cabbage on 0.25ha from April to mid June (2.5 months) would pay (10 × 12) + (40 × 0.5 × 4) + (40 × 0.25 × 2.5) = 225 rand per year. If the DWAF wants to promote farming, it is suggested that any subsidy on operation and maintenance costs only covers the variable component. After a five-year period (subsidy phased out), serious farmers should easily be able to cover the whole fee from their farming income.

This idea of a mixed charging system has been well received by the farmers, but the DWAF strongly rejected it on the grounds that it penalizes those who actually farm. The DWAF supports the idea of a system with a unique fixed fee. It has been suggested that all pay 17 rand per month per hectare. In the likely case that the DWAF introduces subsidies on operation and maintenance costs, however, such a system will provide no incentive to farm whatsoever, since non-farming beneficiaries will also benefit from the subsidy. Conversely, a mixed system allows for targeting subsidies towards farming beneficiaries. The matter should be discussed further, taking account of the farmers' perspectives. Finally, the charging system that the DWAF has developed considers the phasing-in of a water resource management fee (0.0078 rand per m^3 in the Levhuvu-Letaba Water Management Area). This means that ultimately (after 5 years), farmers should pay an additional 22.27 rand per hectare per annum given current water consumption (or 26.17 rand under scenario 2a and 36.30 rand under scenario 2b).

Source: Perret et al (2003)

Table 14.5 *Comparing results from scenarios on land reallocation to commercial farmers in Thabina*

	Base situation	*Reallocation*
Unused area by non-farming occupiers (ha)	98	49
Number of commercial farmers	14	55 (41 newcomers)
Percentage of costs covered	68	59
Average gross margin per irrigated ha (rand)	675	1175
Total costs incurred (rand)	41,400	47,774
Average cost per irrigable ha (rand)	177	204
Total water demand (m^3)	668,150	810,103

Source: Perret et al (2003)

Discussion and Conclusion

The Smile Approach as a Workable Toolbox: Considerations

Although not capturing the actual complexity of a SIS, the model makes it possible for stakeholders to share a common representation of the scheme and its issues, to highlight the issues and to get stakeholders focused on the search for alternative strategies in a very open and flexible manner (i.e. scenario testing). Although requiring accurate and reliable background data, the approach shows interesting potential as it allows more information to flow between stakeholders involved in the rehabilitation and transfer process. It also helps in pointing out where responsibilities, prospects and potential lie, and shows huge potential for training purposes.

The results that are presented here are just examples, limited to certain scenarios, but the Smile simulation platform allows for a large number of possibilities (Perret and Van Schalkwik, 2004).

Recent experience shows that scenarios should be kept simple and sectoral, while farmers often tend to mix up different options and topics into the same question. Sorting out issues and translating them into clear, workable and quantified scenarios proves the most efficient approach in terms of facilitation, participation and collective learning (Perret, 2002b).

The overall idea is not to stick to reality at all costs. Even though coherence and closeness to reality should prevail, it may also prove interesting to test extreme and unrealistic scenarios for the sake of challenging preconceived ideas and raising questions and reactions. Overall, it is all about fuelling discussion and collective learning by local stakeholders (Jiggins and Roling, 1997).

Working in the frame of rehabilitation usually leads to considering two timeframes: the short term, whereby the scheme's immediate future is investigated (testing scenarios based on current trends, for instance, and asking what comes next after rehabilitation), and the longer term, whereby the sustainability of the scheme is investigated.

The first timeframe involves all stakeholders of the rehabilitation/revitalization process, while the second is more local and also more complex as it addresses many different dimensions (Perret, 2003). Nevertheless, it is important to share the scenarios and their outcomes with all stakeholders, since they may have further impacts at different levels (eg production features at the farm level interact with input supply or overall water consumption, and land allocation scenarios only make sense if there's room for action at government and traditional authority levels).

Action-research as a Vehicle to Local Empowerment: Conditions

The scenario approach is not only about imagining what might be the consequences of certain choices or decisions made, but also about starting to think about the conditions that are necessary for these choices to be realistic and workable and for their outcomes to be positive. A simple, cheap and quick simulation helps to shift the discussion towards the necessary conditions for a scenario to become feasible, including the legal, administrative, political and economic requirements.

The whole approach relies on the full and free participation of most local stakeholders (hence the 'workshopping' loops in Figure 14.2). This cannot be taken for granted, especially in the South African context, owing to recent history. Local commitment to discussion followed by collective action requires some form of learning and apprenticeship. Experience shows that the approach helps and supports that learning process, that local participants are increasingly at ease to talk and formulate ideas and scenarios as the process evolves. A renewed stance is also expected from development operators, since their inputs and commitment prove crucial in the sustainability of the changes induced (local reflection and action).

Moreover, a certain amount of learning is also expected on the authority side to achieve certain forms of research and certain outcomes (eg technology transfer, technical prescription, etc). Experience showed that while authorities are seeing researchers as 'accomplices', and farmers as 'abiding beneficiaries', action-research outcomes ultimately may challenge some of their choices.

Technically Speaking, What Have We Learned from the Approach So Far?

Case studies highlight certain recurrent issues facing smallholder irrigation schemes: uncertainty of land tenure does not help, land size does matter (the '1ha or so' basis provides a family with neither a decent farming income, nor food security), subsistence maize is definitely not profitable and farmers tend not to fully use the land at their disposal. More generally, the '1.3ha paradigm' inherited from the Apartheid perspective of the development of black rural areas (Union of South Africa, 1955) must be abandoned in favour of an in-depth reflection as to what is expected from smallholder irrigation schemes, for the benefit of the farmers and for the whole society. All these elements make it difficult to envisage a sustainable and autonomous management of SIS by the farmers at the moment.

However, realistic and well thought-through scenarios involving better yields, some land reallocation, marketing and intensification, for example, show that development and improvement pathways do exist for most SIS. It is important to highlight that these pathways not only involve economic performance, but also equity, food security, social integration and community development (Perret, 2003).

All the experience with the Smile approach indicates that there is probably no such thing as a unique external solution to solve all the problems. Farmers also have to contribute (enhanced collective action and cooperation), as do extension operators (training). Yet these scenarios also show that pathways to success cannot only rely on farmers' mobilization, willingness and action; they also require public investments and support (on capital, extension and training), legal frameworks and proper institutions (especially on land).

The approach often helps to clarify the possible role of the different stakeholders: the authorities (Which support system? Which investments?); the farmers (Which cropping systems? Extensification or intensification?); and the management entity (Which charging system? How to organize collective action? On which issues?).

The approach has not yet been completed. For it to be fully and operationally included in the pre-rehabilitation survey of the Revitalization of Smallholder Irrigation Schemes (RESIS) programme, for instance, some further simplifications and adaptations must take place. This is currently being addressed through work with the Limpopo Department of Agriculture and consultants. Adaptations mainly concern the establishment of a base line situation (diagnosis), which is time consuming and requires specific skills and experienced researchers.

Acknowledgements

The author expresses his gratitude to the DWAF of South Africa and the National Research Foundation of South Africa for their financial support to the research. Stanley Yokwe, Mélanie Lavigne and Nicolas Stirer also contributed significantly to the work.

References

Argyris, C., Putnam, R. and Smith, D. M. (1985) *Action Science: Concepts, Methods and Skills for Research and Intervention*, Jossey-Bass, San Francisco, CA

Bembridge, T. J. (2000) 'Guidelines for rehabilitation of small-scale farmer irrigation schemes in South Africa', WRC report no 891/1/00, Pretoria, South Africa

IWMI (2001) 'Can poor farmers in South Africa shoulder the burden of irrigation management?' IWMI website document, www.cgiar.org/iwmi/home/IMTSAf.htm

Jiggins, J. and Roling, N. (1997) 'Action research in natural resource management: Marginal in the first paradigm, core in the second', *Etudes et Recherches sur les Systemes Agraires et le Developpement*, 1997, no 30, pp151–167

Lamacq, S. (1997) 'Coordination entre l'offre et la demande en eau sur un périmètre irrigué: Des scenarios, des systèmes et des hommes', unpublished doctoral thesis, Ecole Nationale du Génie Rural, des Eaux, et des Forets, Montpellier, France

Le Gal, P. Y., Rieu, T. and Fall, C. (2003) 'Water Pricing and sustainability of self-governing irrigation schemes', *Irrigation and Drainage Systems*, vol 17, pp213–238

Liu, M. (1994) 'Action-research and development dynamics', Systems-Oriented Research in Agriculture and Rural Development, international symposium, Montpellier, France, 21–25 Nov 1994, pp111–116

Moisdon, J. C. (ed) (1997) *Du Mode d'Existence des Outils de Gestion. Les Instruments de Gestion à l'Epreuve de l'Organisation*, Seli Arslan, Paris

Ntsonto, N. (2004) 'Economic performance of smallholder irrigation schemes: A case study in Zanyokwe, Eastern Cape, South Africa', Masters thesis, University of Pretoria, South Africa

Oja, S. N. and Smulyan, L. (1989) *Collaborative Action Research: A Developmental Approach*, Falmer Press, London

Ostrom, E. (1992) *Crafting Institutions for Self-governing Irrigation Systems*, Institute for Contemporary Studies Press, San Francisco, CA

Perret, S. (1999) 'Typological techniques applied to rural households and farming systems: Principles, procedures and case studies', working paper 99/2, Nov 1999, University of Pretoria/Cirad, Pretoria, South Africa

Perret, S. and Le Gal, P. Y. (1999) 'Analyse des pratiques, modélisation et aide à la décision dans le domaine de l'irrigation: Cas de la gestion d'une retenue collinaire collective à la réunion', *Economie Rurale*, vol 254, pp6–11

Perret, S. (2002a) 'Water policies and smallholding irrigation schemes in South Africa: A history and new institutional challenges', *Water Policy*, vol 4, no 3, pp283–300

Perret, S. (2002b) 'Testing scenarios on the viability of smallholding irrigation schemes in South Africa: A participatory and information-based approach', 17th Symposium of the International Farming Systems Association, November 17–20, Lake Buena Vista, FL

Perret, S. (2003) 'Une démarche de modélisation pour accompagner le transfert de gestion de périmètres irrigués en Afrique du Sud', in *Organisation Spatiale et Gestion des Ressources et des Territoires Ruraux*, seminar proceedings, 25–27 February, Cirad, Montpellier, France

Perret, S., Lavigne, M., Stirer, N., Yokwe, S. and Dikgale, K. S. (2003) 'The Thabina irrigation scheme in a context of rehabilitation and management transfer: Prospective analysis and local empowerment', DWAF project 2003-068, final report, Cirad/IWMI/University of Pretoria, Pretoria, South Africa

Perret, S. and Van Schalkwik, P. (2004) *Smile© Version 6: Sustainable Management of Irrigated Land and Environment: Investigating the Viability of Smallholder Irrigation Schemes*, software available for download at www.smile-cirad.co.za

Perry, C. J. (2001) 'Charging for irrigation water: The issues and options with a case study from India', research report no 52, IWMI, Colombo, Sri Lanka

Touchain, E. and Perret, S. (2002) 'A simulation-based approach to access the economic viability of smallholding irrigation schemes in South Africa', University of Pretoria, Cirad-Tera, research report no 02/02, Pretoria, South Africa

Union of South Africa (1955) 'Commission for the socio-economic development of the Bantu areas within the Union of South Africa', summary report, Government Printers, Pretoria, South Africa, UG 61/1955

Vermillion, D. L. (1997) 'Impacts of irrigation management transfer: A review of evidence', IWMI research report no 11, Colombo, Sri Lanka

15

Role-playing Game Development in Irrigation Management: A Social Learning Approach

Anne Chohin-Kuper, Raphaèle Ducrot, Jean-Philippe Tonneau and Edolnice da Rocha Barros

Introduction and Rationale: Users' Participation in Irrigation Management

In the São Francisco river valley in Brazil, irrigation schemes for smallholders were created in the 1960s in order to promote rural development. A little later, in the 1970s, entrepreneurial farms were expected to grow commercial crops, to enable the financial sustainability of these schemes (World Bank, 2002). A regional public development office, Companhia de Desenvolvimento dos Vales do São Francisco e do Parnaíba (CODEVASF, the development company of the São Francisco river valley), was in charge of the operation and maintenance of these schemes from the 1960s until water management transfer started in the 1980s.

Low-cost recovery, a lack of transparency in operation and maintenance (O&M) expenditure and political reluctance to increase water prices resulted in financial deficits and even threatened management sustainability, as has been the case in many other countries. In the past 20 years, several governments have promoted the transfer of irrigation management as a solution to improve the sustainability of their irrigation systems (Johnson, 2002; Shah et al, 2002). In Brazil, this transfer, known as 'emancipation', is implemented through institutional changes with the creation of irrigation districts. These districts are in charge of the O&M of irrigation schemes while the state retains responsibility for investments. The transfer is progressive and the links with the former state organization remain very strong (da Rocha Barros, 2001).

The Maniçoba irrigation scheme in the São Francisco river valley, where this research was conducted, has been managed by the Maniçoba irrigation district since the late 1990s, after more than 20 years of central management by CODEVASF. In the new institutional context, users are expected to participate in the management board and the general assembly of the district. However, the district manager is a former CODEVASF staff member, and links with CODEVASF remain strong. The district is also closely related to Agrovale, the biggest agricultural enterprise of the irrigation scheme, which holds a permanent position on the board.

The institutional reforms and the creation of the water district did not solve budget deficit problems, even though financial reforms, in particular pricing reforms, were initiated. These reforms were expected to improve water management sustainability through higher prices of water and a change in pricing structure. However, users' participation increased the opposition to change. This is partly explained by the fact that users' participation in the pricing reform process in Maniçoba did not respect Ostrom's design principles of self-governing irrigation systems (Ostrom, 1992; Ostrom et al, 1994). First, users' involvement in the definition of rules is questionable. Private firms and CODEVASF managers were in a better position to craft rules than small farmers because of their negotiation power and their relations with the district manager. Second, there was no shared assessment of problems, resulting in two separate visions of irrigation pricing matters (Chohin-Kuper et al, 2004). Stakeholders did not have a shared image of their irrigation system, which is a prerequisite for individuals to reach an agreement (Beuret, 1999). The hypothesis that 'appropriators are more likely to share a common understanding of the rules developed by themselves than those handed down from distant government agencies' (Tang, 1994) was found wanting.

A role-playing game (RPG) was developed to address the issue of capacity building relating to water management decisions. The following question is addressed in this chapter: How can the process of developing an RPG facilitate social learning through improved information, communication and participation?

The RPG focuses on the pricing reforms for three major reasons. First, pricing water is the interface between water users and the district. Second, it is a key issue with respect to irrigation management sustainability. And third, it is a potential source of conflict between water users and district managers.

The first section presents the approach of co-development of an RPG and the underlying concepts of social learning. In the second section, we describe the RPG CAPAGUASF (capacity building in irrigation water management in the São Francisco river valley) and the game sessions. The outcomes of the game are discussed in the third section.

RPG Development Approach

A Social Learning Approach

RPGs have been developed in various contexts of natural resources management (see Daré, 2005 for a comprehensive review). They are used as information and communication tools (ICTs) or as capacity building and negotiation tools, eg the Irrigation Management Game (Burton, 1994) or the forest and pasture management game 'SYLVOPAST' (Etienne, 2002). RPGs are also used in association with computerized models based on multi-agents systems (Barreteau et al, 2001; Barreteau, 2003; Bousquet et al, 2002). They are helpful in terms of transferring information and are now also considered an interactive method that can facilitate social learning (Barreteau, 2003). Social learning refers to the individual learning process based on observation of others and their social interactions within a group (Bandura, 1977; Pahl-Wostl, 2004). The concept of social learning through RPGs also fits into the broader concept of communities of practice described by Wenger (1998a, 1998b): 'Communities of practice are groups of people who share a concern or a passion for something they do and learn how to do it better as they interact regularly'.

Using RPGs in social learning processes is based on two major assumptions. First, 'The game is accepted as a schematic representation of reality'; second, 'The social background of players interferes with role playing in the game' (Daré and Barreteau, 2003; Daré 2005). The use of RPGs as a social learning tool is also based on the principle that what is important is the process and not the results in themselves (Funtowicz and Ravetz, 1993, referring to post-normal science).

The main objective of the RPG in this study is to organize and structure discussions of stakeholders on water management issues, and, more particularly, on the financial sustainability of the irrigation district, including cost recovery and water pricing issues. It also focuses on farmers' understanding of the relationship between water quantity and its price at the farm and district levels. One of the points of the RPG is to simulate and discuss scenarios.

The CAPAGUASF RPG approach presented in this paper is an example of this interactive method and social learning process. It is based on a participatory process and on a co-development approach. This approach differs slightly from self-designed RPGs, where users are involved from the very first steps of game development (d'Aquino, 2002; d'Aquino et al, 2003). In the co-development approach, a simple basic model was developed and this was modified and adapted by users, depending on the results of game sessions. The prototype game CAPAGUASF was elaborated by researchers on the basis of previous research results. The rules were drawn up on the basis of analyses of the pricing system and of the pricing reforms implemented during the last few years (Fernandez, 2001; de Nys, 2004). The underlying hypothesis was that similar outcomes in terms of participation can be obtained from the present approach (co-development of the game) to those of a self-designed RPG approach. To validate this hypothesis, the contributions of the co-development approach of the game to encouraging discussions, sharing problem analysis and improving the collective decision-making process through improved participation, communication and information need to be analysed.

Game Sessions Timing

The game sessions were organized during a six-day workshop and comprised three steps:

1 **First game round** of the basic model. Two irrigation schemes were represented by two groups of players, eight players for each irrigation scheme. In

this first game session, only one game round could be played since it lasted half a day. In the debriefing session, participants could react to the game and suggest changes and improvements. Game participants were farmers, CODEVASF technical staff and district staff. Half a day was devoted to the debriefing session.

2 **Game development**. Suggested improvements and changes were taken into account to further elaborate the game and adapt its presentation. Modifications were made over the weekend.

3 **Second game session**. During this one-day game session it was possible to play two rounds with the modified version. There were still two irrigation schemes represented but with additional players identified during the debriefing session (technical assistant). The game was facilitated by the presence of players who had participated in the first session, although there were a few new participants.

The approach emphasizes interaction, with potential users of the game – farmers, the irrigation district and CODEVASF – involved in the early elaboration and playing steps. The game had not been played before the workshop, but had only been calibrated by the designers of the game to check its coherence.

Game Participants

The interest of the game is to gather the different stakeholders involved in water management decisions. According to Wenger's community of practice approach (1998a, 1998b), the group of players has an 'identity defined by a shared domain of interest'. Therefore, the different types of farmers and district staff were represented (Box 15.1), with a majority formed by small farmers (or 'colonos'), who represent the largest group of users. Although most farmers were water users from Maniçoba, a few farmers from a newly created irrigation scheme were also playing

Box 15.1 *Game participants*

Players:

- 6 small farmers from irrigation schemes ('colonos')
- 1 commercial farmer
- 1 management board (3 players) (role identified in the first test)
- 1 water manager (district)
- 1 technical assistant (identified in the first test)

Other participants:

- Two observers (researchers, consultants) in each virtual irrigation scheme in charge of reporting game outcomes. There was no provision for recording the whole game on video but pictures were taken.
- A game master or facilitator in charge of the explanation of the game.
- A game operator.

Source: chapter authors

the game. The workshop also gave special attention to small farmers, as they are expected to be in a less favourable position in decision-making processes.

CAPAGUASF Design Principles

Building a Simplified Representation of Maniçoba Irrigation System

The first step in the development of the CAPAGUASF RPG was to represent the interface between water users and the irrigation district. The irrigation system is represented by seven farms, six small farms and one large commercial farm, according to the main types of farmers – 'colonos' and private firms – that can be found in the irrigation scheme. Each farm is limited to one mango field, mangoes being the dominant perennial crop in the zone (most farmers moved from annual crops to fruit trees, specializing in mangoes (Ducrot et al, 2001)). Small farmers have a plot of mangoes with a canal water supply. The canal water supply is based on irrigation scheduling. Plots differ with respect to irrigation efficiency depending on the quality of land levelling. Three classes of land levelling correspond to different levels of water loss (0, 5 or 8 hours of canal water). Farmers face diverse debt situations resulting from the planting of the fruit trees. The irrigation district has an electrical pumping plant to supply water to farmers. The district manager is responsible for the pumping plant (operation and maintenance), the payment of the electricity bill to the company (the game operator) and for water bill recovery from farmers.

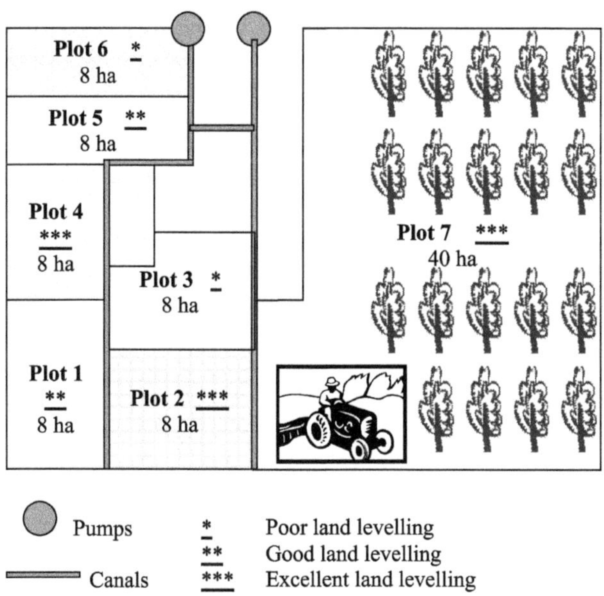

Source: chapter authors

Figure 15.1 *Virtual irrigation system*

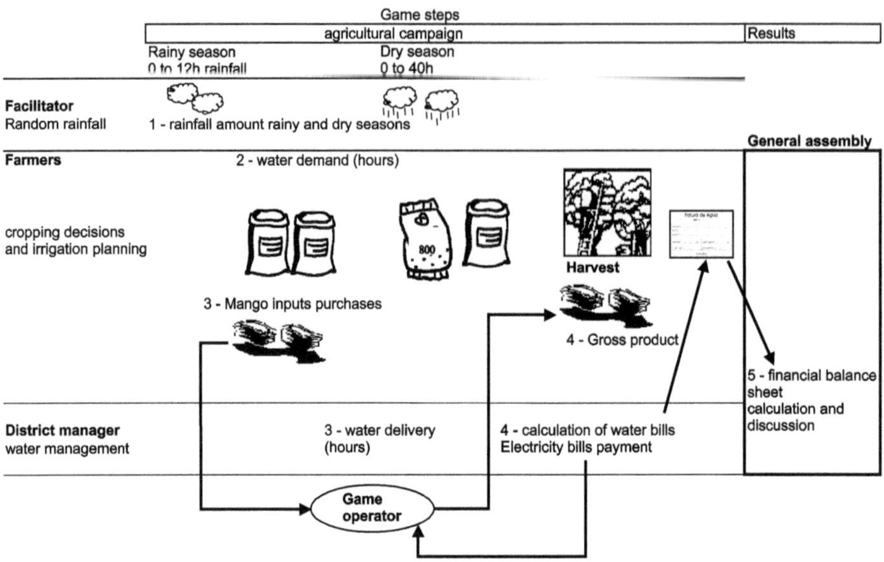

Source: chapter authors

Figure 15.2 *Game steps*

Game Round and Game Steps

During the preparation of the game, small farmers randomly choose their plot and debt situation characteristics while the private firm role is assigned to a commercial farmer. Players also identify a person to play the district manager.

A game round represents an agricultural cycle with two seasons (dry and rainy), with the following steps (Figure 15.2):

1 The total rainfall is given randomly by the game facilitator.
2 Farmers make their cropping and irrigation decisions for each season. Small farmers can choose different technical packages (fertilizer quantity and mango productivity level). They buy fertilizer and other mango chemicals from the game operator. Canal water requirements are determined by farmers. Canal water demand in hours is given to the district manager, who allocates water accordingly. Water requirements should take into account rainfall and water losses due to the land levelling quality. Private firms also decide upon fertilizer quantities and mango productivity. Since they irrigate with drip irrigation from a water reservoir, they can fill the latter on demand.
3 Agricultural cycle expenses are calculated. Each farmer has to pay for harvest costs, based on the mango production. Both small farmers and private firms have to pay for family expenses, although the latter spend more due to their higher standard of living. Finally, private firms have to reimburse a fixed amount every year for the investment in the drip irrigation system.
4 The end of the agricultural cycle. Farmers receive the gross product from their mango plantations from the game operator. The game operator uses a spreadsheet-based model to calculate the gross product for each farmer. The gross

product depends on technical choices (cropping intensity) and on possible water stress that may result from irrigation decisions and plot efficiency. They also receive their water bills from the district manager. The district manager has to pay electricity bills to the game operator.

5 The district manager organizes a general assembly to discuss the results of the cycle. The main issue is the financial balance of the district, which depends on water pricing, cost recovery from farmers and district O&M costs. The district has to recover the O&M costs of its electrical pumping plant, which supplies water to farmers.

6 Analysis of the game and debriefing session. The analysis of the game focuses on players' choices (water and technical choices) and district level decisions in terms of water pricing and cost recovery. The analysis is based on qualitative information recorded by the observers during the game (and debriefing sessions as reported below) and on quantitative indicators. The indicators used are the level of water demand and water users' bills, district costs and receipts, cost recovery level, and water price adjustment to changes in costs.

Key Rules and Principles

Pricing water should be based on a district objective to improve its financial balance. The district objective is to avoid financial deficits during rainy years, when water sales drop, and to improve its financial sustainability.

The pricing system that applies in the game is based on the existing two-part tariff recommended by Brazilian law and applied in Maniçoba. The fixed component of the price is supposed to cover fixed management costs and the variable component should cover variable costs relating to water supplies (mainly pumping costs). However, we use a simplified pricing system with a unique tariff for all farmers. In the 2001 reform, an optional pricing system was implemented with two optional tariffs as indicated in Table 15.1.

Table 15.1 *Changes in water pricing in the irrigation scheme of Maniçoba*

	Water tariff ('K2')			
	Fixed component (reals/ha/month)		*Variable component (reals/1000m³)*	
From 1994 to April 2001	3.94		18.10	
April–September 2001	6		20	
	Colonos Tariff	*Maniçoba Irrigation District Tariff*	*Colonos Tariff*	*Maniçoba Irrigation District Tariff*
September 2001–July 2002	6	17.74	26	8.47
May 2002	6.66	19.66	26	8.47
Since July 2002	6.66		29.34	

Source: chapter authors

Changes in rules, in particular related to water pricing, can be decided by players during the game sessions.

Changes in the structural rules of the game can be discussed and proposed during the debriefing session. Suggestions can then be taken into account, when relevant and feasible, to upgrade the game.

Scenarios

In order to initiate discussions on price adjustment mechanisms, different scenarios were played out during the game session: increase in pumping costs and changes in water supply due to climatic changes. Changes in maintenance costs might also be introduced.

The objective here is to examine price adjustment mechanisms that in reality are never subject to discussion between stakeholders. District water prices were calculated on the basis of average 2000 costs without any anticipation of future costs. In particular, changes in the cost of electricity and inflation may be important in the Brazilian context. In addition, maintenance costs increase regularly with the age of equipment. A good maintenance level is estimated to cost about 2 per cent of the investment costs every year according to Plantey and Blanc (1998). However, maintenance costs are often underestimated by district managers.

Debriefing Sessions to Identify Game Options

Debriefing sessions are very important steps in the co-development approach to RPGs, as discussed below. They can also contribute to identifying game options. During the debriefing sessions, participants identified other important issues they would have liked the game to address. The first issue concerns the organization of the irrigation systems. In the second version of the game, the organization of the irrigation scheme is not pre-defined but is the first step of the game. Participants decide on how decisions are going to be taken, who is going to manage the irrigation system and how the farmers get organized. It is possible to discuss respective roles without limiting the discussion to the stakeholders that were taken into account in the initial game. This was a specific demand from participants of newly created irrigation systems who were interested in using this type of tool in a participatory process.

A second issue relates to the operation and maintenance of the irrigation system. However, incorporating this issue in the game requires modifying the base version, which was not feasible during this workshop.

Game Development Process: What Are the Outcomes?

The Game as a Discussion Tool

The game is a powerful stimulator of discussions and debates, as reported by participants in the debriefing sessions (Box 15.2). The game and debriefing sessions were

Box 15.2 *Outcomes of the debriefing sessions*

The game: a discussion and animation tool

- 'The game facilitates the involvement of stakeholders and creates strong dynamics. It facilitates discussions between stakeholders who do not frequently have them.' (technical staff)
- Players demonstrated a strong interest in the game approach and the dynamics of the game. Social links were developed between the different players, which even impacted on more formal discussions during the workshop.
- Discussions during the game can point out other issues which were originally not addressed. (colono)

The game: a representation of reality

- 'The game facilitates the understanding of farmers' constraints and the identification of problems. It is a way to make a link with reality.' (Colono)
- 'The game is a way to focus on the relationships between farmers and the district but could also focus on other stakeholders.' (district technical staff)

The game: a diagnosis tool to explain problems

- 'To understand the game helps to explain problems and to find solutions: it encourages discussions on possible solutions to solve the problems of the district.' (district manager)

The game: a social learning tool

- 'The strong point of the game was that it was not ready-made.' (private firm, colonos)
- 'What is important is the process and not the results of the game.' (technical staff)
- 'The game focuses on the irrigation scheme's experience and know-how.' (colono)
- The game is a means to 'learn while playing', and can 'reinforce farmers' capacities'. (colono)

Shortcomings of the game

- 'The game cannot address all the issues of interest to farmers and other stakeholders, for instance production and commercialization issues, land problems or irrigation management.' (consultant)
- 'Some problems cannot be identified with the game.' (colono)
- 'The diversity of activities cannot be represented in the game.' (consultant)
- 'The game has to be part of a process and not an objective in itself.' (technical staff)

Source: chapter authors

characterized by sharp discussions and strong debate (Box 15.3). Since the game was played during a workshop, some debates continued during workshop time.

The dynamics and the atmosphere of the game allow for straightforward discussions, roles giving freedom to express opinions and to question other

Box 15.3 *Key issues identified during the debriefing sessions*

Water management and pricing

This is the focal point of the CAPAGUASF game. The main issues that were debated related to questions such as: How are water prices calculated? How are district operation costs defined?

Organization and management of the irrigation scheme

This topic turned out to be a major point of interest to the players, focusing on:

- Collective organization of irrigation management: differences and advantages of the various types of organization (cooperative, association, district, private firm).
- Roles and relationships between the members of an irrigation management organization.
- Design and definition of internal operation rules of the organization.
- Participation mechanisms and methods to reinforce and facilitate involvement of users in the organization.

Operation and maintenance in relation with water management organization

The links between agricultural cycle planning and irrigation district planning (irrigation scheduling) were discussed. The local organization of water management (hydraulic zone: organization, operation and management rules) also came up as a topic.

Profitability of production systems in relation to marketing

Farmers' incomes and the sustainability of the cropping systems were another major theme for discussion:

- Relations between agricultural systems planning, plot management and marketing strategies.
- Organization of producers and commercialization contracts;
- Production costs and credit.
- Market information, the relationship between production quality and commercialization.
- Extension services, information and support to producers.
- Choice of irrigation technology: drip irrigation/surface water.

Source: chapter authors

stakeholders. For instance, small farmers were able to question the former state manager through the game when he was playing his role (the district manager was played by a CODEVASF staff member). Very active discussions between the former state company CODEVASF and farmers could take place. The players debated issues that are not easily discussed in a formal seminar assembly, such as the relation between costs and price of water, and the water management problems of the district.

The Game as a Diagnosis Tool

The discussions during the game and the debriefing sessions helped to validate key issues or to identify other related problems and questions. Problems of organization and management of the irrigation scheme were largely discussed. Issues indirectly related to the game – the profitability of the production systems, marketing issues, and irrigation and soils management – were also debated (Box 15.3).

Co-elaboration and Appropriation

When farmers and stakeholders are given an opportunity to test and play the game and adapt it to their expressed needs they develop a sense of ownership of the process. Hypotheses about roles and rules can be discussed and validated. This makes the game sufficiently realistic and initiates discussions and debates. In this process, players constitute a 'community of practice' where 'members develop among themselves their own understanding of what their practice is about.' (Wenger 1998b).

Promoting a sense of ownership is facilitated by a trial and error process. Some of the changes suggested after the first game session were ultimately not taken into account, some were not relevant and some were rejected outright. What might be seen as a loss of time is in fact useful even if the suggested changes are sometimes abandoned. 'Users' participation in the definition of rules often goes through substantial trial and error.' (Tang, 1994).

Changes in roles, rules and game structure

First, changes in roles were discussed. For instance, it was suggested to remove the role of the CODEVASF but to introduce a technical assistant to help farmers in their decision-making process. It was also proposed to designate a management board of three farmers at the beginning of the game, the three farmers also playing their farmer roles. Despite discussions about the difficulty of playing two roles this change was made prior to the second game session, however, the change did not contribute to any improvement and players recognized that it was not easy for them to change roles during a game session. Other changes appeared not to be very useful in the first place, not so much because they were inappropriate but because they not really taken into account by the players. For instance, the possibility of borrowing money from a bank was given but hardly anybody used it even when they were in debt. However, this does not necessarily mean that this possibility should not be kept, since after a second round of the game people started to study the new option.

Second, changes in rules were debated. Initially, the objective of the game was to discuss different pricing systems that were introduced during the reform, but this proved too complicated in the first game session and participants were in favour of a simplification to a single pricing system. Irrigation management rules and planning were also modified; irrigation efficiency and the possibility of credit were added.

Third, the structure of the RPG appears to be critical for player understanding and the game's functioning. The game was adapted according to participants' image of their system. Participants stressed, for instance, the need for simple record sheets that are adapted to farmers. Game materials – explanations of rules or roles, playing cards, recording sheets and calculation sheets – were then designed accordingly. During the game sessions, participants could choose how they were going to organize themselves. A virtual irrigation system was designed in order to help farmers identify their plot (Figure 15.1). The observers of the game – researchers or consultants – largely contributed to the elaboration of the new materials, which greatly improved the understanding of farmers. In particular, when a translation was required, using vocabulary specific to the region or even to a particular situation was imperative. With respect to the game's organization, it was suggested that the computer interface used by the game operator be abandoned. The interface helps to compute gross products and also to record farmers' operations, but it was viewed as a black box by farmers.

Trade-off between the complexity of the game and reality

In this type of approach there is always a trade-off between the complexity of the tool and the quality of the representation of reality. Although the game is obviously a much simplified representation of reality, it needs to be accepted as a good model of reality in order to be useful for discussion and for social learning. Participants in the game opted for a relatively simple tool that may not be a perfect representation of their reality ('the game is not our field reality') but was still relevant enough to provoke discussion on key water management issues. The basic model, although relatively simple, was somewhat too complicated for farmers and other stakeholders. One of the most important changes after the first test was to make the game even simpler. The number of decision-making steps through the agricultural cycle was reduced. Another significant change was to simplify the water pricing system.

The Game as a Participative and Social Learning Tool

The approach discussed in this chapter facilitates the participation and involvement of the various stakeholders in two ways. First, since the basic version of the game was not a finished product, participants were willing to interact and contribute to develop the game. Participants were very keen in the debriefing sessions to express their points of view, make suggestions or raise other issues. This was nicely summarized by participants during the debriefing: 'The strong point of the game was that it was not ready-made'. Second, during the game sessions themselves, the game was very effective in terms of stimulating participation and interactions between players. Relationships established during the game session are facilitating conditions for emerging collective action, this is an example of the importance of past successful experiences and social capital for sustainable governance of common pool resources (Agrawal, 2001).

The capacity building potential of the game is specific to the issues addressed. For instance, in this game players can learn about pricing methods and about the relationship between O&M costs and pricing design. The social learning capacity depends on game dynamics. For instance, there were more discussions around price design and financial management when the district manager had no experience with this task. Other players would then help him to establish the budget balance, to calculate the water bills. After this learning phase, the district manager would be able to do it alone and other players who participated in the discussions would also understand better how their water bills are calculated.

Participants' demand for training relating to other topics using this type of approach would require developing other modules, new games or other training support tools. RPGs can indeed be part of more comprehensive training programmes where they are used together with other tools. For instance, there was a demand from district managers to help them better anticipate the relationship between crop mix, water demand and the financial balance of the irrigation district. The idea was to use a complementary tool like Smile (Perret, 2003) which is a decision support tool for irrigation district management based on farm and district data. However, this could not be tested during the seminar because of time constraints. Indeed, one drawback of the RPG is the limited number of participants that can play at the same time. In addition, it is relatively time consuming. Even when the game seems relatively simple, the time needed to play a round tends to be underestimated. To use an RPG as a training tool requires taking these factors into account.

The game could also be used as a monitoring or experimental tool. For instance, it can be used as a monitoring tool in a management process, such as the pricing reform that was implemented in the Maniçoba irrigation scheme. In particular, the Baland and Platteau (1996) condition for sustainable management institutions, 'simple rules that are easy to understand' (Agrawal, 2001), can be tested with the game. The game could be used as an experimental tool to test solutions or to test the understanding of new rules. Indeed, playing this type of game before the pricing reform could have been useful to test its feasibility. The game confirmed the complexity of the different pricing structures. Not only farmers, but also technical staff, found it difficult to understand the advantages of the different pricing systems proposed in the game, although the game was trying to keep as close as possible to what happened during the pricing reform. For instance, it was difficult for small farmers to choose between two pricing methods. During group discussions about water pricing (step 5), adjusting the pricing system to changes in costs was complicated, for instance. The fact that people had been through a pricing reform did not make them more comfortable with pricing issues.

Finally, the game should not be seen as an objective in itself but as part of a process. The dynamics that emerged from the game session during the workshop are also of interest for monitoring a participatory process. The game has a capacity to make people interact and establish relationships that will go beyond the game itself; it forms a basis for the kind of regular interactions that characterize communities of practice.

Conclusion

The CAPAGUASF RPG was developed in a context of transition from centrally managed schemes to participatory management. However, this transition is characterized by poor participation on the part of irrigation users and a lack of communication that prevents the development of a common diagnosis. The objective of developing an RPG was to address the issue of capacity building for water management through improved sharing of information, communication and user participation. CAPAGUASF focuses on irrigation district management. It represents a simplified irrigation scheme and the interface between the irrigation district and water users. Four main findings arise from the role playing game process. First, the process stimulates discussions and initiates debates. Stakeholders involved in water management issues at different levels – farmers, district managers and CODEVASF regional managers – debate more easily in a gaming context than in reality. Second, the role playing game is a useful tool for problem identification and diagnosis, during both the game sessions and the debriefing sessions after the game. The game becomes an interesting tool to address the need for multi-stakeholder diagnosis, a requirement for sustainable institutions. Third, the approach offers promising opportunities for better user participation and social learning. A group of players forms an embryonic community of practice where people share a concern and start to learn how to do things better as they interact. Fourth, the RPG should not be seen as an objective in itself but as part of a process. The interaction initiated during the game does not stop with the end of the game but should contribute to emerging collective action. Additionally it can be pointed out that games could be linked to other tools.

In short, if implemented during management reforms, games could contribute to an increased involvement of users and to a dynamic of collective action or social learning.

> *Seria bom que esta usina fosse feita no projeto, pois lá tem gente que está quase parado, não acredita em mais nada, se o produtor fizer o jogo ele vai cair na real.*
>
> *['It would be good to implement this workshop in the irrigation scheme, because there are some people there who do not progress anymore, do not believe in anything, and if farmers were to play this game they would fall back into reality'.]*
>
> João Alexandre, producer, Maniçoba irrigation scheme

Acknowledgements

We would like to thank our Brazilian partners who made this work possible: Companhia de Desenvolvimento dos Vales do São Francisco e Panaíba (Codevasf); Empresa Baiana de Desenvolvimento Agrícola (EBDA); Universidade do Estado da Bahia (UNEB); Empresa Brasileira de Pesquisa Agropecuária (Embrapa); and Associação de Desenvolvimento e Ação Comunitária (ADAC). We also gratefully acknowledge financial support from the Irrigated Systems Common Program (PCSI).

We would also like to thank all the anonymous reviewers for their helpful comments and suggestions and are particularly grateful to Valerie Kelly and Marcel Kuper for their contributions.

References

Agrawal, A. (2001) 'Common property institutions and sustainable governance of resources', *World Development*, vol 29, no 10, pp1649–1672

Baland, J. M. and Platteau, J. P. (1996) *Halting Degradation of Natural Resources: Is there a Role for Local Communities?*, Clarendon Press, Oxford

Bandura, A. (1977) *Social Learning Theory*, Prentice-Hall, Eaglewood Cliffs, NJ

Barreteau, O. (2003) 'The joint use of role-playing games and models regarding negotiation processes: Characterization of associations', *Journal of Artificial Societies and Social Simulation*, vol 6, no 2, http://jasss.soc.surrey.ac.uk/, accessed September 2005

Barreteau, O., Bousquet, F. and Attonaty, J. M. (2001) 'Role-playing games for opening the black box of multi-agent systems: Method and lessons of its application to Senegal River Valley irrigated systems', *Journal of Artificial Societies and Social Simulation*, vol 4, no 2, http://jasss.soc.surrey.ac.uk/, accessed September 2005

Beuret, J-E. (1999) 'Petits arrangements entre acteurs: Les voies d'une gestion concertée de l'espace rural', *Nature Sciences Sociétés*, vol 7, no 1, pp21–30

Bousquet, F., Barreteau, O., d'Aquino, P., Etienne, M., Boissau, S., Aubert, S., Le Page, C., Babin, D. and Castella, J. C. (2002) 'Multi-agent systems and role games: An approach for ecosystem co-management', in Janssen, M. (ed) *Complexity and Ecosystem Management: The Theory and Practice of Multi-agent Approaches,* Elgar Publishers, Northampton, UK, pp248–285

Burton, M. A. (1994) 'The irrigation management game: A role-playing exercise for training in irrigation management', *Irrigation and Drainage Systems,* vol 7, pp305–318

Chohin-Kuper, A., Garin, P., Ducrot, R. and Tonneau, J. P. (2004) 'Irrigation management transfer in the Nordeste, Brazil: The challenge of users' participation', in Cemagref, Cirad, IRD (eds) *Coordinations Hydrauliques et Justices Sociales*, Montpellier, France

d'Aquino, P. (2002) 'Some novel information systems for the empowerment of a decision-making process on a territory: Outcomes from four years of participatory modelling in Senegal' in A. E. Rizzoli and A. J. Jakeman (eds) *Integrated Assessment and Decision Support. Proceedings of the 1st Biennial Meeting of the International Environmental Modelling and Software Society*, proceedings of International Environmental Modelling and Software Society conference, Lugano, Switzerland, 27 June, vol 1, pp132–137

d'Aquino, P., Le Page, C., Bousquet, F. and Bah, A. (2003) 'Using self-designed role-playing games and a multi-agent system to empower a local decision-making process for land use management: The SelfCormas experiment in Senegal', *Journal of Artificial Societies and Social Simulation,* vol 6, no 3, http://jasss.soc.surrey.ac.uk/, accessed September 2005

Daré, W. (2005) 'Comportements des acteurs dans le jeu et dans la réalité: Indépendance ou correspondance? Analyse sociologique de l'utilisation des jeux de rôle en aide à la concertation', PhD Thesis, Engref, Paris

Daré, W. and Barreteau, O. (2003) 'A Role-playing game in irrigated system negotiation: Between play and reality', *Journal of Artificial Societies and Social Simulation*, vol 6, no 3, http://jasss.soc.surrey.ac.uk/ (accessed September 2005)

Da Rocha Barros, E. (2001) 'Evaluation sociologique des organisations avec les petits producteurs', in *Séminaire sur la Gestion des Périmètres Irrigués*, CODEVASF-EMBRAPA-Cirad, Juazeiro, Brazil, 5–7 December

de Nys, E. (2004) 'Interaction between water supply and demand in two collective irrigation schemes in North-East Brazil', PhD Thesis, Katholieke Universiteit Leuven, Leuven, Belgium

Ducrot, R., Le Gal, P. Y., Morardet, S., Jehan, C. and de Nys, E. (2001) 'Transitions institutionnelles et agricoles dans les périmètres irrigués du pôle Petrolina-Juazeiro (Brésil): D'une logique sociale vers une logique manageriale', in *Séminaire PCSI. La gestion des Périmètres Irrigués Collectifs à lAaube du 21ème Siècle: Enjeux, Problèmes, Démarches*, Cemagref, Montpellier, France

Etienne, M. (2002) 'SYLVOPAST, a multiple target role game to assess negotiation processes in sylvopastoral management planning' in *7th Biennial Conference of the International Society for Ecological Economic*, Sousse, Tunisia

Fernandez, S. (2001) 'Emancipation des périmètres irrigués du nordeste au Brésil: Quels outils et règles de gestion de leeau d'irrigation?', Diplôme d'Etudes Approfondies thesis, Engref/Cemagref, Paris X University, Nanterre, France

Fernandez, S., Chohin-Kuper, A., Rieu, T. and De Nys, E. (2001) 'L'évolution de la Tarification dans le Périmètre Irrigué de Maniçoba: Vers une Gestion Durable du Système?' in *Séminaire sur la Gestion des Périmètres Irrigués*, CODEVASF-EMBRAPA-Cirad, Juazeiro, Brazil, 5–7 Dec

Funtowicz, S.O. and Ravetz, J. R. (1993) 'Science for the post-normal age', *Futures*, vol 25, no 7, pp739–755

Gleyses, G., Loubier, S. and Terreau J-P. (2001) 'Evaluation du coût des infrastructures d'irrigation', *Ingénieries*, vol 27, pp59–67

Imperial, M. T. (1999) 'Institutional analysis and ecosystem-based management: The institutional analysis and development framework', *Environmental Management*, vol 24, no 4, pp449–465

Johnson, S. H. III (2002) 'Irrigation Management Transfer: Decentralizing public irrigation in Mexico', in Saleth, R. M. (ed) *Water Resources and Economic Development*, Edward Elgar, Cheltenham, UK, pp437–445

Mucchielli, A. (1983) *Les Jeux de Rôles*. Presses Universitaires de France, Paris

Ostrom, E. (1992) *Crafting Institutions for Self-governing Irrigation System*, ICS Press, Institute for Contemporary Studies, San Francisco, US

Ostrom, E., Gardner, R. and Walker, J. (1994) *Rules, Games and Common Pool Resources*, University of Michigan Press, Ann Arbor, MI

Pahl-Wostl, C. (2004) 'The implications of complexity for integrated resources management', in Pahl-Wostl, C. Schmidt, S. and Jakeman, T. (eds) *Complexity and Integrated Resources Management*, International Environmental Modelling and Software Society, Osnabrück, Germany

Perret, S. (2003) 'Une démarche de modélisation pour accompagner le transfert de gestion de périmètres irrigués en Afrique du Sud', in *Organisation Spatiale et Gestion des Ressources et des Territoires Ruraux*, UMR Sagert-Cirad, Montpellier, France

Plantey, J. and Blanc, J. (1998) 'La maintenance des ouvrages et équipements' in J. R. Tiercelin (coordinator) *Traité d'Irrigation*, Lavoisier, Paris, pp845–849

Shah, T., van Koppen, B., Merrey, D., de Lange, M. and Samad, M. (2002) 'Institutional alternatives in African smallholder irrigation', research report 60, IWMI, Colombo, Sri Lanka

Tang, S. Y. (1994) 'Institutions and performance in irrigation systems', in Ostrom, E. Gardner, R. and Walker, J. (eds) *Rules, Games and Common Pool Resources*, University of Michigan Press, Ann Arbor, MI, pp225–245

Wenger, E. (1998a) *'Communities of Practice: Learning, Meaning and Identity'*, Cambridge University Press, New York

Wenger, E. (1998b) 'Communities of practice, learning as a social system', *The Thinker*, vol 9, no 5, available at www.ewenger.com, accessed September 2005

World Bank (2002) 'Institutional reform in irrigation and drainage', in 'Proceedings of a World Bank Workshop', technical paper 524, World Bank, Washington, DC

16

Support to Stakeholder Involvement in Water Management Circumventing Some Participation Pitfalls

O. Barreteau, G. Abrami, S. Chennit and P. Garin

Introduction

Public participation is increasingly involved in collective decision-making processes at the local level, although what actually is meant by participation is rarely specified and varies widely. In Europe, the Water Framework Directive contains a specific article regarding participation. In Africa, new national policies have attempted to involve stakeholders and citizens in collective decision-making processes and in the implementation of their outputs. Participation is currently promoted by international institutions.

This has led to the widespread acknowledgement that concerned people should be more involved in decision-making processes, at least in the field of natural resource management. This new trend towards participation is generating new issues: How should the involvement of stakeholders be implemented? What pitfalls need to be avoided and what side effects need to be managed? What can be considered as involvement of concerned people, and how diverse can such involvement be?

Agent-based models (ABMs) and role playing games (RPGs) are increasingly being used, separately or together, to facilitate dialogue in territory planning and natural resource management (Bousquet et al, 2002). This is due to the capacity of such tools to simulate and handle complexity.

This paper introduces such a tool, PIEPLUE, as a support for dialogue on water sharing issues. Before presenting PIEPLUE, we discuss some of the diversity in what is considered participation according to two dimensions: the level of involvement and the stage of involvement in the decision-making process. We also identify the pitfalls that PIEPLUE aims to circumvent, such as a disproportionate focus on private interests, the emergence of new lead groups which might not be legitimate and interference in existing conflicts.

Issues Generated by the Implementation of Participation Principles

Member States shall encourage the active involvement of all interested parties in the implementation of this Directive, in particular in the production, review and updating of the River Basin Management Plans. (Water Framework Directive, Article 14, EU, 23-10-2000).

Need for Specific Tools

Institutions in charge of water management, at whatever level of organization – basin institutions, local delegations of concerned departments of state, regions or municipalities – all share the objective of concerted water management noted in the European Water Framework Directive. Their concern, however, is over how to make the concept of participation operational (Richard, 2000). They are more accustomed to dealing with technical consultancy groups, who provide them with technical support in reaching right decisions, than organizing participatory decision-making processes in a way well accepted by stakeholders. They therefore need to know, for example, whether there are certain stages to go through, what kind of interface is available to focus interactions, etc.

In addition, they need to define what they mean by participation and to identify the kind of participation they wish to implement; usually this is only implicit and may range, for example, from asking people to volunteer their labour to build a dike, to asking them whether they would prefer a dike or a channel to protect them from flood, to allowing them to propose scenarios to be explored. Institutions are asking for tools, but the context in which these tools are to be used needs to be specified.

Diversity of Participation

The call for more involvement of stakeholders is specific in the case of the Water Framework Directive quoted above. But it is found in many other places too, from the slogans of grassroots organizations to national policies and international institutions' recommendations. This is evidence of the current consensus that stakeholders should be involved in water management processes. But why? To what extent? And at which stage of the process? Neither is there consensus on these questions nor are there explicit answers.

One reason behind such calls for participation may be an acknowledgement of citizens' fundamental right to be involved in decision processes concerning them. Another may be that participation is seen as a means to convince people to accept a decision that has already been taken. Participation may also be viewed as a legal constraint or a way to share the burden of responsibility and to prevent further criticisms in a context of uncertainty and complexity.

A classic presentation of the range of forms that participation may take was Arnstein's ladder, reaching from information up to co-decision through consultation (Arnstein, 1969). The question regarding the stage in the decision-making process participation is best suited to has been less addressed. A comparison of tools developed to support participatory water management across Europe suggests that some participatory processes aim to encourage stakeholders to choose from a few predefined solutions, while others aim to help stakeholders elaborate common objectives or issues at stake in their area. However, the definition of the issues open to dialogue, like the definition of the set of solutions open for selection, provides power to those given the task (Marengo and Pasquali, 2003). Thus the actual empowerment of stakeholders through participation will differ slightly according to at which stage in the decision-making process they are involved.

In the following discussion, we focus on supports for participatory, collective decision-making as a pragmatic means to reach decisions which are implemented at the level of co-decision and cover all stages, from the elaboration of scenarios to the choice of scenarios. This implies an important collective empowerment of the stakeholders involved in the process.

Pitfalls Encountered in Participatory Processes

No matter what the circumstances, the implementation of a participatory process faces many pitfalls, notably due to the complexity of the societies involved (Eversole, 2003). In many cases, the implementation of participation becomes bogged down by internal conflicts or disagreements among the people involved, and it can even go so far as to foster conflicts by awakening disagreements.

One common pitfall in this context is that power formerly attributed to the administration or some other 'top' institution is taken over by other local social hierarchies. If the purpose of implementing participation is to increase democracy, there is no guarantee that the newly empowered stakeholders have more democratic legitimacy than the former institutions. The possibility that new lead groups will emerge or that local leaderships will be reinforced increases when there is a lack of organizational training of the stakeholders involved. This is particularly true when so-called rapid approaches are used (Platteau and Gaspart, 2003).

Depending on the pre-existing social relations within the group of stakeholders involved, collective decision-making processes may lead to initiating more conflicts than they resolve (Parent and Gallupe, 2001). The evolution of a participatory process is thus rather uncertain and depends on the dynamics within the group and on the professional and personal characteristics of the facilitator of the process. Any approach supporting facilitation thus should provide the possibility of tuning tools and methodologies to the local context.

The NIMBY (not in my back yard) effect is a classic form of the dead ends encountered in participatory processes. It results from the overly strong involvement of private interests: stakeholders agree to collective rules addressing the multiple uses of natural resources provided that their own interests are not affected. Individuals expect the community to provide and manage common goods such as quality water resources but only to the extent that their own interests are not affected. The actors taking part in such collective decision-making processes would be motivated first by selfish reasons. The setting of a new railway or a dam is a common example of this phenomenon. With the implementation of the public participatory processes required by the EU Water Framework Directive, such behavioural patterns might be exacerbated if narrow interests are put to the forefront of the debate. As far as water management is concerned, the interplay between private and collective interests can be found in various issues: flow levels, access to river banks and water quality, for example, and in the public image of sectors such as farming.

Consequences for the Development of Tools to Support Participatory Processes

Information must be available to enable this new trend of power redistribution in decision making. However, there is currently no equality in terms of access to information. Social networks in which stakeholders are involved are a means to obtain information or to control its diffusion. Education levels, the involvement of local NGOs and the size and revenue of a farm all influence relationships with extension services and thus the amount of information available to a stakeholder (Glendinning et al, 2001), but providing explicit and legitimate information is a key feature of any method to support participatory collective decision-making. By 'legitimate information' we mean information that stakeholders accept as being representative of reality and which they are willing to have disseminated. The format of this information also has to fit the actors' cognitive filters. To achieve genuine participation, there is thus a need to improve the management of participatory processes to avoid de facto hijacking by some stakeholders whose social status or specific social networks provide them with better access to information.

The potential pitfall of the nimby effect might be tempered: the actors most involved are often associations or NGOs, ie institutions working to raise awareness and provide education on environmental issues which also feel affected by agreements on general principles, such as the relations between public decisions and environmental issues (Lafaye and Thévenot, 1993). This is often true in cases of civil engineering design where there has been strong public protest and de facto involvement. An example here was the new railway, TGV Méditerranée, in the south-east of France, where NGOs and citizens' associations were able to address very generic issues in discussing the objectives of designing the new railway and ended up by being involved in a re-definition of the issue (Lolive, 1997). Some institutions focus their efforts more on getting pertinent questions which emerge debated than on pursuing pre-determined privately interested ends.

NGOs and self-established institutions aiming to represent certain actors or interests are very heterogeneous. They differ in their attitude towards dialogue, in the interests they hold and in their representativeness. They may already be related to each another, with previous interactions leading either to easier coalitions and positive dialogue or to stronger conflicts. These institutions should be considered not only as single entities but also as being made up of representatives who are involved in a given dialogue process. And such representatives usually have a plurality of interests – even if they are supposed to champion one, they often cannot let go of others (Innes and Booher, 1999). Inviting relevant institutions into a collective decision-making process is consequently difficult and the protocol for their involvement should recognize these issues.

No matter how open to dialogue the stakeholders are, one objective of any tool designed to facilitate participatory approaches is to ensure that generic issues are addressed so that the dialogue process does not simply become a defence of particular interests.

Using ABMs and RPGs to Enhance Stakeholder Involvement

Models, whether they be computer based, drawings, maps or parables, have always been used as supports in the design of public policies (Saunders-Newton and Scott, 2001). For over half a century, policy making advisors have focused on these tools, with an increasing use of science, particularly computer science, in order to rationalize decision-making processes. This is especially true for participatory management where a diversity of viewpoints is involved. Models also are a means of delivering information and of increasing stakeholder awareness. This section presents a specific category of such models, RPGs, which are used jointly with computer ABMs as tools to facilitate participation.

Joint Use of RPGs and ABMs

According to several studies which have used these tools for natural resources management issues (D'aquino et al, 2003; Etienne et al, 2003; Hare and Pahl-Wostl, 2003), ABMs and RPGs are well adapted for work on complex situations and can induce distance from immediate private interests. They are thus viable candidates to be tested as tools to support participatory decision making.

ABMs and RPGs were developed separately to address group decision and dialogue issues (Barreteau, 2003). RPGs, notably including policy exercises, were developed first and have been used to understand collective decision-making processes and to provide training for and solicit support from the stakeholders involved in them. From an analytical viewpoint, they stimulate the emergence of misunderstandings by dividing the decision process into several decision centres (Schelling, 1961) and are similar to the tools developed by experimental economists. They are effective in empowering stakeholders in decision-making processes and in facilitating the sharing of information (Tsuchiya, 1998). However, they are rather complicated to design and the repetition of experiments with control of parameters is difficult (Piveteau, 1995).

More recently, ABMs have been used to simulate complex systems, the idea being to use them not only to represent but also to support collective decision-making processes by broadening the field of information available to participants (Benbasat and Lim, 2000). Providing stakeholders with the potential consequences of various choices involved in an on-going group decision-making process reportedly mobilizes them to be more active (Driessen et al, 2001). The objective of the ABM here is to represent the interests at the centre of the collective decision-making process to lead stakeholders to better formulate the problems or to give them a way of sharing viewpoints. However, as ABMs usually are embedded within a computer tool, they are usually perceived by stakeholders as black boxes, which raises issues of their legitimacy and acceptability.

Formally, ABMs and RPGs have the same structure: autonomous entities situated in an environment and interacting dynamically. This helps to overcome the limits of either one used on its own: an RPG might be used to translate an ABM more clearly to stakeholders while an ABM can be used to repeat and simulate game sessions (Barreteau et al, 2001).

Creating Distance and Involving Complexity

The first feature of these tools that allows them to meet an initial requirement is their ability to deal with complexity; this was mentioned earlier as a reason for using games for policy issue research and education (Schelling, 1961), and ABMs have also been developed to work on other complex issues (Ferber, 1995). The key feature that enables both tools to handle complex issues is the distribution of decision centres, in computer agents in the case of the ABM, in players in the case of the RPG.

RPG design and facilitation is a craft, and even in very closed games a degree of freedom always remains in which players can interpret their roles in various ways. This helps to meet the contingency requirement through the adaptation of the tool by the players themselves.

As is true for many tools, suitability for a specific objective depends on the protocol of use. Many experiments jointly using ABMs and RPGs refer to a companion modelling approach (Bousquet et al, 1999; Bousquet et al, 2002). Significantly, this iterative approach aims at involving stakeholders in the modelling process itself, which provides information to participating stakeholders not only at a factual level but also at the meta-level of the tools being used to deliver the information.

Finally, ABMs and, even more so, RPGs, are very simplified representations of the world. Even though some elements of the real world are present, they are both only thought support tools and cannot identify all individual real-life consequences, leading to a focus on generic issues rather than on the individual's own backyard. Moreover, the gaming atmosphere provides a distance from real issues, providing the possibility to explore scenarios that may not be socially acceptable by leaving a door open to rejecting the consequences of the simulations.

PIEPLUE: A Hybrid RPG–ABM Tool

PIEPLUE is a tool we developed based on the association of RPGs and ABMs; it is used in a companion modelling approach. The tool specifically aims to reorient zero-sum game interactions among farmers sharing water towards win–win situations and has been developed to fit an on-going dialogue process in the south-east of France (the Drôme river basin development plan) and the negotiation of water-sharing rules among irrigating farmers.

The Drôme River Valley Case Study

A local water management plan was signed by the local state representative for the Drôme river valley (a major tributary of the Rhone in the south-east of France) at the end of 1997, building upon an agreement between representatives of local elected bodies, water users' representatives and state representatives. The plan addresses, among other issues, the question of minimum flows to be maintained in downstream areas, where the main use is irrigation, through an agreement among stakeholders on a minimum threshold for downstream flow of $2.4m^3/s$ throughout the year. We have been involved in facilitating dialogue in the farming sector on how to reach this objective through collective control of irrigation uptakes. A set of collective rules were agreed upon in early 2003 but they have not yet been tested due to the exceptional drought that year (the water flow in the upstream part of the irrigated area was below the minimum level for most of the normal irrigation season). This set of collective rules is based upon the use of complementary resources from outside the river basin and the definition of allocation rules among farmers. The RPG and ABM presented below aim to provide an interactive setting for possible revisions of the agreement in the future.

Farmers are the major consumers of irrigation water, mainly for corn fields: irrigation is the main reason for water pumping, with 80 per cent of the irrigated fields (3000ha) on the downstream part of the river. The total pumping capacity for the downstream part is $2m^3/s$. Farmers are partially organized within three irrigation systems managed by users' associations; three-quarters of the irrigated area falls within one of these three irrigation systems, and 85 per cent of farmers belong to at least one of the three associations. The remaining irrigated areas are served by wells in the alluvial aquifer. The whole context is evolving with:

- the occurrence of droughts, causing individual expectations for critically dry years to evolve;
- political interests, such as local elections, which cause new scenarios to appear and others to become unwelcome; and
- national and European agricultural policies, which cause interest in specific crops depending more or less on water to evolve.

In this case study, participation is occurring at a rather late stage as the issues and objectives have already been defined; however, it aims to design some rules within these constraints. Participants cannot change the objectives but they still have full scope to design rules among themselves so that they may collectively cope with the constraints of the agreement with other users on a minimum flow level as well as their own needs. These rules should concern only their activities, i.e. farming; therefore the setting should gather farmers and lead them to discuss only farming practices.

Coordination of the RPG and ABM

Dealing with dialogue support for collective decision-making processes for irrigation management issues raised the need to tackle a large sample of time scales:

from the day, when hydraulic balance and water level in the river is computed or observed and practical decisions for cropping and irrigating are made, to the year, when irrigation investments are made. PIEPLUE thus has to be able to deal with:

- a short-term time scale, typically a day, as the time for the farmer to choose the plot to irrigate;
- a medium-term time scale corresponding to the evolution of priorities among crops, typically a month, for the farmers to update their irrigation patterns; and
- a long-term time scale, typically a year, for cropping pattern choices and collective discussion on rules on sharing water.

The short-term time scale is simulated by the ABM according to the choices made at the other two time scales. The medium- and long-term time scales are simulated in the game and benefit from the simulation results of the short-term simulation.

The RPG constitutes the basis of the interactive setting. As applies to many games using computer tools, the ABM is embedded in the RPG. All players take on the roles of farmers. Two game facilitators are required, one for the gaming part and one to use the ABM. A sequence of several stages is then played out:

- an initialization stage, with the assignment of roles and farm characteristics to players;
- a cropping season starting stage with a choice of cropping patterns for the players;
- a month time step, with a choice of irrigation patterns for the players (four times); and
- a cropping season end stage with the allocation of cropping season results and a discussion of the collective rules.

In the initialization stage, players are each allocated six fields, each characterized by a soil water capacity (superficial, medium or deep), and two water supply facilities that are characterized by location (an individual well or an outlet on a collective irrigation network) and a capacity. These allocations are assigned randomly by the computer to each player. Each player also receives an objective to help in assuming their role. The basic setting involves two collective networks, one with a pumping station in the upstream part of the irrigated area and the other in the downstream part. The collective irrigation networks are initiated, with players with outlets in a network choosing a president and with the definition of water pricing for outlet holders.

During the cropping season starting stage, players choose their cropping pattern based on an allocation of a crop to each field. Players can choose from wheat, maize, tomatoes and garlic, the most common crops in the area.

They then choose an irrigation pattern for each monthly time step. This step is repeated four times as an irrigation season lasts four months in the area (June, July, August and September). For each time step, players fill in a form specifying

Table 16.1 *Table to be filled in by each player at each month time step for each water source*

	Field 1	Field 2	Field 3	Field 4	Field 5	Field 6
Monday						
Tuesday						
Wednesday						
Thursday						
Friday						
Saturday						
Sunday						

Source: chapter authors

a weekly irrigation pattern for their two water sources as shown in Table 16.1. These forms are then entered as parameters for the ABM by a game facilitator on specific computer interfaces.

Players can base their choice of irrigation patterns on information about the previous month's time step. This information deals with the evolution of the private and public indicators detailed below.

Private information systematically concerns the evolution of the state of the crops according to soil-water availability as shown in Figure 16.1. It is printed out and given privately to players for their own fields. Other private information – water consumption for each of their water sources per day, irrigation amount per field and per day – is available and provided to players on request.

Public information is projected directly from the computer. It concerns the series of downstream flows, which appears on a computer interface as presented in Figure 16.2, and is projected systematically. Other indicators – climatic data

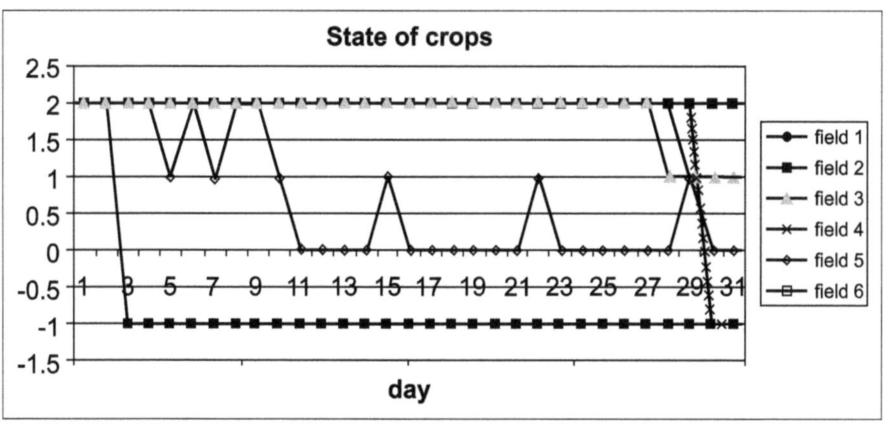

Note: 2 stands for good, 1 stands for thirsty, 0 stands for very thirsty, –1 stands for the absence of crops, usually due to the harvest having been completed
Source: chapter authors

Figure 16.1 *Evolution of the state of crops on a player's fields during a month*

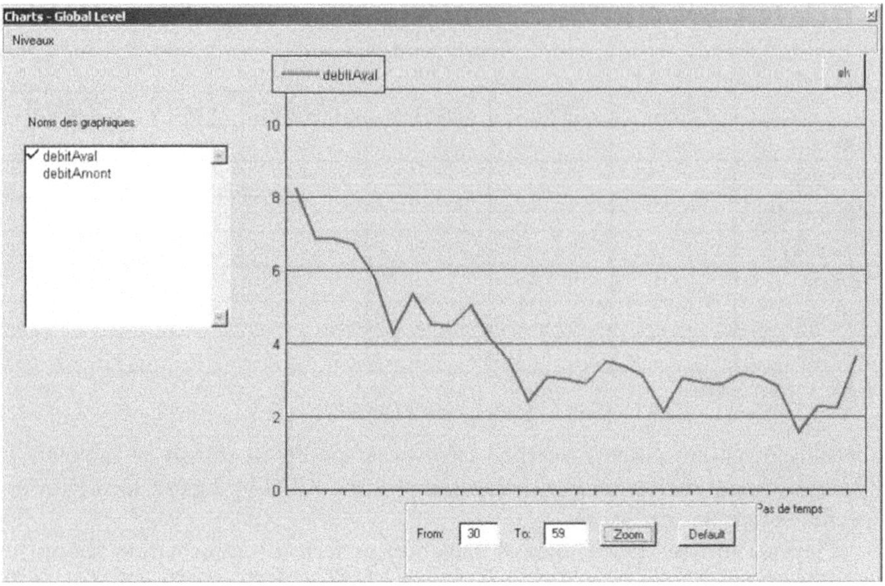

Note: The vertical axis is downstream flow in m^3/s and the horizontal axis is time in days.
Source: chapter authors

Figure 16.2 *Evolution of downstream flow during a month time step*

for the previous month (rain and potential evapotranspiration), the evolution of crisis levels and upstream flow – are computed and can be projected on request from players.

All these indicators are computed between two time steps by the computer ABM. The ABM has the same structure as the RPG:

- Agents are farmers (with one farmer for each player), water users associations (one for each collective irrigated network), and one local water commission (implementing the collective rules and played in the RPG by the RPG facilitator).
- Objects are the fields, the crops, the outlets, the pumps (collective or individual) and the river.

The ABM implements the irrigation patterns provided by each player at the day time step. The resulting downstream water flow is then compared daily to the objective and the collective rules are implemented, possibly generating crises and decreases in water pumping. Expenses for each cropping activity and incomes from yields are also computed, updating the cash level of each player.

At the end of the cropping season, each player is privately provided with his own results: the yield for each crop and current cash level. Current collective rules are then reviewed and possible modifications to these collective rules for a new cropping season are discussed.

Involvement of Stakeholders Through the Use of PIEPLUE

Two Test Sessions

Two test sessions have already been undertaken, one with scientists in the field of irrigation water management, another with employees of the Communauté de Communes du Val de Drôme (CCVD), an association of communes in charge of the implementation of the water management plan. The first test was aimed at calibrating the game and validating the relevance of the ABM simulations from an expert's point of view.

The second test session was held with the institution in charge of the implementation of the water management plan and thus of the facilitation of the collective decision-making processes which are induced thereby. It gathered ten players. The objective was to provide this institution with knowledge of PIEPLUE before a session with real farmers and for the participants to take part in its design on the basis of the prototype. In this second test session, some players were slightly overwhelmed by the amount of rules needed to be learned in the beginning. Significantly, these difficulties of understanding were due to the participation in the game of employees of the CCVD who were not familiar with irrigation issues, which was necessary to get a minimum number of players. The supply of the series of indicators was useful in helping players understand these rules during the play of the first cropping season. Players were able to understand the relations between their choices and their results. However, these explanations were time consuming. Consequently, only one cropping season could be played out, which did not allow for the testing of the capacity of PIEPLUE to foster generations of new collective rules.

The repetition of tasks was well accepted since they were associated with the direct simulation of the consequences of various actions made by each player. These simulations, with the private information provided to players, generated comparisons between neighbours during the session.

PIEPLUE as a Tool to Support Participation

The use of PIEPLUE is suitable for the consultation level of participation. According to the follow-up of a session's outcomes, the tool may even be useful at a higher level of participation; however, this would raise concerns regarding the representativeness of the participants in game sessions generating rules to be implemented in the real world. Due to game facilitation issues, the structure of such exercises limits the number of participants to about 10 or 12.

From a dynamic point of view, PIEPLUE is designed to be used on a long-term time scale. It is meant to prevent crises by making stakeholders discuss calmly the rules to be implemented in case of a crisis. In that sense the choices made by participants are not overly constrained by the given situation to be managed. However, the dynamic framework of the participation process is constrained: the minimum water-flow level objective is not open to discussion and the initial set of rules is given, computed before the formulation of a 'real' set of rules taking into account the simulated area of the game session.

PIEPLUE effectively promotes participation, initially by generating interactions among players which do not exist at the same level in a real setting (i.e. among farmers). The players interacted on several topics: the choice of cropping patterns, the understanding of the dynamics of crop growing and water needs, and the sharing of water capacity within the collective networks. In the debriefing, they also proposed the organization of a crisis management panel within the game. This crisis management panel exists in the current Drôme river management protocol, but it consists of officials and representatives rather than the farmers themselves. Since it was only a test session, no decisions were made at the end of the game; however, the generation of discussion is a requirement of the first stage of participation, at the level of consultation and beyond.

PIEPLUE also demonstrated an ability to deal with the pitfalls of participation identified above. There has been a partial reset of previous relations among players: they did interact as farmers in a same basin, without importing totally their relations in the game. The frame of the test session makes it however difficult to analyse. Players profited from others' knowledge of irrigation and agriculture and asked for advice, but they did not take into account any relation to the real world in their way of playing. The discussion tended to focus on generic issues without any reference to specific ones.

Potential Uses of PIEPLUE

As a test, it was the uses of PIEPLUE itself that generated the most participation, leading to a rich discussion on the tool's possibilities. Whereas it had been designed as a collective decision-making support tool, participants closely related to the farming sector (extension services from the local water institution or water users' association) expressed an interest in it as a professional training tool outside of any real decision-making processes. These players acknowledged the tool's power to make farmers understand that dialogue is an important part of water management. This understanding should lead them to design water management scenarios with parameters other than only pumping rights. However, they did not feel that the tool was a potential aid to collective decision-making support in an operational context. They argued that negotiations already take place within the farming sector without this kind of interactive session. Biases, such as the hijacking of decision-making processes by a few leaders of the farming sector, which is observed in real negotiation patterns, were not relevant to them since their aim is to provide advice to the farming sector and to lead them to good practices. They therefore asked for more control input on simulation scenarios played during game sessions so that the lessons learned from the game would be those they need: representation of climatic diversity and the proposal of a few cropping patterns to players in order to reach contrasted outputs. However, such use would require a deeper validation of the model because it assumes less knowledge and less experience from players and thus less capacity to sort out game outputs. With this type of tool and players, it is easy to pass from training to manipulating.

The aim of the representatives of extension services and of the river basin institution is to make water users' practices evolve towards what they view as new and better practices. Their viewpoint, however, may not be shared by the water users themselves. While in the setting of PIEPLUE, farmers are considered as actors to be involved in the decision-making process, institutional stakeholders consider them as final users who need to be convinced to adopt good practices. The rationale for involving these final users in interactive sessions is to lead them to take 'better decisions' defined by others and according to objectives which go beyond their own sphere. From the perspective of the institutional stakeholders, there is no real purpose in involving farmers in the decision-making process itself, which would imply that there would be no a priori expected decision.

The whole session, including the game and the debriefing, thus led to a reformulation of the initial question and objectives, which is characteristic of a participatory process. However, the direction it took in terms of redesigning the question and objectives raises two new concerns: the acceptability of a participatory process that leads to a lowering of the level of participation itself and the legitimacy of a participatory process that draws conclusions concerning people who were not represented in the process.

Conclusion

The development of PIEPLUE follows a companion modelling approach (Bousquet et al, 2002) which implements a succession of versions that are confronted with the field at various levels. Here, we moved from experts external to the field closer and closer to actors who are more directly involved in the use of water resources.

The tool is open to new rules despite the constraint on the topic of discussion. There must be prior agreement that sticking to cropping patterns and the management of irrigation patterns is suitable and sufficient to deal with water sharing issues among farmers. There is an empowerment of the stakeholders who happen to play in the game. However, this raises the issue of the power they eventually gain over those who do not participate. Since the game cannot be played by more than a dozen players, it has to be repeated so that this technical feature does not induce more changes in power relations than those generated by the implementation of any participatory process. The protocol for these repetitions and the synthesis of conclusions of sessions are still to be designed.

In this case study, relations within the farming sector are not too difficult. Playing randomly attributed roles also leads players to avoid involving actual relational issues, for example between individual and collective water uses, since they are not personally responsible for their played behavioural patterns. The randomly attributed roles and the generic spatial support of not keeping to any one part of the basin leads participants to tackle more generic issues as it is impossible to assess through the game what may be the particular consequences of the proposed rules to any one individual.

Acknowledgements

We are grateful to the CCVD for the quality of interactions in the case study and to the French Ministry of the Environment's *Concertation et décision en environ-*

nement (dialogue and decisions on the environment) programme for financing this research. We also wish to thank our colleagues from the ComMod group for the rich discussions on companion modelling and the use of RPGs. The group's charter can be found at http://cormas.cirad.fr/en/reseaux/ComMod/index.htm.

References

Arnstein, S. (1969) 'A ladder of citizen participation', *Journal of the American Planning Association,* vol 35, pp216–224

Barreteau, O. (2003) 'The joint use of role-playing games and models regarding negotiation processes: Characterization of associations', *Journal of Artificial Societies and Social Simulations,* vol 6, http://jasss.soc.surrey.ac.uk/6/2/3.html

Barreteau, O., Bousquet, F. and Attonaty, J-M. (2001) 'Role-playing games for opening the black box of multi-agent systems: Method and teachings of its application to Senegal River Valley irrigated systems', *Journal of Artificial Societies and Social Simulations,* vol 4, http://jasss.soc.surrey.ac.uk/4/2/5.html

Benbasat, I. and Lim, J. (2000) 'Information technology support for debiasing group judgments: An empirical evaluation', *Organizational Behavior and Human Decision Processes,* vol 83, pp167–183

Bousquet, F., Barreteau, O., d'Aquino, P., Etienne, M., Boissau, S., Aubert, S., Le Page, C., Babin, D. and Castella, J-C. (2002) 'Multi-agent systems and role games: An approach for ecosystem co-management', in Janssen, M. (ed) *Complexity and Ecosystem Management: The Theory and Practice of Multi-agent Approaches,* Edward Elgar Publishers, Cheltenham, UK

Bousquet, F., Barreteau, O., Le Page, C., Mullon, C. and Weber, J. (1999) 'An environmental modelling approach: The use of multi-agent simulations', in Blasco, F. and Weill, A. (eds) *Advances in Environmental and Ecological Modelling,* Elsevier, Paris

d'Aquino, P., Le Page, C., Bousquet, F. and Bah, A. (2003) 'Using self-designed role-playing games and a multi-agent system to empower a local decision-making process for land use management: The SelfCormas experiment in Senegal', *Journal of Artificial Societies and Social Simulations,* vol 6, http://jasss.soc.surrey.ac.uk/6/3/5.html

Driessen, P. P. J., Glasbergen, P. and Verdaas, C. (2001) 'Interactive policy making: A model of management for public works', *European Journal of Operational Research,* vol 128, pp322–337

Etienne, M., Le Page, C. and Cohen, M. (2003) 'A step by step approach to build up land management scenarios based on multiple viewpoints on multi-agent systems simulations', *Journal of Artificial Societies and Social Simulations,* vol 6, http://jasss.soc.surrey.ac.uk/6/2/2.html

Eversole, R. (2003) 'Managing the pitfalls of participatory development: Some insight from Australia', *World Development,* vol 31, pp781–795

Ferber, J. (1995) *Les Systèmes Multi-agents, vers une Intelligence Collective.* InterEditions, Paris

Glendinning, A., Mahapatra, A. and Mitchell, C. P. (2001) 'Modes of communication and effectiveness of agroforestry extension in Eastern India', *Human Ecology,* vol 29, pp283–305

Hare, M. P. and Pahl-Wostl, C. (2003) 'Stakeholder categorisation in participatory integrated assessment processes', *Integrated Assessment,* vol 3, pp50–62

Innes, J. E. and Booher, D. E. (1999) 'Consensus building as role playing and bricolage: Toward a theory of collaborative planning', *Journal of the American Planning Association,* vol 65, pp9–26

Lafaye, C. and Thévenot, L. (1993) 'Une justification ecologique? Conflits dans l'aménagement de la nature', *Revue Française de Sociologie,* vol 34, pp495–524

Lolive, J. (1997) 'La montée en généralité pour sortir du NIMBY. La mobilisation associative contre le TGV Méditerranée', *Politix,* vol 39, pp109–130

Marengo, L. and Pasquali, C. (2003) 'How to construct and share a meaning for social interactions?' in conference proceedings *Conventions et Institutions: Approfondissements Théoriques et Contributions au Débat Politique,* Paris

Parent, M. and Gallupe, R. B. (2001) 'The role of leadership in group support systems failure', *Group Decision and Negotiation,* vol 10, pp405–422

Piveteau, V. (1995) *Prospective et Territoire: Apports d'une Réflexion sur le Jeu.* Cemagref éditions, Antony, France
Platteau, J-P. and Gaspart, F. (2003) 'The risk of resource misappropriation in community driven development', *World Development,* vol 31, pp1687–1703
Richard, A. (2000) *Analyse comparée de l'acceptabilité des SAGE et contrats de milieu,* Cemagref/ Ecole Polytechnique, Palaiseau, Montpellier, France
Saunders-Newton, D. and Scott, H. (2001) '"But the computer said!": Credible uses of computational modeling in public sector decision making', *Social Science Computer Review,* vol 19, pp47–65
Schelling, T. C. (1961) 'Experimental games and bargaining theory', *World Politics,* vol 14, pp47–68
Tsuchiya, S. (1998) 'Simulation/gaming as an essential enabler of organizational change', *Simulation and Gaming,* vol 29, pp400–408

Index

For Product Safety Concerns and Information please contact our EU representative GPSR@taylorandfrancis.com
Taylor & Francis Verlag GmbH, Kaufingerstraße 24, 80331 München, Germany

www.ingramcontent.com/pod-product-compliance
Ingram Content Group UK Ltd.
Pitfield, Milton Keynes, MK11 3LW, UK
UKHW020018200926
475702UK00014B/1946
9780415852029